Explorations

in

Complex Analysis

© *2012 by the Mathematical Association of America, Inc.*

Library of Congress Catalog Card Number 2012943816

Print edition ISBN 978-0-88385-778-6

Electronic edition ISBN 978-1-61444-108-3

Printed in the United States of America

Current Printing (last digit):

10 9 8 7 6 5 4 3 2 1

Explorations in Complex Analysis

Michael A. Brilleslyper
United States Air Force Academy

Michael J. Dorff
Brigham Young University

Jane M. McDougall
Colorado College

James S. Rolf
Yale University

Lisbeth E. Schaubroeck
United States Air Force Academy

Richard L. Stankewitz
Ball State University

Kenneth Stephenson
University of Tennessee, Knoxville

Published and Distributed by
The Mathematical Association of America

CLASSROOM RESOURCE MATERIALS

Classroom Resource Materials is intended to provide supplementary classroom material for students—laboratory exercises, projects, historical information, textbooks with unusual approaches for presenting mathematical ideas, career information, etc.

101 Careers in Mathematics, 2nd edition edited by Andrew Sterrett

Archimedes: What Did He Do Besides Cry Eureka?, Sherman Stein

Calculus: An Active Approach with Projects, Stephen Hilbert, Diane Driscoll Schwartz, Stan Seltzer, John Maceli, and Eric Robinson

Calculus Mysteries and Thrillers, R. Grant Woods

Conjecture and Proof, Miklós Laczkovich

Counterexamples in Calculus, Sergiy Klymchuk

Creative Mathematics, H. S. Wall

Environmental Mathematics in the Classroom, edited by B. A. Fusaro and P. C. Kenschaft

Excursions in Classical Analysis: Pathways to Advanced Problem Solving and Undergraduate Research, by Hongwei Chen

Explorations in Complex Analysis, Michael A. Brilleslyper, Michael J. Dorff, Jane M. McDougall, James S. Rolf, Lisbeth E. Schaubroeck, Richard L. Stankewitz, and Kenneth Stephenson

Exploratory Examples for Real Analysis, Joanne E. Snow and Kirk E. Weller

Geometry From Africa: Mathematical and Educational Explorations, Paulus Gerdes

Historical Modules for the Teaching and Learning of Mathematics (CD), edited by Victor Katz and Karen Dee Michalowicz

Identification Numbers and Check Digit Schemes, Joseph Kirtland

Interdisciplinary Lively Application Projects, edited by Chris Arney

Inverse Problems: Activities for Undergraduates, Charles W. Groetsch

Keeping it R.E.A.L.: Research Experiences for All Learners, Carla D. Martin and Anthony Tongen

Laboratory Experiences in Group Theory, Ellen Maycock Parker

Learn from the Masters, Frank Swetz, John Fauvel, Otto Bekken, Bengt Johansson, and Victor Katz

Math Made Visual: Creating Images for Understanding Mathematics, Claudi Alsina and Roger B. Nelsen

Mathematics Galore!: The First Five Years of the St. Marks Institute of Mathematics, James Tanton

Methods for Euclidean Geometry, Owen Byer, Felix Lazebnik, Deirdre L. Smeltzer

Ordinary Differential Equations: A Brief Eclectic Tour, David A. Sánchez

Oval Track and Other Permutation Puzzles, John O. Kiltinen

A Primer of Abstract Mathematics, Robert B. Ash

MAA Service Center
P.O. Box 91112
Washington, DC 20090-1112
1-800-331-1MAA   FAX: 1-301-206-9789

Contents

3 Applications to Flow Problems   **161**
Michael Brilleslyper (text)
James S. Rolf (software)

4 Anamorphosis, Mapping Problems, and Harmonic Univalent Functions   **197**
Michael J. Dorff (text)
James S. Rolf (software)

5 Mappings to Polygonal Domains   **271**
Jane McDougall and Lisbeth Schaubroeck (text)
James S. Rolf (software)

Introduction

This book is written for undergraduate students who have studied some complex analysis and want to explore additional topics in the field. It could be used as

- a supplement for an undergraduate complex analysis course allowing students to explore a research topic;

- a guide for undergraduate research projects for advanced students;

- a resource for senior capstone courses; or

- a portal for the mathematically curious, a hands-on introduction to the beauties of complex analysis.

This book differs from other mathematics texts. It focuses on discovery, self-driven investigation, and creative problem posing. The goal is to inspire students to investigate, explore, form conjectures, and pursue mathematical ideas. Students are taken on a guided tour of the topics and are given many opportunities to pursue their own investigations.

Interlaced in the reading are exercises, explorations using computer applets, and projects. They are an essential part of the learning process. For this reason, most of them end with the phrase *Try it out!* to remind the student that the activity needs to be done before going on. Activities include:

Examples—Students should be sure that they can follow the arguments and provide small details when needed.

Exercises—These have a well-defined goal and should be done before going on to the next paragraph in the text. Skipping them would result in the reader missing a fundamental skill or idea.

Explorations—These also should be done before going on. Generally, they do not have a well-defined problem to solve. Some may include undirected investigating or playing with applets. There is no specific outcome expected, but much will be gained from engaging with the material. Such activities are at the heart of what this book is for: getting students to explore mathematics on their own.

Small Projects—These are optional activities that may take up to a few weeks to complete.

Large Projects—These could be a semester-long project, a capstone project, or an honors thesis.

Additional Exercises—Additional exercises may include any of the previous activities. They appear at the end of chapters and are optional.

During these activities, students should consider such questions as: *Why was this problem posed? Why is it interesting? If I changed the problem slightly, does it make it easier? harder? impossible? What does it say about the general theory?* Thinking about such questions is part of mathematics research and investigating the unknown. This can make for slow reading. Progress should be measured not by the number of pages read but by the amount of independent thought given to the material. If students read just a few pages of a chapter and then become motivated to work on a problem or set of problems devised on their own, the authors of this book would be delighted. As Albert Einstein said, quoting from Philipp Frank, *Einstein: His Life and Times*, Knopf, New York, 1947:

> It is not so very important for a person to learn facts. For that he does not really need a college. He can learn them from books. The value of an education in a liberal arts college is not learning of many facts but the training of the mind to think something that cannot be learned from textbooks.

This book delves into six current research topics, providing numerous directions for student investigation. Each topic is presented in a self-contained chapter with mathematical background, new material, exercises, explorations, problems suitable for projects, and computer software for exploring the topic. Two appendices provide background. Appendix A recaps key definitions and results from undergraduate complex variables, while Appendix B discusses the Riemann sphere. Results from appendices will be cited in various chapters, but students can take these as known and continue reading, or look to the appendix for further details. There is also an Index of Notation on page 365 for quick reference to symbols employed throughout.

Here are descriptions of the chapters in this book:

Chapter 1: Complex Dynamics   This chapter investigates chaos and fractals as they relate to dynamical systems that come from iterating complex-valued functions, i.e., given an initial value z_0 we consider the values $z_1 = f(z_0), z_2 = f(z_1) = f(f(z_0)), z_3 = f(z_2) = f(f(f(z_0))), \ldots$, and how the sequence $\{z_n\}$ behaves. This kind of iteration arises in Newton's method for approximating roots, and so our chapter begins by asking: *Which initial values will work for Newton's method (i.e., converge to a root)? If the initial value z_0 is changed slightly, will similar or drastically different behavior result?* The questions are considered computationally, visually, and experimentally with the aid of computer applets. We then consider the iteration of any complex analytic map, which leads to the mathematics behind the Mandelbrot set, and much more.

Chapter2: Soap Films, Differential Geometry, and Minimal Surfaces   Dipping a wire frame into soapy water produces an iridescent soap film that clings to the frame. Such soap films are related to minimal surfaces, which are beautiful geometric objects that minimize surface area locally. Visually, minimal surfaces can be thought of as saddle surfaces that bend upward in one direction in the same amount that they bend downward in the perpendicular direction. In this chapter, we present the necessary background from differential geometry, a field of mathematics in which the ideas and techniques of calculus are applied to geometric shapes, to give an introduction to minimal surfaces. Then we use ideas from complex analysis to present a nice way to describe minimal surfaces and to relate the geometry of the surface with this description. This allows us to begin investigating some of the interesting properties and new research questions that can be explored with the help of the applets.

Chapter 3: Applications to Flow Problems   Two dimensional vector fields are used to model and study a wide range of phenomena. Of particular interest are those that are irrotational and incompressible, which can be used to model the velocity of an ideal fluid flowing in a region or the electric field in a region free of charges. Modeling two dimensional fluid flow is a standard application of conformal mappings in complex variables. This chapter takes a geometric and visual approach to this topic and then extends it to several other applications. Fields of interest include those generated by combinations of sources and sinks that add or remove fluid from the flow. Throughout the chapter, we use examples, theory, and exercises to develop methods that allow these fields and many others to be modeled and analyzed. Also, the included applet *FlowTool* displays the streamlines and equipotential lines for a wide variety of fields, and permits real-time dynamical experimentation with sources and sinks in several pre-selected regions. Students with an interest in using technology to visualize mathematical objects will find many opportunities to explore their own ideas in extending this material.

Chapter 4: Anamorphosis, Mapping Problems, and Harmonic Univalent Functions Anamorphosis is a method of distorting an image that appears normal when viewed from a different perspective. Think of a warped mirror that makes you look taller and thinner or shorter and fatter. This method has been used in the past in painting and is still used today for dramatic effect. In complex analysis, analytic maps such as Möbius transformations distort images in a systematic way. We begin this chapter by discussing geometric properties of Möbius transformations and other analytic functions. But analytic maps have been around for a long time. Is there anything new to investigate? Yes, there is. The real and imaginary parts of analytic functions are harmonic and satisfy the Cauchy-Riemann equations. What happens if we remove the condition of satisfying the Cauchy-Riemann equations? We then create a new collection of functions known as complex-valued harmonic functions. Recently, there has been a resurgence in the study of one-to-one complex-valued harmonic functions also known as harmonic univalent functions. We will explore geometric properties of these functions and see some of the bizarre behaviors they have that form the basis of an exciting new area of research.

Chapter 5: Mappings to Polygonal Domains   A rich source of problems in analysis is determining when, and how, we can create a univalent (one-to-one) function from one region onto another. In this chapter, we consider the problem of mapping the unit disk onto a polygonal domain by two classes of functions, analytic and harmonic functions. For analytic functions we give an overview of the Schwarz-Christoffel transformation, that leads to some rich mathematics, the study of special functions. We then use the Poisson Integral Formula to find harmonic functions that map onto polygonal domains. Proving that they are univalent requires us to explore the theory of harmonic functions and uses some new techniques.

Chapter 6: Circle Packing   *Circle packings* are configurations of circles with prescribed patterns of tangency. They exist in quite amazing and often visually stunning variety, but what are they doing in a book on complex analysis? Complex analysis is, at its heart, a geometric topic. The reader will see this in the global geometry on display in Chapters 1–5, but the foundation lies at the local level where, as the saying goes, "analytic functions map infinitesimal circles to infinitesimal circles." In Chapter 6 this geometry will come to life in the theory of discrete analytic functions when we map actual circles to circles. Using the Java application *CirclePack*, we will create, manipulate, and display maps between circle packings that are the discrete analogues of familiar maps between plane domains, including some of those encountered earlier in the book. Direct access to the underlying geometry gives insight into fundamental topics like harmonic measure, extremal length, and branching. Moreover, we will see that our discrete functions not only mimic their continuous counterparts, but also converge to them under refinement. In short, Chapter 6 is about *quantum complex analysis*.

Acknowledgements

We sincerely thank all those colleagues and their students who read through early drafts of the manuscript and provided very valuable feedback. We will list these people individually with the chapters they reviewed. We also thank Melissa Mitchell who provided valuable comments on the Introduction and Appendices. Of course, any errors that remain are the full responsibility of the authors.

Further, we thank the National Science Foundation for their support of this project through grants No. DUE-0632969, No. DUE-0632976, No. DUE-0633125, and No. DUE-0633266.

We offer special acknowledgement and sincere gratitude to Jim Rolf. Jim has managed to design, code, and maintain fourteen separate applets for the first five chapters of this text. His work to keep things well organized while dealing with a seemingly unlimited stream of requests from so many authors has been astounding. His work turned out to be well beyond what any of us initially thought we would need or want from him, yet he adapted and produced professional applets tailored to our specific desires. Since active use of these applets is critical to this text, we especially thank Jim for his tremendous efforts which were most critical to improving the overall project.

Chapter 1: Complex Dynamics   As a text that hopes to inspire students by showing them the beauty of complex variables research, I would like to dedicate this chapter to the professors who have inspired me, namely, Juha Heinonen, Joe Miles, and Aimo Hinkkanen.

Also, I thank Neal Coleman, Stephanie Edwards, Dan Lithio, Kevin Pilgrim, and Irina Popovici, who provided very useful feedback after reading through early drafts of this chapter. I also thank Bob Devaney and Phil Rippon for helpful conversations where they shared ideas and insights, and gave encouragement.

Richard L. Stankewitz

Chapter 2: Soap Films, Differential Geometry, and Minimal Surfaces
Chapter 4: Anamorphosis, Mapping Problems, and Harmonic Univalent Functions
I would like to thank my colleagues Casey Douglas and Michelle Hackman who read through the chapter on minimal surfaces and provided me with very helpful feedback. Also, I would like to thank my 2009 and 2010 REU students who worked through both

chapters and gave me lots of advice from the students' perspective. These students include Valmir Bucaj, Sarah Cannon, Amanda Curtis, Sam Ferguson, Laura Graham, Jamal Lawson, Rachel Messick, Jessica Spicer, Ryan Viertel, and Melissa Yeung. Finally, I would like to thank Lonette Stoddard who used Adobe Acrobat to draw many of the images in the chapters.

Michael Dorff

Chapter 3: Applications to Flow Problems   I am particularly indebted to Harrison Potter who used ideas from an early draft of this work in writing his undergraduate honors thesis at Marietta College. He inspired several of the chapter's exercises and projects. I would also like to acknowledge Professor Tristan Needham and his beautiful text *Visual Complex Analysis,* Oxford University Press, 1997, for giving me a greater appreciation of the geometric aspects of complex analysis. I offer a special thank you to my colleagues Beth Schaubroeck and Jim Rolf, who agreed to participate in this crazy scheme. Finally, I wish to thank my co-authors for sticking with a project that lasted far longer than any of us thought it would.

Michael Brilleslyper

Chapter 5: Mappings to Polygonal Domains   We gratefully acknowledge the numerous students, colleagues, and reviewers who gave us insight into the material and the writing of this chapter. We especially thank Peter Duren, both for the inspiration through his work in harmonic functions, and also for detailed and helpful feedback on an earlier draft of the chapter. We would like to dedicate this work to our husbands and children, whose support and love made our efforts possible.

Jane McDougall and Lisbeth Schaubroeck

Chapter 6: Circle Packing   I would first thank William Thurston for bringing this wonderful topic into existence. Next, my PhD students, Tomasz Dubejko, G. Brock Williams, Matt Cathey, Gerald Orick, Chris Sass, and James Ashe, along with the undergraduate and REU students who have shared in circle packing discoveries. Finally, thanks to those who helped in programming *CirclePack*, especially Fedor Andreev, Sam Reynold, Benjamin Pack, Chris Sellers, and Alex Fawkes. I gratefully acknowledge support from the National Science Foundation over many years. I dedicate this chapter to my lovely wife Dee: thanks for your patience, dear.

Kenneth Stephenson

Using Java Applets

Software is essential in using this text. Java applets are provided to explore complex functions in Chapters 1–5, while the Java application *CirclePack* is provided to explore Chapter 6. All software provided for use with this text may be found at www.maa.org/ebooks/ EXCA/applets.html. The applets can be run over the internet in a web browser or downloaded onto your computer and run locally, while *CirclePack* must be downloaded and run locally.

We now demonstrate how to use the applet *ComplexTool*, which is used in several chapters and which shares many features with the other applets. We can graph the image of a variety of standard domains under numerous familiar functions. We begin by examining how to draw the image of the unit disk $\mathbb{D}$ under the function $f(z) = (1 + z)^2$. Open *ComplexTool* (see Figure 0.1).

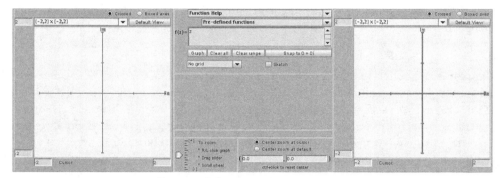

FIGURE 0.1. The applet *ComplexTool*

In the middle section near the top there is a box that has **f(z) =** before it. In this box, enter **(1+z)** ∧ **2**. (In general, you will enter functions into *ComplexTool* using the same syntax you would in a standard computer algebra system.) Below this, there is a window that reads **No grid**. Click on the down arrow ▼ and choose the option **Circular grid**; an image of a circular grid appears on the left to show the domain being graphed. Next, click on the button **Graph** which is in the middle section below the function you entered earlier. The image of the circular grid under the action of the function appears on the right. To change the size of the image window, click on the down arrow ▼ above the image and choose a different size, such as **Re: [−3,3] Im: [−3,3]**. You can also zoom in and out by left- or right-clicking either image, by dragging the slider in the bottom left of the center panel, or by turning the scroll wheel on the mouse. Also, you can move the axes so that the image is centered by positioning the cursor over the image and depressing the mouse button while dragging the image (see Figure 0.2). To go back to the original region you graphed, push the **Default View** button above either the domain or range. If you want to change the domain that is being graphed, you can do so manually by changing **Center**, **Interior circles**, **Rays**, the boxes surrounding θ, and the boxes surrounding **radius**. Alternatively, you can check the box that says **Vary** θ or **Vary radius** to see sliders that dynamically change the θ-range and radius range. (You may have to push the **Graph** button to regraph after you make changes.) You

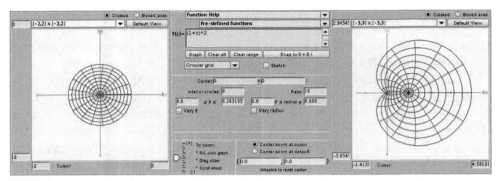

FIGURE 0.2. The image of the unit disk under the map $f(z) = (1 + z)^2$.

can also position the cursor over the domain and drag it around while watching the range dynamically change. This will not change the size of the domain, but will change its center. To go back to where you had originally centered the object, you can push the **Snap to** button.

To figure out exactly which points in the domain correspond to points in the range, you can check the box in front of **Sketch**. Then you can scribble in the domain window and see its image appear in the range. When the **Sketch** function is turned on, you cannot dynamically move the domain, as a mouse-click in the domain window is interpreted as sketching instead.

Exploration 0.1. Use *ComplexTool* to graph several functions. Some functions to try are:

(a) Graph the image of $\mathbb{D}$ under the function $f(z) = \frac{z}{1-z}$.

(b) Consider the two analytic functions $f_1(z) = 2z^2 - z$ and $f_2(z) = \frac{1}{2}z^2 - z$ each defined on $\mathbb{D}$. One is one-to-one while the other is not. Graph the image of $\mathbb{D}$ under the function $f_1(z)$ and under $f_2(z)$. Determine how you can tell by looking at the applet which function is one-to-one.

(c) Determine the smallest set of values that you can use for **theta** in a circular wedge domain to get an image of a complete disk under $f(z) = z^3$. (The **Vary** θ option is good to use for this exercise.)

(d) For the function $f(z) = e^z$, plot the images of vertical and horizontal lines. Explain your result mathematically.

(e) Determine if there exists a domain D so that the image of D under $f(z) = \log(z)$ covers the plane.

(f) Demonstrate the periodicity of $\sin(z)$ using *ComplexTool*. (Find a region to graph and drag it horizontally to watch its image change.)

Try it out!

In addition to the action of the applet, you can use the menus across the top to change settings. Perhaps most helpful is the option to export screen shots from the applet. The **Export Settings** menu allows you to choose which file type to export, and then you can choose to export the domain, range, or the whole screen. As you do the exercises and explorations in this text, you will find it helpful to document the work that you have done by taking hand-written or electronic notes.

Using Links in the Electronic Book

The electronic version of this book comes with links to assist with its use. Each applet or computer application is colored red and is linked directly to a web page where the software can be found. To illustrate, click on the red link *ComplexTool*. In addition to these links to software, there are links to each numbered referenced item, e.g., Definition A.1, Theorem A.8, and Bibliography item [1] but these links are not colored. Simply click on the number of the reference, e.g., "A.1." in the above Definition A.1., to be taken to the item.

When using Adobe Reader X, use the **Page Navigation** buttons called **Previous View** and **Next View** to return to the location in the text where the link was clicked. If you right-click on the tool bar, under **Page Navigation** you can select **Previous View** and **Next View** for these buttons to appear on the tool bar. Alternatively, click on the **VIEW** drop down menu, then click **Page Navigation**, then select **Previous View**. A backward arrow will then appear in the toolbar at the top. Repeat for **Next View** to add a forward arrow.

PDF readers other than Adobe Reader X will likely have buttons with similar functions. We encourage you to use these to make the best use of the links in this text.

Adobe Reader X can be downloaded for free from get.adobe.com/reader/.

Complex Dynamics: Chaos, Fractals, the Mandelbrot Set, and More

Richard L. Stankewitz (text and software design),
James S. Rolf (software coding and design)

1.1 Introduction

This chapter introduces *complex dynamics*, an area of mathematics that has inspired and continues to inspire research and experimentation. Its goal is not to give a comprehensive description of the topic, but to engage you with the general notions, questions, and techniques of the area and to encourage you to actively pose as well as pursue your own questions.

Dynamics, in general, is the study of mathematical systems that change over time, i.e., *dynamical* systems. For example, consider a Newtonian model for the motion of the planets in our solar system. Here the mathematical system is a collection of variables corresponding to the location and velocity of the planets relative to the sun, and this system changes over time according to Newton's laws of motion. We can describe the process by which the system evolves (i.e., the rules of how the system changes over time) by differential equations relating the system variables to each other and to Newton's laws of motion.

Many dynamical systems can be described similarly, for example the population of bacteria in a petri dish, the weather in Ann Arbor, Michigan (temperature, pressure, and wind velocity, to be more precise), the average temperature of Earth's atmosphere and oceans, and the flight of a paper airplane that you might toss across the room. The models have a set of system variables, and some rule or set of equations that describes how they change over time. Their values at time t is called the *state* of the system at time t.

Knowing the initial state of the system (e.g., today's location and velocity of each planet), we often try to analyze the equations that describe how the system variables change over time in order to answer such questions as: *What will the state of the system be tomorrow? next week? next year? one hundred years from now? Will the system in the long run settle into some sort of equilibrium? Will small changes or errors in our knowledge of the initial state only lead to small changes or errors in the system at some future time, or could they lead to huge changes?* Such natural questions about the future states of a dynamical system have led to results of practical importance, and to some beautiful mathematics.

1

The famous physicist Richard P. Feynman said ([15, p. 9]): "Physicists like to think that all you have to do is say, these are the conditions, now what happens next?" Unfortunately there are many systems which we cannot adequately solve for the purposes of making useful predictions, and so *What happens next?* is a question that we cannot answer. Frequently this is due to our inability to find the right pattern, or to solve some differential equation. Perhaps someone (maybe you!) will come along and find a clever solution, so we can then predict the future of these systems. But for a large class of systems there is no hope of ever being able to predict, with any useful accuracy, how it will behave. What's astounding is that, for these systems, it is not a matter of finding the right solution. In fact, we sometimes even have what we thought was a great solution, a formula even, but there is a problem with applying it.

The heart of the problem is that we have only solved a model, an approximation to the real world, and so there will be some error built into the model, which we hope is small. The problem is that: (1) we can never pinpoint the initial conditions EXACTLY, and (2) ANY approximation (or error) to the initial conditions leads to errors so large that we cannot have any confidence at all in applying our prediction to the real world. Such systems are called *chaotic*, a term you will explore and even be asked to mathematically define in this text. Though chaotic systems do not allow for precise answers to some of the questions scientists like to ask, much can be gained from studying them.

In this chapter we will study simple chaotic systems that can be analyzed using complex analysis. We do this because they are of interest in their own right, and because it will lead to understanding fundamental principles of complicated systems, like those mentioned above. The dynamical systems we consider are *discrete iterative* systems, which are in some sense the easiest. For them time is represented by a natural number n. Also, there is no need to solve a differential equation to determine the system's state; we just repeatedly apply a function. Furthermore, the states of our systems are described not by a large number of variables (as are needed to represent the positions and velocities of the planets), but just a single complex variable. The system is called iterative because the state z_n of the system at time n will evolve according to the rule $z_{n+1} = f(z_n)$ where f is a complex-valued function. Thus, given an initial state, computing the future state z_n is just a simple matter of computing the values $z_1 = f(z_0), z_2 = f(z_1) = f(f(z_0)), z_3 = f(z_2) = f(f(f(z_0)))$, and so on. However, predicting the behavior of the sequence of states $\{z_n\}$ (e.g., deciding if $\{z_n\}$ converges or not), an altogether different problem, is by no means simple!

This chapter uses tools from complex function theory to investigate several types of discrete iterative systems, including Newton's method, polynomial iteration, exponential iteration, and trigonometric iteration. We also consider what happens when these systems are perturbed by changing a parameter, leading us into bifurcation theory. There are many ways to perturb a system and in the concluding section we describe two more: perturbation with a pole and random dynamics. We begin with perhaps the most familiar discrete dynamical system, Newton's method.

Although we are not studying real world dynamical systems directly, we should keep in mind that they exhibit many of the same behaviors as the systems we do study.

How to Use this Chapter

The sections of this chapter can be worked through in order. However, to proceed to Section 1.3 quickly, Section 1.2 may be skipped, with the exceptions of Sections 1.2.3 and 1.2.6. Also, Section 1.6 may be skipped by anyone not wishing to investigate the dynamics of transcendental entire maps. Reading the entire chapter and working on several additional exercises as well as projects would suit a three credit hour 15-week semester reading course. A 2-3 week group or individual project for the end of the semester in an undergraduate complex variables course is provided by either Section 1.2 or Sections 1.2.3 and 1.2.6 together with Section 1.3.

Because the natural setting for this chapter is the Riemann sphere all of Appendix B should be worked through, though it is not necessary that it be completed in full before starting. There are also three chapter appendices that begin on page 78 providing added information relevant to the chapter. Results from the appendices, when referenced, may be accepted as stated without interrupting your reading, or their details may be investigated in the appendix.

In addition to *ComplexTool*, introduced on page xvi, this chapter will utilize the following applets which can be accessed at www.maa.org/ebooks/EXCA/applets.html.

1. *Real Newton Method Applet*, to visualize the real-valued Newton's method.

2. *Complex Newton Method Applet*, to visualize the complex-valued Newton's method.

3. *Real Function Iterator Applet*, to iterate a real function, and see its orbit displayed as a numerical list and as points on a number line.

4. *Complex Function Iterator Applet*, to iterate any complex function, and see its orbit displayed as a numerical list and as points in the complex plane.

5. *Cubic Polynomial Complex Newton Method Applet*, to explore the attracting basins for Newton's method applied to the cubic polynomials $p_\rho(z) = z(z - 1)(z - \rho)$.

6. *Global Complex Iteration Applet for Polynomials*, to draw the basin of infinity for a polynomial.

7. *Mandelbrot Set Builder Applet*, to illustrate how the Mandelbrot set is constructed.

8. *Parameter Plane and Julia Set Applet*, to investigate the parameter plane and dynamic plane pictures for the families of functions $z^2 + c$, $z^d + c$, ce^z, $c\sin(z)$, $c\cos(z)$, and $z^d + c/z^m$.

1.2 Newton's Method

Solving equations, finding solutions to ordinary differential equations, finding eigenvalues of a matrix: all are important mathematical procedures. However, they cannot always be done *exactly* (by computing a value exactly we mean being able to express it in terms of standard mathematical operations and functions, e.g., $\sqrt{\cos(\pi/12)}$). When exactness is not possible, often the solution can be approximated with an iterative numerical method.

For example, consider the problem of finding a *root* of a complex-valued function $f(z)$, i.e., a value α such that $f(\alpha) = 0$. If the function is the quadratic $f(z) = az^2 + bz + c$ where $a \neq 0$, then there are two roots given by the quadratic formula $\alpha = (-b \pm \sqrt{b^2 - 4ac})/2a$. If the function is a cubic or quartic polynomial, then there exists a formula (or, more precisely, procedure) for finding the roots exactly. However, if $f(z)$ is a quintic polynomial, then there is not, in general, a procedure that will exactly find any of its roots. (Even though any polynomial of degree n has, by the fundamental theorem of algebra, n roots in the complex plane, this theorem does not help us to actually find them.) The same is true for many transcendental functions such as $h(z) = \cos z - z$ and $g(z) = e^z - 4z$. In such cases we must give up on finding exact roots and resort to approximation.

For real-valued functions of a real variable, there are approximation methods for root finding based on the intermediate value theorem, such as the bisection method. When considering a complex-valued function f, we can sometimes approximate a root of f using Rouche's theorem (see [1, p. 294]). Specifically, if f and g are analytic on and inside a simple closed curve C with $|f| > |g|$ on C, then f and $f + g$ have the same number of zeroes (counting multiplicities) inside C. Hence, if g can be chosen such that $f + g$ has a root inside of C, then so must f. However, even if f is a nice function (e.g., a polynomial), this can be a difficult method to implement. So we look for a better method. One of the best is Newton's method, for finding both real and complex roots. Often it allows us to quickly approximate roots with extreme accuracy, if we have access to a computer. In this section we examine Newton's method for both real and complex functions, with the goal of understanding when it succeeds and when it fails.

1.2.1   Real Newton's Method

If we seek to find a root α of a differentiable real-valued function $f(x)$ defined for a real variable x, then we can apply Newton's method as follows. We start with an initial guess x_0 close to α and define

$$x_1 = x_0 - \frac{f(x_0)}{f'(x_0)}, \qquad x_2 = x_1 - \frac{f(x_1)}{f'(x_1)}, \qquad x_3 = x_2 - \frac{f(x_2)}{f'(x_2)},$$

and, in general,

$$x_{n+1} = x_n - \frac{f(x_n)}{f'(x_n)}.$$

The geometry behind Newton's method, as illustrated in Figure 1.1, is as follows: Given an approximation x_0 to the root α, we considers a linear function $\tilde{f}$ that approximates f near x_0. The best linear approximation is the first order approximation $\tilde{f}(x) = f(x_0) + f'(x_0)(x - x_0)$ whose graph is the tangent line L to the graph of f at x_0. The root of $\tilde{f}(x)$, i.e., the x-intercept of L, is then the definition of x_1.

Exercise 1.1. Before reading further, compute $\tilde{f}(x)$ for $f(x) = x^3 - 2x$ and $x_0 = 2$ as in Figure 1.1. Then use $\tilde{f}(x)$ to find the equation of the tangent line L and also check that x_1, given by the above formula, is the x-intercept of L. ***Try it out!***

Exercise 1.2. Verify that for general $f(x)$, the function $\tilde{f}(x)$ has a root at x_1 where $\tilde{f}$ and x_1 are as given above. ***Try it out!***

In general (as illustrated in Figure 1.1), we expect that the root x_1 of $\tilde{f}(x)$ will be a better approximation to α, the sought-after root of $f(x)$, than the initial guess x_0. This process is repeated using x_1 as the initial guess to generate a new approximation x_2. We

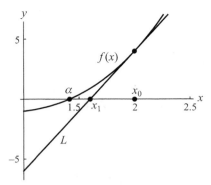

FIGURE 1.1. An illustration of the first step in Newton's method where $f(x) = x^3 - 2x$ and $x_0 = 2$. Newton's method can be described by saying from x_0 move to the graph of $f(x)$, slide along the tangent to the x-axis, and repeat.

then iterate to generate successive approximations x_n, for $n = 0, 1, 2, \ldots$. We express this in terms of iteration of the following function.

Definition 1.3. The function $F(x) = x - \dfrac{f(x)}{f'(x)}$ is called the *Newton map for* f.

Thus, the *orbit*, i.e., sequence of iterates $F(x_0), F(F(x_0)), F(F(F(x_0))), \ldots$, is the same as the sequence $\{x_n\}$ generated above, and it will be proven (in Proposition 1.20) that x_n converges to the root α whenever our original guess x_0 is close enough to α. What it means to be close enough is important, and we will come back to this in Section 1.2.4, but we first explore Newton's method with some examples.

Example 1.4. Consider $f(x) = x^2 - 3x + 2 = (x - 1)(x - 2)$, which has roots at 1 and at 2. Let's apply Newton's method. The Newton Map for f is

$$F(x) = x - \frac{x^2 - 3x + 2}{2x - 3} = \frac{x^2 - 2}{2x - 3}.$$

Making an initial guess $x_0 = 0.5$ and using your calculator (do this now) you can compute $x_1 = F(x_0) = 0.875, x_2 = F(F(x_0)) = F(x_1) = 0.9875$, and $x_3 = F(F(F(x_0))) = F(x_2) = 0.99984756$. Since using a calculator is drudgery and computers are efficient at such tasks, we have created the *Real Newton Method Applet* for you to use. Use this now to confirm the calculations above and then, by taking many iterates of the Newton map using this applet, convince yourself that with the *initial value* (a term we use interchangeably with *seed value* and *starting point*) $x_0 = 0.5$, we have $x_n \to 1$. *Try it out!*

Now use the *Real Newton Method Applet* to determine the behavior using an initial guess $x_0 = 3$. *Try it out!*

What you learned above is that some initial guesses for x_0 find the root 1 (i.e., have the corresponding $\{x_n\}$ converge to 1) while others find the root 2. This suggests the question: *Given an initial guess x_0, how do we know which root it will find?*

Exploration 1.5. Make a prediction about which seed values for x_0 in Example 1.4 will find the root 1 and which will find the root 2. Are there seed values for which Newton's method fails to find either root? Experiment with the *Real Newton Method Applet* to test your prediction. *Try it out!*

Exploration 1.6. Analyze Newton's method for $f(x) = x(x - 1)(x + 1)$ using different initial guesses for x_0 in the *Real Newton Method Applet*. Describe (as best as you can) the set of seed values for x_0 which find the root -1, and then do the same for the roots 0 and 1. *Try it out!*

1.2.2   Global Picture of Real Newton Method Dynamics

Identifying how the orbits of all seed values behave, as you attempted in Explorations 1.5 and 1.6, is what we mean by looking at the global dynamics of Newton's method. In some cases this can be done without too much work, but in other cases it can be complicated. To help with this problem, we use the **Graph basins of attraction** feature in the *Real Newton Method Applet* to illustrate which initial guesses will find which roots of $f(x)$ when Newton's method is applied.

Exercise 1.7. For each of Explorations 1.5 and 1.6, use the *Real Newton Method Applet* to get a snapshot of what the dynamics of Newton's method are for all starting values. *Try it out!*

What you see when using the *Real Newton Method Applet* with $f(x) = (x - 1)(x - 2)$ is what you might expect. The initial values that are closer to the root at 1 will find 1, and those that are closer to the root at 2 will find 2. One should also check that Newton's method fails for the initial value $x_0 = 1.5$. Analytically we see this in the formula because $f'(1.5) = 0$ leads to a zero in the denominator in the calculation of x_1. Geometrically we see this because the tangent line at $x_0 = 1.5$ is horizontal, never crosses the x-axis, and thus does not determine a value for x_1. We can also understand this dynamically, at least in a heuristic way. The value 1.5 separates those values that are pulled or attracted to 1 and those that are attracted to 2, and so by an informal use of symmetry, it would seem that something must fail to work out at $x_0 = 1.5$.

Though this reasoning is informal, it captures an important idea. As a rule, we encourage you to use and create your own heuristic ideas to explain or describe mathematics. Sometimes it's hard to be formal with all of your mathematical ideas. But don't let that stop you from thinking of and sharing great mathematical thoughts, even if you can't make them precise or formal. Some of the best mathematics, if not all mathematics, starts off as raw unformed ideas with no foundation in formalism. Later, you can (and should) try to be formal with you ideas.

The case when $f(x) = x(x - 1)(x + 1)$ is much different from the quadratic case. A seed for Newton's method will not always find the root it is nearest to. For example, $x_0 = 0.55$ will find the root at -1, even though it is closer to the roots at 0 and 1. Indeed, in this case the set of initial guesses on the real line is divided into intricate regions that find the roots. In fact, if you zoom in near the point $x = 0.4472135951871958$ you will see a cascade of ever shrinking and alternating colored intervals of blue and turquoise (see Figure 1.2). It turns out that this pattern goes on forever. Use the **Zoom** feature of the *Real Newton Method Applet* to observe this.

Small Project 1.8. Prove the existence of the infinite cascade of ever shrinking and alternating colored intervals of blue and turquoise found in Figure 1.2. Hint: First use the applet to understand what each of the colored intervals means dynamically, and then try to give a proof for what you observe.

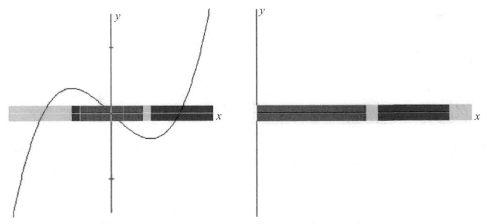

FIGURE 1.2. On the left is a cascade of intervals in the global picture of Newton method dynamics for $f(x) = x(x - 1)(x + 1)$, and on the right is a magnification centered at $x = 0.4472135951871958$.

We have seen that Newton's method gets really complicated to understand globally when we switch from a quadratic to a cubic, as from $f(x) = (x - 1)(x - 2)$ to $f(x) = x(x - 1)(x + 1)$. This suggests the questions: *Why? How? Is there a way to know ahead of time when a system will be simple or complicated?*

Answers will come from taking the advice of Jacques Hadamard who said, "The shortest path between two truths in the real domain passes through the complex domain." So let's look at Newton's method applied to complex-valued functions of a complex variable. There are many wonderful theorems at our disposal when we consider complex analytic maps instead of mere real-valued differentiable maps. Let's take advantage of them.

Before we investigate Newton's method applied to complex-valued functions, however, we first develop some concepts that we will need to understand any iterative dynamics.

1.2.3   Orbits, Examples, and Fixed Points

The general questions we consider in iterative dynamical systems concern describing and predicting orbits. Let g be a function mapping its domain set domain(g) into itself (the *domain set* of a function g is the set of its possible inputs), which we take to be a subset of the Riemann sphere $\overline{\mathbb{C}}$. (The Riemann sphere is the natural setting for this chapter and so familiarity with the material in Appendix B starting on page 357 is crucial.) We denote the nth iterate of g by g^n. Thus $g^n(z) = (g \circ \cdots \circ g)(z)$ where g is applied n times, e.g., $g^3(z) = g(g(g(z)))$. In this chapter $g^3(z)$ does not denote the value $g(z)$ raised to the third power, which would instead be denoted $[g(z)]^3$. We define g^0 to be the identity map, i.e., $g^0(z) = z$. Furthermore, for any starting (seed) value $z_0 \in$ domain(g), the sequence of values $z_n = g^n(z_0)$, for $n = 1, 2, 3, \ldots$, is called the *orbit* of z_0 under the map g.

When trying to predict the behavior of the evolution of a seed value, we ask: *Does the orbit converge, fall into a repeating pattern, or show no signs of following any pattern? What happens for different starting values z_0? Do we get the same (or even similar) behavior if we choose starting values near z_0?* Keep these questions in mind as you consider the following examples.

Example 1.9. Let $f(x) = e^x$ for all $x \in \mathbb{R}$. Thus the iterates are $f^0(x) = x$, $f^1(x) = e^x$, $f^2(x) = e^{e^x}$, $f^3(x) = e^{e^{e^x}}$, and so on. Experiment with the *Real Function Iterator Applet* to convince yourself that $f^n(x) \to +\infty$ no matter what x we start with. Additional Exercise 1.157 asks for a formal proof.

Example 1.10. Let $f(x) = \sin x$ where $x \in \mathbb{R}$ is given in radians. Experiment with the *Real Function Iterator Applet* to convince yourself that $f^n(x) \to 0$ for any real number x. Additional Exercise 1.158 asks for a formal proof.

Example 1.11. Let $f(x) = \cos x$ where $x \in \mathbb{R}$ is given in radians. Experiment with the *Real Function Iterator Applet* to convince yourself that $f^n(x) \to 0.739085\ldots$. for any real number x. Additional Exercise 1.159 asks for a formal proof.

Example 1.12. Let $f(x) = x^2 - 1$ and $x_0 = 0.9$. Use the *Real Function Iterator Applet* to convince yourself that the tail end of the orbit $x_n = f^n(0.9)$ appears to oscillate back and forth between 0 and -1.

Example 1.13. Let $f(x) = 4x(1 - x)$ and $x_0 = 0.2$. Use the *Real Function Iterator Applet* to see that the orbit $x_n = f^n(0.2)$ appears to have no pattern to it at all, even after the first 25,000 orbit points are plotted. Zoom in on the orbit points to see how it appears that they are *dense* in the interval $[0, 1]$, that is, for every open interval (a, b) that meets $[0, 1]$, there is some orbit point $x_n \in (a, b)$.

Though many types of behavior can be exhibited in orbits, we focus on one for the moment. In Examples 1.9, 1.10, and 1.11, we see that there is a point that attracts the orbits of many seeds. This motivates the following fundamental definition.

Definition 1.14 (Attracting Basin). Let $w \in \overline{\mathbb{C}}$. For a complex-valued function g mapping its domain set into itself, we define the *basin of attraction* of w (also called the *attracting basin*) under the function g to be the set $A_g(w)$ of seed values whose orbits approach the point w, i.e., $A_g(w) = \{z \in \mathrm{domain}(g) : g^n(z) \to w\}$.

The point w in this definition does not have to lie in $\mathrm{domain}(g)$ (as in Example 1.9). However, if $w \in \mathrm{domain}(g)$ and g is continuous, as in Examples 1.10 and 1.11, then Theorem 1.15 shows that w must necessarily be a *fixed point* of g, i.e., $g(w) = w$, whenever $A_g(w)$ is non-empty.

Theorem 1.15. *Let $f : domain(f) \to domain(f)$ be a continuous map where $domain(f)$ is contained in $\overline{\mathbb{C}}$. Suppose a and x_0 are in $domain(f)$ and $f^n(x_0) \to a$. Then $f(a) = a$.*

Proof. Since the sequence $f^n(x_0) \to a$ and f is continuous at a, we have $f(a) = f(\lim_{n \to \infty} f^n(x_0)) = \lim_{n \to \infty} f(f^n(x_0)) = \lim_{n \to \infty} f^{n+1}(x_0) = a$. $\square$

Fixed points play a major role in dynamical systems and so we will pay special attention to them whenever they arise. In our previous examples the roots of f appear to be fixed points of the corresponding Newton map $F(z)$, a fact we prove in Proposition 1.20. Additional Exercise 1.161 will shed some light on the extent to which the converse holds.

We call the fixed points in Examples 1.10 and 1.11 *attracting* fixed points because the orbits of seed values near the respective fixed points will converge to them. To be more precise we give a formal definition, but first we give a reminder about a key relationship between the Euclidean metric on $\mathbb{C}$ and spherical metric σ on $\overline{\mathbb{C}}$, which is stated in Appendix Proposition B.7 on page 361. Namely, for points $z, w \in \mathbb{C}$, we have $|z| > |w|$ if and only if $\sigma(z, \infty) < \sigma(w, \infty)$.

Definition 1.16 (Attracting Fixed Point). Let f be a map from its domain set $\Omega \subseteq \overline{\mathbb{C}}$ into itself (Ω could be a subset of $\mathbb{R}$).

(a) We call a (finite) fixed point $a \in \mathbb{C}$ an *attracting fixed point* of f if there exists a neighborhood U of a such that for any point $z \in \Omega \cap U \setminus \{a\}$, we have $|f(z) - a| < |z - a|$, i.e., the action of f is to move each point in $\Omega \cap U \setminus \{a\}$ closer to a.

(b) Suppose $f(\infty) = \infty$. We call ∞ an *attracting fixed point* of f if there exists a neighborhood $U \subseteq \overline{\mathbb{C}}$ of ∞ such that for any point $z \in \Omega \cap U \setminus \{\infty\}$, we have $|f(z)| > |z|$, i.e., the action of f is to move each point in $\Omega \cap U \setminus \{\infty\}$ closer to ∞, as measured by the spherical metric.

In Definition 1.16(a) we used the Euclidean metric to describe when the action of f moves points closer to a, but we could have equivalently used the spherical metric by writing $\sigma(f(z), a) < \sigma(z, a)$. It is important to become comfortable with understanding when a particular metric used in a definition or result could equivalently be changed to another. We often use the simplest metric at our disposal, assuming it is clear when another metric could also be used. A good way to try to become comfortable with this is to consider how sets appear when visualized in the flat plane $\mathbb{C}$ and in the sphere $\overline{\mathbb{C}}$. An exposition of this is in Appendix B on page 357.

If a is an attracting fixed point of a continuous map f, then there exists a neighborhood U of a such that $U \subseteq A_f(a)$. The proof of this does not require that f be (real or complex) differentiable at a, but without a differentiability condition the proof is more technical. Since we are interested only in specific differentiable functions in this chapter, we only provide the following.

Theorem 1.17. *Suppose $\Omega \subseteq \mathbb{R}$ or $\Omega \subseteq \mathbb{C}$. Let $f : \Omega \to \Omega$ be such that $f(a) = a$ and $|f'(a)| < 1$. Then a is an attracting fixed point of f. Furthermore, there exists some $\varepsilon > 0$ such that $\triangle(a, \varepsilon) \cap \Omega \subseteq A_f(a)$, where $\triangle(z_0, r) = \{z \in \mathbb{C} : |z - z_0| < r\}$ is the open Euclidean disk with center z_0 and radius r.*

The proof applies to both cases $\Omega \subseteq \mathbb{R}$ or $\Omega \subseteq \mathbb{C}$, where f' denotes, respectively, the real or complex derivative.

Proof. Since $|f'(a)| < 1$ we may select β such that $|f'(a)| < \beta < 1$. Since, by definition, $f'(a) = \lim_{z \to a} \frac{f(z) - f(a)}{z - a}$, there exists $\varepsilon > 0$ such that for any $z \in \Omega \setminus \{a\}$ for which $|z - a| < \varepsilon$, we have

$$\left| \frac{f(z) - a}{z - a} \right| = \left| \frac{f(z) - f(a)}{z - a} \right| < \beta.$$

Hence for $z \in \Omega$ with $|z - a| < \varepsilon$, we know that $|f(z) - a| \le \beta |z - a| < \varepsilon$. This says that for points z near a (within a distance of ε), the function f moves z closer to a by a factor of at least $\beta < 1$. Hence a is an attracting fixed point by Definition 1.16. If we iterate the map f at z, generating the orbit of z, we know that each application of f takes each orbit point a step closer to a, by a factor of β. Hence induction shows that $|f^n(z) - a| \leq \beta^n |z - a| \leq \beta^n \varepsilon \to 0$ whenever $z \in \Omega$ with $|z - a| < \varepsilon$. Thus $\triangle(a, \varepsilon) \cap \Omega \subseteq A_f(a)$. □

Remark 1.18. The smaller the value of $|f'(a)|$ is (and hence the smaller the value of β may be chosen) in the theorem, the faster the convergence is. In particular, if $f(a) = a$ and $f'(a) = 0$, then β can be taken to be extremely small leading to very fast convergence, and so such a fixed point a is called *super attracting*.

Example 1.19. Use the *Real Function Iterator Applet* to compare the rates of convergence given by the following maps. For $f(x) = \sin x$ consider the rate at which $f^n(\frac{1}{2})$ converges to 0. For $g(x) = x^2$ consider the rate at which $g^n(\frac{1}{2})$ converges to 0. For $h(x) = \frac{1}{2}x$ consider the rate at which $h^n(\frac{1}{2})$ converges to 0. Compare the absolute value of the derivative at each function's fixed point and note the correspondence with Remark 1.18. ***Try it out!***

By Theorem 1.17, to check if a fixed point is attracting, you can calculate the absolute value (or modulus) of the derivative at the fixed point. In Example 1.11, we see that $|f'(0.739085....)| < 1$, which proves that the fixed point is attracting. When considering the Newton's method dynamics, it appears in the cases we explored experimentally, that the roots of f are attracting fixed points for the Newton map F. This is the case, and we will prove it in Proposition 1.20 by showing that $|F'| < 1$ at each root of f. But before we do, we give a word of caution.

We must be careful using Theorem 1.17 since it is not an if and only if statement. Consider the fixed point $a = 0$ for the map $f(x) = \sin x$ in Example 1.10. Here $f'(0) = 1$, but this function as a real map, defined only for all real numbers x, does have a genuine attracting fixed point at $a = 0$. However, if we consider the complex map $g(z) = \sin z$, defined for all complex numbers z, the fixed point $a = 0$ is no longer attracting. Indeed, as you are asked to prove in Additional Exercise 1.160, $g^n(\pm i\varepsilon) \to \infty$ for any $\varepsilon > 0$, which you can illustrate using the *Complex Function Iterator Applet*. ***Try it out!***

1.2.4   Complex Newton's Method

We now investigate Newton's method when we allow our variables and output values to be complex. As we saw in the examples, our experimentation with the applets suggest that the Newton map always has an attracting fixed point at each root of f. We can now state and prove this fact in the real and complex cases.

Recall that a real-valued function of a real variable is said to be *real analytic* if it possesses derivatives of all orders and agrees with its Taylor series in a neighborhood of every point in its domain set.

Proposition 1.20 (Attracting Property of Newton's Method). *Given a non-constant real or complex analytic function f with a root at $\alpha \in \mathbb{C}$, the point α is an attracting fixed point of the Newton map F and thus there exists $r > 0$ such that all points within a distance r of α are in the attracting basin $A_F(\alpha)$.*

Put another way, the proposition states that for initial values z_0 that are close enough to α (within a distance r) the successive approximations $F^n(z_0)$ converge to α. That is, starting from such a z_0 and defining $z_{n+1} = z_n - \frac{f(z_n)}{f'(z_n)}$, we have $z_n \to \alpha$.

Proof. We consider the case that $f(z)$ is complex analytic. The same proof applies when f is real analytic. From Theorem 1.17, it suffices to show that for $F(z) = z - \frac{f(z)}{f'(z)}$ we have $F(\alpha) = \alpha$ and $|F'(\alpha)| < 1$. We note that $F'(z) = \frac{f''(z)f(z)}{(f'(z))^2}$.

Since $f(\alpha) = 0$, we apply Appendix A Lemma A.22 on page 354 to express $f(z) = (z - \alpha)^k h(z)$ where $h(\alpha) \neq 0$ and $k \in \mathbb{N}$ is the multiplicity of the root. If $f'(\alpha) = 0$, the quotient $\frac{f(z)}{f'(z)}$ appearing in the definition of $F(z)$ is not formally defined at α. However, $\frac{f(z)}{f'(z)}$ has a removable singularity at α because

$$\frac{f(z)}{f'(z)} = \frac{(z-\alpha)^k h(z)}{k(z-\alpha)^{k-1}h(z) + (z-\alpha)^k h'(z)} = \frac{(z-\alpha)h(z)}{kh(z) + (z-\alpha)h'(z)},$$

which equals 0 for $z = \alpha$. This lets us define $F(\alpha) = \alpha$ whether $f'(\alpha) = 0$ or not.

Exercise 1.21 asks you to show that $F'(z)$, which has a removable singularity at α when $f'(\alpha) = 0$, gives $F'(\alpha) = \frac{k-1}{k}$. Thus $|F'(\alpha)| < 1$, which completes the proof. $\square$

Exercise 1.21. Provide the details in the proof that $F'(\alpha) = \frac{k-1}{k}$ when f has a root of order k. Also, note that α is a super attracting fixed point of the Newton map F if and only if $k = 1$, i.e., α is a simple root of f. *Try it out!*

It is interesting to note that the Newton map of $f(z)/f'(z)$ (as opposed to the Newton map of $f(z)$) always has super attracting fixed points at the roots of f regardless of the order of the root of f. You are asked to show this in Additional Exercise 1.162.

Exploration 1.22 (Convergence Rates for Newton's method). Let $f(z)$ and $g(z)$ be analytic, with Newton maps $F_f(z) = z - \frac{f(z)}{f'(z)}$ and $F_g(z) = z - \frac{g(z)}{g'(z)}$. Suppose f has a root at α of order k and g has a root at α of order m, where $k < m$. The rate of convergence of F_f^n near α is faster than the rate of convergence of F_g^n near β since $|F_f'(\alpha)| = \frac{k-1}{k} < \frac{m-1}{m} = |F_g'(\alpha)|$ (Remark 1.18 relates the rate of convergence to the derivative at the attracting fixed point). Let's explore this in the real variable case with $f(x) = x^k$ and $g(x) = x^m$. Use the *Real Newton Method Applet* to visualize how the order of the root of f (respectively, g) influences the tangent lines used in Newton's method. Note the effect the order of the root has on the curvature of the graph near the root and how this provides a visual way to understand the relative rates of convergence of F_f^n and F_g^n near α. *Try it out!*

Remark 1.23. We see that the value r in Proposition 1.20 gives us a lower bound on how close an initial guess z_0 must be to a root α for Newton's method to be guaranteed to find α. Because of this, we call any such r a *radius of convergence* (for F and α) and call the corresponding circle $C(\alpha, r) = \{z : |z - \alpha| = r\}$ a *circle of convergence*. In practice, it is useful to gauge r so that we can guarantee the success of Newton's method. In Additional Exercise 1.163 you are asked to show that there is no universal estimate for r that always works since it very much depends on the particular map f. However, for certain classes of maps f, useable estimates for r can be found and you can explore these in Additional Exercise 1.164 and Small Projects 1.165 and 1.166.

Let us now consider some examples using complex functions.

Example 1.24. For constants $\alpha, \beta \in \mathbb{C}$, let $f(z) = (z - \alpha)(z - \beta)$, with roots at α and β. The Newton map for f is $F(z) = \frac{z^2 - \alpha\beta}{2z - (\alpha + \beta)}$.

Exploration 1.25. Set $\alpha = 0$ and $\beta = 1 + i$ in Example 1.24 and iterate the Newton map F with starting values $z_0 = 2$, $z_0 = -3 - 2i$, and $z_0 = i + 1$, which you can compute using the *Complex Newton Method Applet*. As expected, different starting values for z_0 find different roots of $f(z)$. Can you make a guess as to which seed values will find which root? Try to determine if there are any seed values z_0 for which Newton's method fails to find either root. Experiment with the *Complex Newton Method Applet* to test your predictions. ***Try it out!***

Exploration 1.26. Using $f(z) = z^3 - 1$, determine the Newton map for f and then analyze Newton's method using different initial guesses in the *Complex Newton Method Applet*. Describe (as best as you can) which seed values will find which of the roots 1, $e^{2\pi i/3}$, and $e^{-2\pi i/3}$. We are asking you to describe (as best as you can) the basins of attraction $A_F(1)$, $A_F(e^{2\pi i/3})$, and $A_F(e^{-2\pi i/3})$. ***Try it out!***

1.2.5 Global Picture of Complex Newton Method Dynamics

As with the real-valued maps, we wish to understand the global dynamics of Newton's method, i.e., how the orbits behave for all seed values. We employ the **Graph basins of attraction** feature of the *Complex Newton Method Applet*, which uses different colors to display which initial guesses will find which roots.

Exercise 1.27. Use the *Complex Newton Method Applet* to view the basins of attraction for the Newton maps in Explorations 1.25 and 1.26. ***Try it out!***

Looking at the picture of the two attracting basins corresponding to Exploration 1.25, it appears that the boundary between them is a straight line; points on one side look closer to the root of f on that side than to the other root. That is, this boundary appears to be the perpendicular bisector (denoted by L in Figure 1.3) of the line segment from α to β. This is true (pictures, however, can sometimes be misleading) and we can prove it using the technique of *global conjugation*.

1.2.6 Global Conjugation

We learn in linear algebra that a change of basis can facilitate calculations and procedures, and lead to better understanding. Similarity of matrices plays a key role (matrices A and B are called *similar* when there is an invertible matrix P such that $A = PBP^{-1}$). An analog often used in dynamics is a type of change of coordinates provided by what we call conjugation.

Definition 1.28. Let ϕ be a Möbius map, i.e., $\phi(z) = \frac{az+b}{cz+d}$ where $ad - bc \neq 0$, noting it is a bijection of $\overline{\mathbb{C}}$ onto itself. When f and g are rational maps (quotients of two polynomials) such that $g = \phi \circ f \circ \phi^{-1}$, we say f and g are *globally conjugate* by the map ϕ.

Often the point of conjugating a map f to a map g is that g is easier to work with than f. And, as we will see, the information we usually want from f can be obtained from the simpler map g. Iterates are related as such, $g^n = (\phi \circ f \circ \phi^{-1}) \circ (\phi \circ f \circ \phi^{-1}) \circ \cdots \circ (\phi \circ f \circ \phi^{-1}) \circ (\phi \circ f \circ \phi^{-1}) = \phi \circ f^n \circ \phi^{-1}$. Hence ϕ transfers information between the iterates of f and the iterates of g. In particular, fixed points and their derivatives are transferred in the following way.

Exercise 1.29. Suppose maps f and g are globally conjugate by the Möbius map ϕ, i.e., $g = \phi \circ f \circ \phi^{-1}$. Prove that for $a \in \overline{\mathbb{C}}$, $f(a) = a$ if and only if $g(\phi(a)) = \phi(a)$. Furthermore prove that if $a, \phi(a) \in \mathbb{C}$ and $f(a) = a$, then we have $f'(a) = g'(\phi(a))$. *Try it out!*

Additional Exercises 1.167–1.170 explore several examples of global conjugation.

Remark 1.30. There is also a very useful technique called *local conjugation* that can simplify calculations. In fact, an important problem is determining for which maps of the form $f(z) = a_1 z + a_2 z^2 + \ldots$, having a fixed point at 0, can there be found be an analytic map ϕ defined near 0 such that $\phi \circ f \circ \phi^{-1}$ is simply $z \mapsto a_1 z$. This is called *linearizing* the map f near 0. It can always be done when $0 < |a_1| \neq 1$, but only sometimes when $|a_1| = 1$. Such results and their proofs can be found in the literature (e.g., [1, 3, 24]).

1.2.7 Newton Map of a Quadratic Polynomial

We now use global conjugation to simplify the analysis of the Newton map $F(z) = \frac{z^2 - \alpha\beta}{2z - (\alpha + \beta)}$ in Example 1.24. We first choose a Möbius map that sends α and β to 0 and ∞, respectively, and then analyze the simpler map obtained by conjugation. In particular, the map $\phi(z) = \frac{z - \alpha}{z - \beta}$ conjugates F to $g(z) = \phi \circ F \circ \phi^{-1}(z) = z^2$. We leave it to you to show this, as well as to show that $g^n(z) = z^{2^n}$.

Our goal is to show that if $z \in \mathbb{C}$ is closer to α than to β, then $F^n(z)$ iterates to α, i.e., $|z - \alpha| < |z - \beta|$ implies $F^n(z) \to \alpha$. Let $|z - \alpha| < |z - \beta|$, which implies $|\phi(z)| < 1$. Since $|g^n(\phi(z))| = |\phi(z)^{2^n}| = |\phi(z)|^{2^n} \to 0$, we have $g^n(\phi(z)) \to 0$. By the conjugation property we have $F^n(z) = \phi^{-1}(g^n(\phi(z))) \to \phi^{-1}(0) = \alpha$. Thus we have shown that the points $z \in \mathbb{C}$ which are closer to α than to β are in $A_F(\alpha)$.

Exercise 1.31. Show that the points $z \in \mathbb{C}$ that are closer to β than to α are in $A_F(\beta)$. *Try it out!*

We illustrate this conjugation in Figure 1.3, called a *commutative diagram* because the maps $\phi \circ F$ and $g \circ \phi$ from the upper left to the bottom right are equal.

The points α and β are moved by ϕ to the points 0 and ∞, and the line $L \cup \{\infty\}$ in the top pictures is transformed to the unit circle $C(0, 1)$ in the bottom pictures. Using this conjugation, we are able to analyze the relatively simple dynamics of g to get information about the dynamics of F, in particular, $A_F(\alpha) = \phi^{-1}(A_g(0))$ and $A_F(\beta) = \phi^{-1}(A_g(\infty))$.

Let's return to the question of whether Newton's method can fail in this example. Are there initial values for which the Newton's method orbit never finds a root of f? In Exploration 1.5, we saw that Newton's method fails when an initial value x_0 is such that $f'(x_0) = 0$. This, what we term an analytic obstruction of having a zero in the denominator, however, is overcome when we allow ∞ to take its equal place with the other values in the Riemann Sphere $\overline{\mathbb{C}}$. Even though in Example 1.24 we have $f'(\frac{\alpha + \beta}{2}) = 0$, the Newton

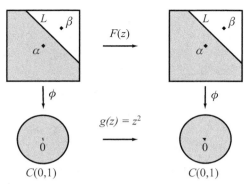

FIGURE 1.3. Commutative diagram for global conjugation of the Newton map $F(z)$ of $f(z) = (z - \alpha)(z - \beta)$.

map gives $F(\frac{\alpha+\beta}{2}) = \infty$ (see Appendix Section B.4 on page 361 for a review of a discussion on functions defined at ∞). Also, since $F(\infty) = \infty$, we see that an initial value $z_0 = \frac{\alpha+\beta}{2}$ will never find either root α or β because it generates the following sequence of Newton approximations $\frac{\alpha+\beta}{2}, \infty, \infty, \infty, \ldots$ The obstruction for the success of Newton's method starting with seed value z_0 is not that F cannot be appropriately defined at z_0, but that such a definition forces F to never iterate z_0 to a root of $f(z)$ (since $F^n(z_0) = \infty$ for all $n \geq 1$). The seed $\frac{\alpha+\beta}{2}$ is not the only complex number that fails to find a root of $f(z)$ though. We can show, again with the help of conjugation, that the boundary line L that divides $A_F(\alpha)$ from $A_F(\beta)$ consists of exactly those points in $\mathbb{C}$ for which Newton's method fails to find either root. As in the real variable examples, we can understand this dynamically since it divides the points that are pulled or attracted to α from those that are attracted to β. We would think that something must fail to work out for points on this line. We have not proven anything carefully yet. We have only looked at compelling computer-generated evidence, which is always to be viewed with a healthy bit of skepticism.

However, a formal argument can be made. For a point $z \in L$, we have $|\phi(z)| = \frac{|z-\alpha|}{|z-\beta|} = 1$ and thus by the conjugation property we have $|g^n(\phi(z))| = |\phi(z)^{2^n}| = |\phi(z)|^{2^n} = 1$ for all $n \in \mathbb{N}$. Hence for all $n \in \mathbb{N}$, we see that $F^n(z) = \phi^{-1}(g^n(\phi(z))) \in \phi^{-1}(C(0,1)) = L \cup \{\infty\}$. In particular, $F^n(z)$ converges to neither α nor to β.

Exercise 1.32. The analysis applies to any monic quadratic polynomial $p(z)$ with distinct roots, but what if the leading coefficient is not 1? What if the quadratic polynomial $p(z)$ has a double root instead of two distinct roots? Analyze what happens in these situations. *Try it out!*

1.2.8   Newton Map of a Cubic Polynomial

The behavior found in Exploration 1.26 with the cubic function $f(z) = z^3 - 1$ is far more complicated than the quadratic case. A starting point under Newton's method will not always find the root to which is is nearest. If it did, then the picture of the global dynamics would look like Figure 1.4. However, the better picture that represents the dynamics is Figure 1.5, which shows that the set of initial guesses in the complex plane is divided into very intricate regions.

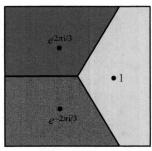

FIGURE 1.4. A reasonable (but false) guess for the description of the global dynamics of $F(z)$, the Newton map for $f(z) = z^3 - 1$.

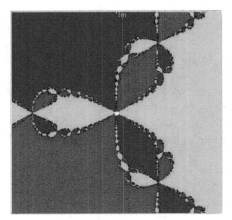 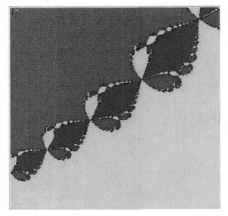

FIGURE 1.5. A more accurate picture of the global dynamics of $F(z)$, the Newton map for $f(z) = z^3 - 1$ (with magnification on the right). Here the turquoise regions represent $A_F(1)$, the blue regions represent $A_F(e^{2\pi i/3})$ and the red regions represent $A_F(e^{-2\pi i/3})$.

Let's experiment with the zoom feature of the *Complex Newton Method Applet* to investigate this. When you zoom in on a point that is on the boundary of one colored region (attracting basin), you always find tiny bulbs of the other two colors (attracting basins) nearby. This happens no matter how much you zoom in! This shocking feature is why we call such sets *fractals* (a set which when you zoom in reveals new features not seen from the coarse larger scale picture). In this case, at a large scale there are extremely tiny bulbs (smaller than a single pixel) that are not revealed in the picture unless one zooms in far enough to see them.

So again we see that Newton's method became very complicated when $f(z)$ changed from a quadratic to a cubic map. Earlier we posed the questions: *Why? How? Is there a way to know when a system will necessarily be simple or complicated ahead of time?* With the depth of complex analysis knowledge to aid us we can give some good reasons why the pictures, and hence the dynamics they represent, must be so complicated. We first remind ourselves of some key concepts.

Definition 1.33. The *boundary of a set* E in $\overline{\mathbb{C}}$ is $\partial E = \overline{E} \cap \overline{\overline{\mathbb{C} \setminus E}}$, which is the set of points $z \in \overline{\mathbb{C}}$ that have the property that every open disk in the spherical metric $\triangle_\sigma(z, r)$ intersects both E and the complement of E no matter how small $r > 0$ is.

We can now describe the fractal features we observed in Figure 1.5 by saying $\partial A_F(1) = \partial A_F(e^{2\pi i/3}) = \partial A_F(e^{-2\pi i/3})$. We see this phenomenon in Example 1.4 and Explorations 1.6 and 1.25 as well when we consider the complex versions of these maps. It turns out that, in general, all attracting basins of a Newton map F share the same set of boundary points. Specifically, we have the following Common Boundary Condition. The proof, however, uses Proposition 1.20 and the forthcoming Theorem 1.59 and so cannot be presented until we develop a few more concepts.

Theorem 1.34 (Common Boundary Condition). *Let $f(z)$ be an analytic function such that its Newton map $F(z) = z - \frac{f(z)}{f'(z)}$ is a rational map. If w_1 and w_2 are roots of $f(z)$, then $\partial A_F(w_1) = \partial A_F(w_2)$.*

Let's examine how Theorem 1.34 forces the dynamics in Figure 1.5 to be complicated. By Proposition 1.20 there exists $r > 0$ such that $\triangle(1, r) \subseteq A_F(1)$, $\triangle(e^{2\pi i/3}, r) \subseteq A_F(e^{2\pi i/3})$, and $\triangle(e^{-2\pi i/3}, r) \subseteq A_F(e^{-2\pi i/3})$. So, starting from Figure 1.6, we need to consider how to color the rest of the points in $A_F(1)$, $A_F(e^{2\pi i/3})$, and $A_F(e^{-2\pi i/3})$ with turquoise, blue, and red, respectively, while being certain to make sure that the boundary of each color matches the boundary of the other two colors. Reflecting on what this means we see that this Common Boundary Condition forces the picture to be very complicated, and also rules out having the global dynamics in Figure 1.4 since, for example, a point on the negative real axis lies on the boundary of only two of the three color basins.

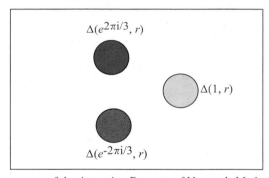

FIGURE 1.6. Consequence of the Attracting Property of Newton's Method (Proposition 1.20).

This Common Boundary Condition is also at the heart of what we call *chaos* in the dynamics of Newton's method. Consider a point z on the boundary of any basin and then draw a tiny disk B around it. According to the Common Boundary Condition, it contains all three colors. So if we wish to determine the orbit of z with a computer, what we will find is that tiny errors, such as roundoff error, in the coordinates of z could lead to drastically different results. By changing z even by the slightest amount, we can change the orbit of z tremendously as z could slip into any of the red, turquoise, or blue regions. This is the essence of what we call chaos. Although a more formal and more thorough understanding of chaos will come later in the text, use our current understanding to explain why each point on the line L that appeared in the analysis of the quadratic Newton method case has this behavior. Also, explain why each point in one of the attracting basins (which has a tiny disk of all the same color) is not such a point. ***Try it out!***

Remark 1.35. We mention here a remarkable fact about sets that share a complicated boundary. The sets $A_F(1)$, $A_F(e^{2\pi i/3})$, and $A_F(e^{-2\pi i/3})$ share the same boundary because each is broken up into an infinite number of pieces that are then arranged in the complicated fractal pattern in Figure 1.5. However, it is true that three open sets (or even n open sets for any fixed $n \in \mathbb{N}$) can all share the same boundary set and also have the property that each is connected (an open set A is called *connected* when given any two points $z, w \in A$ there exists a polygonal line in A which connects z to w). Such sets are complicated indeed! You can read about the Lakes of Wada in [16, p. 143].

Let us now consider the question of whether there are any seed values for which the Newton map for $f(z) = z^3 - 1$ fails to find a root of $f(z)$. In the previous examples there always existed such points; however, they were, in some sense, relatively few. You might guess that each point in $\partial A_F(1) = \partial A_F(e^{2\pi i/3}) = \partial A_F(e^{-2\pi i/3})$ is such a point, and you would be right. We can use our intuition to imagine that such points fail to find any root because they are being attracted by the three different roots with equal force. You are asked to prove this in Additional Exercise 1.171. However, assuming that each point in $\partial A_F(1) = \partial A_F(e^{2\pi i/3}) = \partial A_F(e^{-2\pi i/3})$ fails to iterate under $F(z)$ to any of the roots of $f(z)$ does not necessarily give the fate of all starting points. Is it true that all points in $\overline{\mathbb{C}}$ lie in either one of the attracting basins or on the common boundary of these sets? We ask

1. Is there a point not in $\partial A_F(1) = \partial A_F(e^{2\pi i/3}) = \partial A_F(e^{-2\pi i/3})$ that fails to find any root of $f(z)$?

2. Can there be an open disk of such points?

When considering $f(z) = z^3 - 1$ (or any of the maps f mentioned thus far in this chapter) and its related Newton map $F(z)$, the answer to both questions is no, as you are asked to explore in Additional Exercise 1.182. However, it is possible for the answer to the second question, and hence the first, to be yes.

Exploration 1.36. Consider the map $f(z) = z(z-1)(z-.909-.416i)$ and its related Newton map $F(z)$. Using the *Complex Newton Method Applet* you can find regions of seed values colored black that fail to find a root of $f(z)$. For example, there is one near $0.64 + 0.14i$. Use the applet to select such a point and then iterate $F(z)$ to explore the behavior. Experiment with many seed values chosen from the black regions. Zoom to see whether or not it appears that the boundary of the black region matches the boundary of the attracting basins for the roots of $f(z)$. *Try it out!*

You can investigate Newton's method applied to other cubic polynomials in Additional Exercise 1.172.

1.2.9 Newton's Method Applied to a Family of Cubic Polynomials

In this section we apply Newton's method to an arbitrary cubic polynomial. Let $\mathcal{F}$ denote the collection of all polynomials $p_\rho(z) = z(z-1)(z-\rho)$ where ρ is in $D = \{\rho : \mathrm{Im}\,\rho \geq 0, |\rho| \leq 1, |\rho - 1| \leq 1\}$.

This class of polynomials $\mathcal{F}$ is representative of all polynomials with three distinct roots in the following sense.

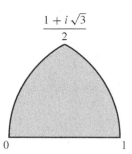

FIGURE 1.7. Region D of ρ values corresponding to the maps p_ρ in $\mathcal{F}$.

Proposition 1.37. *For a cubic polynomial $p(z)$ with three distinct roots, there exists $\rho \in D$ such that the Newton map F_p is globally conjugate by a linear map T to the Newton map F_{p_ρ} of $p_\rho(z)$, i.e., $T \circ F_p \circ T^{-1} = F_{p_\rho}$.*

In order to prove it we require the next proposition, which applies to polynomials of any degree.

Proposition 1.38. *Let $p(z)$ be a polynomial and let $T(z) = az + b$ for $a \neq 0$ where $a, b \in \mathbb{C}$. Then for the polynomial $q(z) = p(T(z))$, we have*

$$T \circ F_q \circ T^{-1} = F_p$$

where F_q and F_p are the Newton maps of q and p respectively.

We leave the proof of Proposition 1.38 to the reader. However, we say a few words about its meaning and usefulness. The polynomial q (which is sometimes called the "rescaling" of p) has the same degree as p. Furthermore, if p has roots at $r_1, \ldots, r_d$, then q will have roots at $T^{-1}(r_1), \ldots, T^{-1}(r_d)$. So the proposition says that we can move the roots of p by choosing T^{-1} appropriately and generating a new polynomial q. Studying the dynamics of F_q will be essentially the same as studying the dynamics of F_p since they are globally conjugate to each other.

Example 1.39. Let $p(z) = (z + i/2)(z - 1)(z + 1)$. We illustrate Proposition 1.37 by finding $\rho \in D$ so that F_{p_ρ} is conjugate to F_p. A triangle is formed by the roots of $p(z)$ at $-i/2, 1$, and -1. We construct a linear map T^{-1} to transform it into a triangle whose longest side is the interval $[0, 1]$ and whose third vertex (which will be our choice of ρ) is in the upper half plane $\{\operatorname{Im} z \geq 0\}$. Then $T^{-1}(1) = 0$ and $T^{-1}(-1) = 1$, which determine that $T^{-1}(z) = \frac{-1}{2}z + \frac{1}{2}$ and $\rho = T^{-1}(-i/2) = 1/2 + i/4$ which is in D.

Let's verify that this choice of ρ works. Because $q(z) = p(T(z))$ has roots at $0, 1$, and ρ and q and p_ρ share the same roots, they agree up to a multiplicative constant, i.e., we have $q(z) = cp_\rho(z)$ for some constant $c \neq 0$. This implies that $F_q = F_{p_\rho}$ (why?). Hence, by Proposition 1.38, we have that $T \circ F_{p_\rho} \circ T^{-1} = T \circ F_q \circ T^{-1} = F_p$ as desired.

Use the *Complex Newton Method Applet* to compare the pictures of the attracting basins for F_p and F_{p_ρ} noting the similarities because they are globally conjugate. Try to see the effects of T when you compare the two pictures. ***Try it out!***

Exercise 1.40. Find $\rho \in D$ so that for $p(z) = (z - 4)(z + i)(z + 4)$ the map F_{p_ρ} is globally conjugate to F_p. ***Try it out!***

Proposition 1.37 can be proven as in Example 1.39 and so we omit the details. We leave it to the reader to investigate what can be said about the dynamics of F_p when $p(z)$ is a cubic polynomial with a double or triple root. In addition, we invite the reader to consider generalizations such as the following.

Small Project 1.41. Consider how Proposition 1.38 is used to move roots by a linear transformation T^{-1} in the proof of Proposition 1.37. With this in mind, can you justify the statement that all cubic polynomials whose roots form an equilateral triangle have globally conjugate Newton maps (and hence essentially the same dynamics)? What about quartic polynomials whose roots form a square? Does this generalize to higher degree polynomials? Do the roots need to form a regular n-gon? What can you say about generalizing the class $\mathcal{F}$ (and the set D) when considering polynomials of fixed higher degree?

Aside from the exceptional cases mentioned above, we can study the Newton's method dynamics of all cubic polynomials by studying the maps p_ρ where $\rho \in D$. We have created the *Cubic Polynomial Complex Newton Method Applet* to help. It allows the user to generate the pictures of the attracting basins for Newton's method applied to any p_ρ, not just $\rho \in D$. Also, it allows the user to investigate the parameter plane of ρ values since each such value will be colored according to the dynamics of F_{p_ρ}.

We present a note about the coloring of the parameter plane of ρ values. You will see in Section 1.4 that the orbit of points where $F'_{p_\rho}(z) = 0$ are critically important to understanding the dynamics of F_{p_ρ}. Since $F'_{p_\rho}(z) = \frac{p_\rho(z) p''_\rho(z)}{(p'_\rho(z))^2}$, we see that $F'_{p_\rho}(z) = 0$ only at the roots of $p_\rho(z)$ or when $p''_\rho(z) = 0$. Since the roots of $p_\rho(z)$ are attracting fixed points for F_{p_ρ}, their orbits are understood. However, since $p''_\rho(z) = 0$ for $z = (1 + \rho)/3$ this, so-called, *free critical point* is important. In the *Cubic Polynomial Complex Newton Method Applet*, we track, for each ρ, the *critical orbit* of $z_0 = (1 + \rho)/3$. When it is attracted to one of the roots of p_ρ, we color ρ in the parameter plane the corresponding color of the attracting basin. To illustrate, click on the **Show Critical Orbit** checkbox to see the critical orbit appear as white dots in the right picture (dynamical plane). Thus, if the critical orbit converges to, say, the red root of p_ρ, then ρ (marked by a +) is colored red in the left picture (parameter plane). If this orbit is not attracted to any of the roots of p_ρ, we color ρ black. Try clicking in the parameter plane to a select different colored ρ value, and then observe the convergence of the critical orbit to a different colored root of p_ρ.

Exploration 1.42. Use the *Cubic Polynomial Complex Newton Method Applet* to investigate the dynamics of Newton's method applied to any p_ρ. Look for symmetries and experiment with the dynamical behavior you find. Make conjectures, and see if you can prove them. *Try it out!*

1.3 Iteration of an Analytic Function

In this section we study the dynamics of analytic maps that do not necessarily arise as Newton maps of polynomials. One goal is to understand which dynamical features of Newton's method extend to such a wider class of maps. We pay particular attention to the iteration of polynomial maps of the form $z^2 + c$ where c is a complex parameter since these are the simplest maps that produce a rich variety of dynamical behaviors. (The iteration of Möbius maps is simpler to study because they have relatively simple dynamical behavior. This is shown in Appendix 1.B.) Additional Exercise 1.169 asks you to use the global conjugation technique of Section 1.2.6 to show that every quadratic map is globally conjugate, and in some sense dynamically equivalent, to exactly one map of the form $z^2 + c$. Hence, in this way the dynamics of maps $z^2 + c$ represent the dynamics of all quadratic maps.

Sections 1.2.3 and 1.2.6 must be read before proceeding with this section, but not all of Section 1.2 is necessary.

Though quadratic maps of the form $z^2 + c$ are genuinely very simple and well understood as functions, we will see that their dynamics can be complicated. We will look at the dynamics of a map $z^2 + c$ noting how the dynamics change as c changes, thus leading us into *bifurcation theory*.

In Section 1.6, we consider the dynamics of more exotic transcendental entire complex analytic maps such as ce^z, $c \sin z$, and $c \cos z$.

1.3.1   Classification of Fixed Points for Analytic Maps

In Section 1.2.3 we defined attracting fixed points. Now we extend our discussion to two more types of fixed points, *repelling* and *indifferent*. We begin with an example.

Example 1.43. Let $f(x) = x^2$ and $g(x) = \sqrt{x}$, restricted so that $f, g : \mathbb{R}^+ \to \mathbb{R}^+$ where $\mathbb{R}^+ = \{x \geq 0\}$. Clearly 0 and 1 are fixed points of both f and g. We have $f^n(x) \to 0$ if $0 < x < 1$, $f^n(x) \to \infty$ if $x > 1$, and $g^n(x) \to 1$ for $x \in \mathbb{R}^+ \setminus \{0\}$. Note that 0 is an attracting fixed point for f (why?). If x is close to, but not equal to 1, the orbit $f^n(x)$ moves away from 1. We then call 1 a *repelling* fixed point for f. The function $g(x)$ has an attracting fixed point at 1 and a repelling fixed point at 0. We illustrate the dynamics in Figure 1.8.

FIGURE 1.8. Dynamics of f and $g = f^{-1}$.

Because f and g are inverses of each other, it is reasonable that an attracting fixed point for f must be a repelling fixed point for g, and vice versa. This becomes clear by considering the following definition.

Definition 1.44 (Repelling Fixed Point). Let f be a map with domain set $\Omega \subseteq \overline{\mathbb{C}}$, which could be a subset of $\mathbb{R}$.

(a) We call a (finite) fixed point $a \in \mathbb{C}$ a *repelling fixed point* of f if there exists a neighborhood U of a such that for $z \in \Omega \cap U \setminus \{a\}$, we have $|f(z) - a| > |z - a|$, i.e., the action of f is to move each point in $\Omega \cap U \setminus \{a\}$ farther from a.

(b) Suppose $f(\infty) = \infty$. We call ∞ a *repelling fixed point* of f if there exists a neighborhood $U \subseteq \overline{\mathbb{C}}$ of ∞ such that for $z \in \Omega \cap U \setminus \{\infty\}$, we have $|f(z)| < |z|$, i.e., the action of f is to move each point in $\Omega \cap U \setminus \{\infty\}$ farther from ∞, as measured by the spherical metric.

Theorem 1.45. *Suppose $\Omega \subseteq \mathbb{R}$ or $\Omega \subseteq \mathbb{C}$. Let $f : \Omega \to \Omega$ be such that $f(a) = a$ and $|f'(a)| > 1$. Then a is a repelling fixed point of f. Also, there exists $\varepsilon > 0$ such that for* *each $z \in \Omega \cap \triangle(a, \varepsilon) \setminus \{a\}$ the orbit $f^n(z)$ eventually leaves $\triangle(a, \varepsilon)$, i.e., there exists N such that $f^N(z) \notin \triangle(a, \varepsilon)$.*

Remark 1.46. The proof is a modification of the proof of Theorem 1.17, and so we leave the details to the reader. Additional Exercise 1.173 illustrates the fact that the theorem does not preclude the possibility that the orbit of z, after leaving $\triangle(a, \varepsilon)$, might reenter $\triangle(a, \varepsilon)$.

Remark 1.47. Theorem 1.45 is not an if and only if result. Find a real-valued function $f : \mathbb{R} \to \mathbb{R}$ such that 0 is a repelling fixed point that has $|f'(0)| = 1$. Hint: Consider maps of the form $x \mapsto x \pm x^n$.

We provided examples of real-valued maps which show that the converses of Theorems 1.17 and 1.45 do not hold. However, the next theorem shows that such examples cannot be found amongst the complex analytic maps.

Theorem 1.48. *Let $f(z)$ be an analytic map on an open set $\Omega \subseteq \mathbb{C}$ such that $f(a) = a$. Then,*

(a) the point a is an attracting fixed point if and only if $|f'(a)| < 1$, and

(b) the point a is a repelling fixed point if and only if $|f'(a)| > 1$.

Theorems 1.17 and 1.45 provide two of the four implications. The proofs of the remaining two, outlined in Additional Exercise 1.174, use special properties of complex analytic maps.

Theorem 1.48 applies only to finite fixed points and so we ask if there is a corresponding result when ∞ is fixed. The following examples prepare for handling this.

Example 1.49. The map $h(z) = z + 1$ on $\overline{\mathbb{C}}$ fixes ∞ and shifts a point in the complex plane one unit to the right. For large and positive real z, we have $|h(z)| > |z|$, which makes it appear that ∞ is attracting. However, for large and negative real z, we see that $|h(z)| < |z|$, which makes it appear that ∞ is repelling. We conclude that the fixed point at ∞ is neither attracting nor repelling. The nth iterate $h^n(z) = z + n$ and so for z in $\overline{\mathbb{C}}$ we have $h^n(z) \to \infty$, always moving parallel to the x-axis. Thus $A_h(\infty)$ is the entire Riemann Sphere $\overline{\mathbb{C}}$. This is somewhat surprising seeing that ∞ is not an attracting fixed point. We represent the dynamics on $\mathbb{C}$ and $\overline{\mathbb{C}}$ in Figure 1.9.

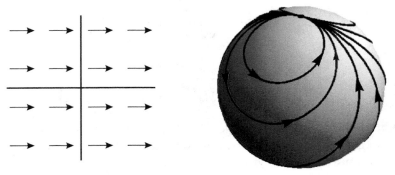

FIGURE 1.9. Graphical representations of the dynamics of $h(z) = z + 1$ on the plane $\mathbb{C}$ (left) and on the Riemann sphere $\overline{\mathbb{C}}$ (right).

Example 1.50. The map $g(z) = z/2$ on $\overline{\mathbb{C}}$ takes $z \in \mathbb{C}$ to a point with one half the modulus, but with the same argument. The origin is an attracting fixed point and $A_g(0) = \mathbb{C}$. Also, $g(\infty) = \infty$ and if $z \in \mathbb{C}$ is near ∞ then $g(z)$ moves away from ∞, i.e., $|g(z)| < |z|$. Thus ∞ is a repelling fixed point. Note also that $g^n(z) = z/2^n$. We represent the dynamics on $\mathbb{C}$ and $\overline{\mathbb{C}}$ in Figure 1.10.

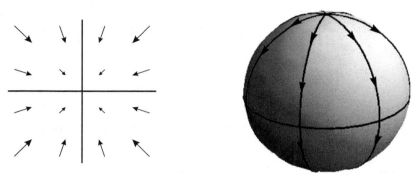

FIGURE 1.10. Graphical representations of the dynamics of $g(z) = z/2$ on the plane $\mathbb{C}$ (left) and on the Riemann sphere $\overline{\mathbb{C}}$ (right).

Example 1.51. For $f(z) = z^2$ on $\overline{\mathbb{C}}$ we have that $f^n(z) = z^{2^n}$ for all $n = 1, 2, \ldots$. Thus, $|z| < 1$ implies $|f(z)| = |z|^2 < |z|$ and $|f^n(z)| = |z^{2^n}| = |z|^{2^n} \to 0$. Also, $|z| > 1$ implies $|f(z)| = |z|^2 > |z|$ and $|f^n(z)| = |z^{2^n}| = |z|^{2^n} \to +\infty$. Thus, we conclude that 0 and ∞ are attracting fixed points with attracting basins $A_f(0) = \triangle(0, 1)$ and $A_f(\infty) = \overline{\mathbb{C}} \setminus \overline{\triangle(0, 1)}$. We represent the dynamics in Figure 1.11, noting that the angle doubling property of the map is not represented. If $z_0 = re^{i\theta}$ in polar form we have that $f(z_0) = r^2 e^{i2\theta}$. So, starting with seed $z_0 = 0.999e^{i\pi/100}$, the orbit z_n will converge towards 0, doubling the angle at each step. Observe this in the *Complex Function Iterator Applet*, using the **Polar seed form** and the **Polar computation mode**.

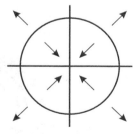

FIGURE 1.11. Graphical representation of the dynamics of $f(z) = z^2$.

The last three examples were chosen to illustrate fixed points at ∞. We were able to calculate a formula for f^n. This is rare. However, having a formula for the iterates f^n is unnecessary because we can analyze the dynamics without appealing to them.

In Examples 1.49, 1.50, and 1.51 we saw that ∞ is a fixed point that we were able to classify by examining the dynamics near ∞. However, we wonder (in light of Theorem 1.48) if we can use the derivative to classify fixed points at ∞. It turns out that we can, but we must be careful. We saw that $|h'(\infty)| = 1$ corresponded to a fixed point that was neither attracting nor repelling, $|g'(\infty)| = 1/2$ corresponded to a repelling fixed point, and $f'(\infty) = \infty$ corresponded to an attracting fixed point. The derivative at ∞ does not play the same role as it does in Theorem 1.17 for finite points. However, as we now see, it is not far off.

Definition 1.52 (Multiplier of a Fixed Point). When f is analytic at ∞ with $f(\infty) = \infty$, we define the *multiplier* λ at ∞ to be $1/f'(\infty) = \lim_{z\to\infty} 1/f'(z)$. When a is a finite fixed point in $\mathbb{C}$, we define the *multiplier* λ at a to be $f'(a)$.

We now classify a fixed point of an analytic map, whether finite or not, based on its multiplier.

Definition 1.53 (Classification of Fixed Points). Suppose $f : \Omega \to \Omega$ is analytic where Ω is an open subset of $\overline{\mathbb{C}}$ and a is a fixed point with multiplier λ. Then a is called
(a) *super attracting* if $\lambda = 0$
(b) *attracting* if $0 < |\lambda| < 1$
(c) *repelling* if $|\lambda| > 1$
(d) *indifferent* if $|\lambda| = 1$.

You can check that this matches what we found in Examples 1.49, 1.50, and 1.51. The motivation for the definition of the multiplier when f has a fixed point at ∞ is that by globally conjugating f by $\phi(z) = 1/z$ we obtain the map $k(z) = 1/f(1/z)$, which has a fixed point at 0. We defined the multiplier of f at ∞ to be the multiplier of k at 0, which is calculated in Appendix B Lemma B.20 on page 363 to be $k'(0) = 1/f'(\infty)$. In Additional Exercise 1.175, you are asked to prove that the multiplier correctly corresponds to the notions of attracting and repelling fixed points as given in Definitions 1.16 and 1.44.

Remark 1.54. Indifferent fixed points can exhibit both a partial attracting nature and a partial repelling nature. For example, we saw in Section 1.2.3 that the indifferent fixed point at the origin for the complex map $g(z) = \sin z$ is attracting for real-valued seeds, but repelling for purely imaginary seeds. Likewise, the indifferent fixed point at ∞ for the map $h(z) = z + 1$ attracts large positive real z, but repels (at least initially) large negative real z. However, sometimes an indifferent fixed point, like the origin under the map $f(z) = e^{\sqrt{2}\pi i} z$, neither attracts nor repels. Indifferent fixed points of analytic maps can exhibit many types of dynamical behavior and their study can be complicated. It is complicated enough that we will not say more about them here, but refer the interested reader to [1, 3, 24].

1.3.2 A Closer Look at the Dynamics of $f(z) = z^2$

Let us return to the dynamics of $f(z) = z^2$. If $|z| = 1$, then $|f^n(z)| = |z^{2^n}| = |z|^{2^n} = 1^{2^n} = 1$, i.e., if a point is on the unit circle $C(0, 1) = \{z : |z| = 1\}$, then so is its entire orbit $\{f^n(z)\}_{n=1}^{\infty}$. From this and from our previous work in Example 1.51 you might conclude that its dynamics has been solved. As we will see, this is not at all close to the truth!

If we know the orbit of one seed value, will nearby seed values have similar orbits? If z lies within the unit disk, then it and points near it in the unit disk have the same behavior, namely, each orbit converges to 0. If z lies outside the closed unit disk, then it and points near it outside of the closed unit disk have the same behavior, namely, each orbit converges to ∞. However, the story is different for seed values on the unit circle. For any seed on $C(0, 1)$ we can find other seed values arbitrarily close to it that have drastically different orbits (namely that limit to either ∞ or 0).

Let's examine this using the computer. For the map $f(z) = z^2$ use the *Complex Function Iterator Applet* to iterate the seed value $z_0 = 1 + i$ as well as seed values close to z_0. You can see that z_0 and the nearby seed values have the same behavior. Repeat by iterating z_0 for the values $-0.4 + 0.5i, 1, i$, and $0.6 + 0.8i$, testing nearby seed values that you choose yourself.

Remark 1.55 (A word of caution about using computers). If we are trying to calculate the orbit under the map $f(z) = z^2$ of a point on $C(0, 1)$, we may have problems getting our computer to provide accurate results. Let's test this on the *Complex Function Iterator Applet*. For the map $f(z) = z^2$ try iterating seed value $z_0 = 0.6 + 0.8i$ (which lies on $C(0, 1)$) by entering the seed value in **Euclidean form** and using the **Euclidean computation mode**. Iterating 20 times, you will see what you would expect. However, after iterating about 55 times you will see the orbit move inside the unit disk where it will then iterate to 0. Why? If the computer truncates, rounds, or approximates any of the values in the orbit (as computers often do), these small errors will likely push the orbit outside or inside the unit circle, causing the computer mistakenly to calculate the orbit as tending to either ∞ or 0. When using a computer it is important to know if the computer will approximate values it uses, and whether the approximations will be significant or not in the end result.

As an illustration of the subtle issues that can wreak havoc on your computations, we show the problems that can arise from the fact that z^2 and $z * z$ are not always equivalent. Of course the expressions z^2 and $z * z$ are mathematically equivalent, but they are not computationally equivalent. The former is evaluated as $e^{2 \text{Log} z}$, where $\text{Log} z$ is the principal logarithm, and the latter is evaluated with complex multiplication. Each incorporates different rounding errors at times. The result of this very subtle difference is quite evident when iterating the seed $0.6 + 0.8i$ in **Euclidean computation mode** under these maps. *Try it out!*

So we see that it is important to identify those seed values where approximations (or errors introduced by using approximations) lead to significant errors in calculations of orbit values. Such seed values are said to be in the *chaotic set*. The chaotic set is called the *Julia set*, in honor of the mathematician Gaston Julia who in 1918, at the age of 25, published his 199-page masterpiece [17] titled "Mémoire sur l'itération des fonctions rationnelles" describing the iteration of complex rational functions.

Remark 1.56. The Julia set is not tied only to errors or approximations that a computer might introduce. We need to know whether a tiny error in the seed value (no matter what caused it) could produce a significant error in an orbit value. For example, the seed $z_0 = i$ would be in the Julia set of $f(z) = z^2$ even though its orbit $i, -1, 1, 1, 1, \ldots$ is computed exactly by a computer. However, we still say it is a chaotic seed value since given any allowable error in the seed value, even one as small as 10^{-631}, we can find another seed z_0' close to $z_0 = i$ (i.e., within the error) such that z_0' has a drastically different orbit from the orbit of z_0. Thus, even if a computer wouldn't introduce an error in its orbit calculation for a seed z_0, the seed value might still be in the Julia set.

The Julia set for $f(z) = z^2$ is $C(0, 1)$ because the dynamics there are *chaotic*, i.e., for any $z_0 \in C(0, 1)$ there is a point z_0' arbitrarily close by that has a drastically different orbit. A computer often fails to be accurate since we would need it to store such values and each point in its orbit with an infinite degree of accuracy. To put it another way, approximation of the starting value (or some future iterate value) ruins our confidence in the calculations of the long term behavior. This is called *sensitive dependence on initial conditions*, and it is the defining feature in what is called *chaos*.

We call the complement $\overline{\mathbb{C}} \setminus C(0, 1)$ the *stable set* for $f(z) = z^2$ because orbits are stable there, i.e., for any $z_0 \in \overline{\mathbb{C}} \setminus C(0, 1)$ its orbit z_n will behave like the orbit z_n' of any seed value z_0' chosen sufficiently close to z_0. The stable set is commonly called the *Fatou set* because of Pierre Fatou's role in developing the theory of complex function iteration in 1917 (see [12, 13, 14]). In honor of their pioneering work in the field, the dynamics of complex analytic functions is often called the Fatou-Julia theory.

Exercise 1.57. Use your understanding of the meanings of the Fatou and Julia sets to determine these sets for the maps $h(z) = z + 1, g(z) = z/2$, and $k(z) = 3z$. ***Try it out!***

Exercise 1.58. Try to write a precise mathematical definition of the Fatou set and the Julia set that applies to any rational function $g(z)$ defined on $\overline{\mathbb{C}}$. Try to formulate what it means for orbits to be drastically or significantly different. ***Try it out!***

Notation: For a rational function $g(z)$, we employ the notation $F(g)$ for the Fatou set of g and $J(g)$ for the Julia set of g. Formal definitions along with a more advanced development can be found in the chapter Appendix 1.A on page 78. These should be read and compared with your own definitions, but it is not necessary to know these formal definitions well to continue on in the text; an intuitive understanding is enough.

We showed that $J(f) = C(0, 1)$ for the map $f(z) = z^2$ by noting that for $z_0 \in C(0, 1)$ there is a point arbitrarily close to z_0 whose orbit tends to ∞ and is thus drastically different from the orbit of z_0 which must remain on $C(0, 1)$. It is also true that f is chaotic on $C(0, 1)$ if we restrict ourselves to using only seed values from $C(0, 1)$. To see this, consider $z_0 = e^{i\theta}$ written in polar form. For any nearby point $z_0' = e^{i\alpha}$ we must, because f doubles angles, have $|f(z_0) - f(z_0')| > |z_0 - z_0'|$. Indeed, denoting the arclength distance between z_0 and z_0' along the unit circle by $\beta = |\theta - \alpha|$, we see that the distance between $f(z_0)$ and $f(z_0')$ along the unit circle is 2β. By induction, we see that the distance between $f^n(z_0)$ and $f^n(z_0')$ along the unit circle is $2^n \beta$ as long as $2^n \beta < \pi$. So no matter how close z_0 and z_0' start out, i.e., no matter how small β is, corresponding orbit points will eventually be far apart. As soon as $2^n \beta > \pi/3$ we have $|f^n(z_0) - f^n(z_0')| > 1$. Thus arbitrarily close seed values on $C(0, 1)$ do not have corresponding orbit values that forever stay close. Use the *Complex Function Iterator Applet*, using both **Polar computation** mode and **Polar seed form**, to see the sensitive dependence. Simultaneously iterate (one step at a time) **Seed 1** $z_0 = e^{2.18i}$ and **Seed 2** $z_0' = e^{2.19i}$.

We close this section by reflecting on some commonalities between the Newton map dynamics in Section 1.2.4 and the dynamics of $f(z) = z^2$. As we saw in each Newton's method example and for the $f(z) = z^2$ dynamics, any two basins of attraction of attracting fixed points share a common boundary. For $f(z) = z^2$, the basins $A_f(0)$ and $A_f(\infty)$ share a common boundary that is the Julia set $J(f) = C(0, 1)$ where the attractive pull of each attracting fixed point is balanced by the other. If you read Section 1.2, you should now go back and consider each complex Newton's method example to convince yourself that the Julia set of the Newton map is exactly the common boundary of any attracting basin. This is no coincidence and we state the precise result as follows.

Theorem 1.59. *Let $f(z)$ be a rational map. If w is an attracting fixed point of $f(z)$, then $\partial A_f(w) = J(f)$.*

Remark 1.60. Theorem 1.59 together with Proposition 1.20 imply Theorem 1.34.

Remark 1.61. If the fixed point is not attracting in Theorem 1.59, then the conclusion might not follow. We saw this in Example 1.49 where $A_h(\infty) = \overline{\mathbb{C}}$, which has empty boundary, but $J(h) = \{\infty\}$, which we leave the reader to show.

An incomplete sketch of the proof of Theorem 1.59. Showing the full details of the proof requires a precise definition of the Julia set and complex analysis beyond the level of this text. We provide a sketch of some of the arguments. By Additional Exercise 1.171, $A_f(w)$ is an open set. Thus each point in $A_f(w)$ has a neighborhood of points whose orbits act in the same way, showing, informally, $A_f(w) \subseteq F(f)$. Likewise, a point in $\partial A_f(w)$ contains points arbitrarily close to it that iterate to w and points arbitrarily close to it whose orbit points stay far away from w. Such a point must be in the Julia set, i.e., $\partial A_f(w) \subseteq J(f)$. What remains is to show that $\overline{\mathbb{C}} \setminus \overline{A_f(w)}$ is in the Fatou set. By showing $f(\overline{\mathbb{C}} \setminus \overline{A_f(w)}) \subseteq \overline{\mathbb{C}} \setminus \overline{A_f(w)}$, we may use Montel's Theorem (Appendix 1.A Theorem 1.210) to conclude $\overline{\mathbb{C}} \setminus \overline{A_f(w)} \subseteq F(f)$. The details of the proof can be found in [3, p. 58]. □

1.3.3 Dynamics of Maps of the Form $f_c(z) = z^2 + c$

We have investigated the dynamics of the map $f(z) = z^2$, though there is still more to investigate since a finer analysis of the dynamics on $C(0, 1)$ reveals some very interesting behavior. This is pursued in Appendix 1.A Remark 1.214 on page 79 and Additional Exercises 1.173 and 1.208. Now we investigate what happens when we change the map by adding a constant.

A justification for studying this (although the only justification a mathematician requires is that the problem be interesting) is that we want to know what happens when a system we are studying is slightly altered, or has some error causing us to believe that we cannot be certain that our mathematical model is exactly correct as opposed to being a good approximation. In the dynamics of $f(z) = z^2$, we identified which seed values z_0 were stable (i.e., in the Fatou set) or chaotic (i.e., in the Julia set) by studying the effects of allowing arbitrarily small errors in the seed value. Now we investigate what type of stability may or may not be present when we allow for an error or perturbation in the map we are iterating.

We will consider the dynamics of maps of the form $f_c(z) = z^2 + c$ where c is a complex parameter. Specifically, we fix $c = c_0$ and study the dynamics of f_{c_0}, then vary c to see if the dynamics of the maps f_c have similar behavior. Just as we saw the seed values z for a fixed map split into the stable Fatou set and the chaotic Julia set, we will also see the set of parameters c split into stable parameters and what we will call bifurcation parameters.

We begin by noting a common aspect of the dynamics of f_c for all values of c. Each map f_c has a super attracting fixed point at ∞ (verify by checking the multiplier of the fixed point at ∞). Thus, if any orbit point $z_n = f_c^n(z_0)$ of a seed value z_0 should be large enough (by this we mean that $|z_n|$ is large enough), then the orbit will tend to ∞. A calculation in Additional Exercise 1.191 shows that for the maps f_c, we have $|z_k| > \max\{2, |c|\}$ for some $k \in \mathbb{N}$ if and only if $z_n \to \infty$.

We now consider a small perturbation of the map $f_0(z) = z^2$. If we let c be small (by which we mean $|c|$ is small), then we might expect the dynamics of f_c to be similar to those of $f_0(z) = z^2$. This turns out to be true in many respects.

Exploration 1.62. Fix $c = 0.1$ and use the *Complex Function Iterator Applet* to study the dynamics of $f_{0.1}(z) = z^2 + 0.1$. Try many different seed values including $z_0 = 0, 1 + i, -2 - 0.5i, \ldots$ and record the types of behavior you are able to find. ***Try it out!***

Similarly to f_0, it seems that $f_{0.1}$ has only two types of long term behavior. Iterates of $f_{0.1}$ seem to either approach ∞ or the attracting fixed point $p \approx 0.1127$. Our intuition tells us that this cannot be the whole story. If there are two attracting fixed points, then we expect there to be tension between attracting basins, and that points on their boundaries will not be attracted to either fixed point. The map $f_0(z) = z^2$ has two attracting basins each having the chaotic set $J(f_0) = C(0, 1)$ as its boundary. This also occurs with the map $f_{0.1}$, except that the chaotic set $J(f_{0.1})$ is not a circle (though some advanced mathematics can show that it is a simple closed curve, see [3, p. 126]). Let's use the *Global Complex Iteration Applet for Polynomials* to see a picture of $J(f_{0.1})$. The applet will color each seed in the basin of attraction of ∞ for the map $f_{0.1}$ based on how many iterates it takes for the orbit to become strictly larger than $\max\{2, |0.1|\} = 2$, and it will color the remaining points black. By Theorem 1.59, the basin $A_{f_{0.1}}(\infty)$ has boundary set equal to $J(f_{0.1})$. Experiment with iterating seed values (two at a time) near $J(f_{0.1})$ to see the sensitive dependence on initial conditions. ***Try it out!***

The set of points colored in black in the applet has a name that we present here for any polynomial.

Definition 1.63. For a polynomial $g(z)$, we define the *filled-in Julia set* $K(g)$ to be the set of points which do not iterate to ∞, i.e.,

$$K(g) = \{z \in \mathbb{C} : \{g^n(z)\}_{n=0}^\infty \text{ is bounded in } \mathbb{C}\} = \overline{\mathbb{C}} \setminus A_g(\infty).$$

Remark 1.64. For a polynomial $g(z)$ of degree greater than or equal to two (which has a super attracting fixed point at ∞), it is true that $\partial A_g(\infty) = \partial K(g)$, and so by Theorem 1.59 we have that $J(g), \partial A_g(\infty)$, and $\partial K(g)$ are identical sets. You are asked to prove this in Additional Exercise 1.176.

Returning to the dynamics of the map $f_{0.1}$, we ask if we could have predicted that there would be a finite attracting fixed point. The answer is yes, and here is how. We find the two finite fixed points of $f_{0.1}$ by solving $f_{0.1}(z) = z$ (why?). We then show that the two finite fixed points $p = (1 - \sqrt{0.6})/2$ and $q = (1 + \sqrt{0.6})/2$ are attracting and repelling, respectively, since $|f'_{0.1}(p)| < 1$ and $|f'_{0.1}(q)| > 1$.

One way to relate the dynamics of f_0 and $f_{0.1}$ is to say that when c moves from $c = 0$ to $c = 0.1$, the super attracting fixed point at 0 with multiplier $\lambda_0 = 0$ becomes an attracting fixed point at $p \approx 0.1127$ with multiplier $\lambda_p = f'_{0.1}(p)$, the circle $J(f_0)$ becomes a distorted circle $J(f_{0.1})$, and the super attracting fixed point at ∞ persists, i.e., remains a super attracting fixed point for $f_{0.1}$. Thus the small change in the c parameter led to only a small change in the dynamics.

Let us explore other c values to decide which c values have dynamics similar to f_0 and which do not.

Exercise 1.65. Let $c = 0.2 + 0.2i$ and use the *Global Complex Iteration Applet for Polynomials* to study the dynamics of $f_{0.2+0.2i}(z) = z^2 + 0.2 + 0.2i$ and to see the global picture of the attracting basins and the Julia set. Calculate by hand the exact value of the attracting fixed point of this map. You can test your calculation by using the applet to iterate a nearby seed. *Try it out!*

Now let us see what happens if we move c far from 0.

Exercise 1.66. Let $c = -1$ and use the *Global Complex Iteration Applet for Polynomials* to study the dynamics of $f_{-1}(z) = z^2 - 1$. Are they similar to those of f_0? *Try it out!*

Exercise 1.67. By generalizing the calculations for $c = 0.1$, mathematically describe the set K_1 of all c values such that f_c has a finite attracting fixed point in $\mathbb{C}$. Start by solving an equation to find the fixed points, and then consider what conditions need to be met to make one of them attracting. For all c values, f_c has fixed points. Our goal is to determine what c values have a fixed point that is attracting. You can see the picture of K_1, which is called a *cardioid* (a heart shaped region) in Figure 1.13 on page 32. *Try it out!*

By examining the attracting basins and Julia sets for maps f_c where $c \in K_1$ (using the *Global Complex Iteration Applet for Polynomials*) you can see that each function has dynamics similar to the dynamics of f_0. Thus we have seen our first example of *stability in the parameter*, i.e., for any c in this set of parameters, the dynamics does not fundamentally change when you change the parameter slightly. Another way of saying this is that K_1 is an open set of parameter values.

We know that for $c \in K_1$ there is an attracting fixed point p_c of the map f_c. Let's call $\lambda(c)$ the multiplier at p_c and thus, using Theorem 1.48, which says that the multiplier of a (super) attracting fixed point must have modulus strictly less than 1, we may regard λ as a map from K_1 into $\triangle(0, 1)$. If we follow the calculations in Exercise 1.67 carefully, we see that this *multiplier map* $\lambda : K_1 \to \triangle(0, 1)$ is one-to-one, continuous, and onto (onto means that for every $\lambda_0 \in \triangle(0, 1)$ there is a parameter $c_0 \in K_1$ such that $\lambda(c_0) = \lambda_0$). In fact, this map can be extended to be defined and continuous on all of $\overline{K_1}$. (We say a map g defined on its domain set D can be *extended* to a larger set $\tilde{D} \supset D$, if there exists an extension map $\tilde{g}$ on $\tilde{D}$ such that $\tilde{g} = g$ on D. By a slight abuse of notation we use g to denote the extension, which means that we assume g itself is defined on the larger set $\tilde{D}$.) However, for $c \in \partial K_1$ we have only $\lambda(c) \in C(0, 1)$ and the corresponding fixed point p_c is indifferent.

Let us explicitly compute the multiplier map and consider its inverse. By a direct calculation, we see that $p_c = (1 - \sqrt{1 - 4c})/2$ and $q_c = (1 + \sqrt{1 - 4c})/2$ are the fixed points of f_c. Note that p_c will be attracting for an appropriate choice of c, but q_c will always be repelling (why?). Hence, the multiplier map is given by $\lambda = \lambda(c) = f_c'(p_c) = 2p_c = 1 - \sqrt{1 - 4c}$, when $c \in K_1$. The inverse of the multiplier map, as you can readily compute, is $c = c(\lambda) = \frac{\lambda}{2} - \frac{\lambda^2}{4}$, which gives the c value for a fixed point with multiplier λ. Since c values in K_1 correspond to (super) attracting fixed points, the set K_1 is the image of the set of all $\lambda \in \Delta(0, 1)$ under the map $c(\lambda)$. This produces the set K_1 illustrated in Figure 1.13 which you should take a moment to verify using the *ComplexTool* applet.

Exercise 1.68.

(a) Find the c value such that f_c has an attracting fixed point p_c with multiplier $\lambda = 0.7e^{\pi i/4}$, and then use an applet to illustrate the attractiveness of p_c by iterating points near it.

(b) Find the c value such that f_c has an indifferent fixed point p_c with multiplier $\lambda = e^{\pi i/3}$, and then use an applet to illustrate the behavior of orbits of points near p_c.

Additional Exercise 1.177 will assist in gaining a better understanding of the role of the multiplier in the dynamics near a fixed point.

1.3.4 Cycles of $f_c(z) = z^2 + c$

As we saw, the orbit of 0 under the map $f_{-1}(z) = z^2 - 1$ is $-1, 0, -1, 0, -1, 0, \ldots$. We summarize this by saying that f_{-1} has a 2-cycle $\{0, -1\}$. Also, as we saw (for example, by iterating the seed $z_0 = 0.2 - 0.3i$), the 2-cycle seems to be attracting (actually, we see later that it can be called super attracting).

Exercise 1.69. Modify Definition 1.16 of an attracting fixed point to produce your own definition of an attracting 2-cycle. *Try it out!*

Exercise 1.70. Modify Definition 1.14 of an attracting basin of a point to produce your own definition of an attracting basin for a 2-cycle. *Try it out!*

Using an applet to produce the orbit of several seed values, we seem to have only two types of long term behavior for the map f_{-1}. Iterates of f_{-1} seem to either approach ∞ or approach the 2-cycle $\{0, -1\}$. Intuition suggests that this cannot be the whole story. Tension between attracting basins probably leads to points which are neither attracted to ∞ nor to the 2-cycle $\{0, -1\}$. Use the *Global Complex Iteration Applet for Polynomials* to see the basins, and the corresponding Julia set. *Try it out!*

Example 1.71. Let $c = -0.9 - 0.1i$ and use one of the applets to study the dynamics of the map $f_{-0.9-0.1i}(z) = z^2 + (-0.9 - 0.1i)$ to verify that it has an attracting 2-cycle. *Try it out!*

Example 1.72. Let $c = -0.13 + 0.73i$ and use one of the applets to study the dynamics of the map $f_{-0.13+0.73i}(z) = z^2 + (-0.13 + 0.73i)$ to verify that it has what we could call an attracting 3-cycle. *Try it out!*

We would like to calculate which c parameters will lead to $f_c(z) = z^2 + c$ having an attracting 2-cycle, 3-cycle, 4-cycle, and so on. We will be able to make some progress, but it will be limited because our methods require us to find the roots of polynomials of high degree. We begin by defining cycles and then we see how to classify them as attracting, repelling, or indifferent by using the derivative (or more precisely the multiplier).

1.3.5 p-Cycles and their Classification

Definition 1.73 (Cycles). A point $w \in \overline{\mathbb{C}}$ is called *periodic with period p* for the map f if $f^p(w) = w$ and $w, f(w), \ldots, f^{p-1}(w)$ are distinct points. In such a case we call the set $\{w, f(w), \ldots, f^{p-1}(w)\}$ a *p-cycle* for the map f.

Periodic points correspond to fixed points of higher iterates f^p of the map f. For example, the periodic point of period $p = 2$ at $w = 0$ for the map f_{-1} is a fixed point of the second iterate f_{-1}^2. We can then use this to classify a cycle as attracting, repelling, or indifferent based on the multiplier of the corresponding iterate.

Definition 1.74 (Multiplier for Cycles). Suppose $\{w_0, \ldots, w_{p-1}\}$ forms a p-cycle for the map f. We define the *multiplier λ* of this cycle (also called the multiplier of each point $w_0, \ldots, w_{p-1}$ of period p) to be the multiplier of the map f^p at its fixed point w_0. Then the p-cycle $\{w_0, \ldots, w_{p-1}\}$ of the map f is called

(a) *super attracting* if $\lambda = 0$

(b) *attracting* if $0 < |\lambda| < 1$

(c) *repelling* if $|\lambda| > 1$

(d) *indifferent* if $|\lambda| = 1$.

Example 1.75. The 2-cycle $\{0, -1\}$ for f_{-1} is super attracting since $\lambda = (f_{-1}^2)'(0) = f_{-1}'(0) \cdot f_{-1}'(f_{-1}(0)) = f_{-1}'(0) \cdot f_{-1}'(-1) = 0$. Instead of using 0 we could have used the other point in the 2-cycle to calculate $\lambda = (f_{-1}^2)'(-1) = f_{-1}'(-1) \cdot f_{-1}'(f_{-1}(-1)) = f_{-1}'(-1) \cdot f_{-1}'(0) = 0$.

Since the multiplier λ of a p-cycle is defined via the derivative of f^p, the p-fold composition of f with itself, it is important to understand the use of the chain rule in the classification of p-cycles. Let's examine it more closely.

Chain Rule in $\mathbb{C}$: If f and g are analytic functions at finite points z_0 and z_1, respectively, and if $z_0 \overset{f}{\mapsto} z_1 \overset{g}{\mapsto} z_2$, then $(g \circ f)'(z_0) = g'(f(z_0))f'(z_0) = g'(z_1)f'(z_0)$. In other words, to compute the derivative of the composite function, we multiply the derivatives of each function evaluated at the appropriate point along the way.

Suppose that the set of finite points $\{w_0, \ldots, w_{p-1}\}$ forms a p-cycle for the analytic map f as pictured in Figure 1.12. We have $f^p(w_0) = w_0$, and so by the chain rule we compute

$$\lambda = (f^p)'(w_0) = f'(w_0)f'(w_1) \ldots f'(w_{p-1}) = (f^p)'(w_j), \quad j = 0, \ldots, p-1, \quad (1.1)$$

which shows, among other things, that the multiplier in Definition 1.74 is well defined since the derivative of f^p is the same at any point in the cycle.

We now wish to understand the relationship between the classification of the cycle as attracting, repelling, or indifferent (determined by λ) and the dynamics of the map f near

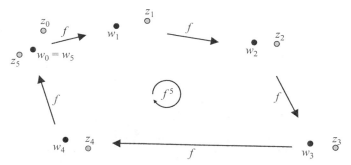

FIGURE 1.12. Illustration of a 5-cycle $\{w_0, w_1, \ldots, w_4\}$ with the partial orbit of a point z_0 chosen near w_0.

the cycle. By the continuity of the map f for any seed z_0 sufficiently close to w_0, the orbit points $z_1, \ldots, z_{p-1}, z_p$ will be close to the points $w_1, \ldots, w_{p-1}, w_0$. Supposing that $|\lambda| < 1$, the map f^p has an attracting fixed point at w_0. Thus, if we choose a seed z_0 sufficiently close to w_0, we have that $f^p(z_0)$ is closer to w_0 than z_0 is (i.e., $|f^p(z_0) - w_0| < |z_0 - w_0|$) by Definition 1.16 applied to the fixed point w_0 of f^p. This argument works for any of $w_1, \ldots, w_{p-1}$, and so one way to describe such an attracting cycle is to say that when you apply the map f a total of p times, points near a w_k will move around the cycle only to return closer to w_k. See Figure 1.12 where the orbit of z_0 exhibits this behavior. In a similar way we justify the classification of repelling cycles.

The calculations require only minor modifications when one of the points w_k in the cycle is ∞. In the spherical metric, where ∞ is no more special than any other point in $\mathbb{C}$, we see that the dynamic behavior (attraction or repulsion) of the cycle behaves in the same way as cycles in the finite plane $\mathbb{C}$.

1.3.6   Attracting Cycles for the Maps $f_c(z) = z^2 + c$

Let us return to investigating the dynamics of the maps f_c. We have seen two examples of maps $f_c(z) = z^2 + c$ with attracting 2-cycles. Let us now determine the set K_2 of all c values such that f_c has an attracting 2-cycle. As 2-cycles generally exist, the issue here is to determine when such a cycle will be attracting. Any point in a 2-cycle must be a fixed point of f_c^2, and not be a fixed point of f_c. Thus we wish to solve the equation A, $(z^2 + c)^2 + c = z$, and exclude the solutions of equation B, $z^2 + c = z$. Since each solution to B, rewritten as $z^2 + c - z = 0$, is a solution to A, rewritten as $(z^2 + c)^2 + c - z = 0$, we have that $z^2 + c - z$ must divide $(z^2 + c)^2 + c - z$. After long division, we rewrite A as $(z^2 + z + 1 + c)(z^2 + c - z) = 0$. Thus a 2-cycle $\{u, v\}$ must be such that both u and v solve $z^2 + z + 1 + c = 0$, i.e., $(z - u)(z - v) = z^2 + z + 1 + c$, which implies $uv = 1 + c$. According to (1.1) the multiplier of the 2-cycle $\{u, v\}$ is $\lambda = f_c'(u) f_c'(v) = 4uv = 4(1 + c)$. Hence the 2-cycle will be attracting exactly when $4|1 + c| = |\lambda| < 1$, i.e., when $|c - (-1)| < 1/4$. Hence $K_2 = \triangle(-1, 1/4)$, which is the disk pictured in Figure 1.13.

Exploration 1.76. Test c values in K_2 by using the *Global Complex Iteration Applet for Polynomials* to see that you get an attracting 2-cycle. Which seed values in the picture produced by the applet seem to find the attracting 2-cycle? ***Try it out!***

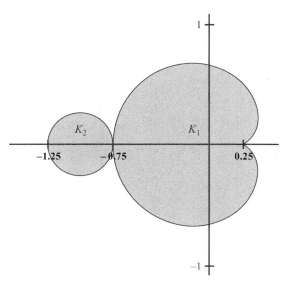

FIGURE 1.13. Parameter space of c values showing the cardioid K_1 with cusp at $c = 1/4$ and the disk $K_2 = \triangle(-1, 1/4)$. The boundaries of K_1 and K_2 meet at $c = -3/4$.

In Additional Exercise 1.178 you are asked to investigate the relationship between the multiplier and the convergence towards the 2-cycle. In Additional Exercise 1.179 you are asked to investigate a multiplier map defined on K_2 and its inverse.

Let us now try to determine the set K_3 of c parameters that lead to attracting 3-cycles for $f_c(z) = z^2 + c$. A point in a 3-cycle must be a fixed point of f_c^3, and not be a fixed point of f_c, and so must satisfy (when substituted for z) the eighth degree polynomial equation $[(z^2 + c)^2 + c]^2 + c - z = 0$, but not be a root of $z^2 + c - z = 0$. After long division we are left with a degree six polynomial to solve if we are to find the points of the 3-cycle. Because we cannot, in general, solve a polynomial of degree five or greater, we will have considerably more difficulty determining the attracting 3-cycles than we did the 2-cycles. However, we were able to locate the c values that correspond to attracting 2-cycles without having to solve for the points of the cycle, although we could have by applying the quadratic formula. We wonder if it is possible to use similar techniques, or devise new ones, to describe as much as we can of this set K_3.

Large Project 1.77. Is it possible to determine the set K_3 precisely, as was done for K_1 and K_2? Though we might not be able to solve the degree six polynomial equation, perhaps we could use other root solving techniques, such as Newton's method, to approximate roots closely enough to be useful. The roots could be tested with the *Global Complex Iteration Applet for Polynomials* to see their role in the dynamics of f_c. Related questions of interest are: Can you show that K_3 is an open set? It turns out that K_3 has more than one connected component (though K_1 and K_2 are connected). Can you determine how many connected components K_3 has and what relationship they have to each other? Can you find the c parameters which lead to super attracting 3-cycles? The roots of the degree six polynomial must include three points which make up the attracting 3-cycle, but what do the other three points represent? What about finding the c parameters that yield attracting $4, 5, 6, \ldots$ cycles?

We present a definition and theorem that may aid in understanding the project because it shows a relationship between multiplier maps and components of K_3. (To recall the definition of a component, see Appendix B.3 on page 359.)

Definition 1.78 (Hyperbolic Components). For each $n \in \mathbb{N}$, we define the set K_n to be the set of parameters c such that f_c has an attracting n-cycle. We call any connected component W of some K_n a *hyperbolic component* of K_n.

Although K_1 and K_2 are connected sets, some K_n are disconnected. However, each connected component of K_n has the following nice property corresponding to its multiplier map. The proof, which is beyond the scope of this text, may be found in [3, p. 134].

Theorem 1.79 (Multiplier Map Theorem). *Let W be a hyperbolic component of some K_n. Let $\lambda : W \to \triangle(0, 1)$ be the multiplier map that takes $c \in W$ to the multiplier $\lambda(c)$ of the associated attracting n-cycle. Then the map λ is one-to-one, analytic, and onto, i.e., it maps W conformally onto $\triangle(0, 1)$. Furthermore, λ extends to a one-to-one continuous map of $\overline{W}$ onto $\overline{\triangle(0, 1)}$.*

Definition 1.80. For a hyperbolic component W of K_n, we call the unique $c \in W$ the *center* of W if $\lambda(c) = 0$, i.e., it is the unique c in W for which f_c has a super attracting n-cycle.

Example 1.81. We have already observed that the center of K_1 is $c = 0$ and the center of K_2 is $c = -1$.

We can prove the theorem for K_1 and K_2 by studying the explicit form of the multiplier map, which you are asked to do in Additional Exercise 1.179. For the other sets K_n, however, it is not so easy. The proof of Theorem 1.79 uses properties of the multiplier map without explicitly constructing it. One important application is that it proves that each hyperbolic component W of K_n is open since it is the conformal image of an open set, namely $W = \lambda^{-1}(\triangle(0, 1))$. Hence, each K_n is also open. Recall, that a map is called *conformal* when it is both one-to-one and analytic.

As we have seen, if we move the parameter c within the cardioid K_1, the dynamics of f_c do not change much. Similarly, if we move the parameter c within the disk K_2, the dynamics of f_c do not change much. Thus we call K_1 and K_2 *stable* regions of parameter space. It is also true that each K_n is a stable region of parameter space, which, given its definition, is another way of saying that K_n is an open set. This follows from Theorem 1.79, but also seems reasonable without appealing to the theorem. If you slightly alter a function with an attracting cycle (keeping in mind the strict inequality condition on the multiplier), then it seems reasonable the new function will have an attracting cycle (i.e., will have a strict inequality condition on the multiplier) in roughly the same place. The details can be shown by working through Additional Exercise 1.180. This is exactly what we see when we move the c parameter by small amounts within the cardioid K_1 or within the disk K_2. We cannot, however, expect this to happen with an indifferent fixed point or cycle since altering the multiplier of an indifferent fixed point could make the modulus be strictly less than one (attracting) or strictly greater than one (repelling), instead of remaining exactly equal to one.

As opposed to the stable parameters found in the sets K_n, we call a parameter c *unstable* if there are parameters arbitrarily close to it for which the maps f_c have fundamentally different dynamics. For example, $c = -3/4$ is an unstable parameter.

Exploration 1.82. Give several reasons why $c = -3/4$ is an unstable parameter. Use the *Global Complex Iteration Applet for Polynomials* to explore the dynamics when c is close to $c = -3/4$, paying special attention to the dynamics for $c_1 = -0.75 + .05, c_2 = -0.75 - .05$, and $c_3 = -0.75 + .05i$. *Try it out!*

Now that you have provided your own reasons, we illustrate the unstable nature of the parameter $c = -3/4$ by describing three particular ways in which the dynamics changes there. We note how the type of attracting cycle, the Julia set, and the orbit of the origin all undergo fundamental changes. We call $c = -3/4$ a *bifurcation point*, since it is the parameter on the boundary of two regions in parameter space where the dynamics undergoes a fundamental change.

1. *Attracting Cycle:* We don't have to know the dynamics exactly at the point $c = -3/4$ to show it is unstable. It is enough to know that there are parameters arbitrarily close to and less than $c = -3/4$ (in K_2) that give rise to an attracting 2-cycle (but no attracting fixed point) and there are parameters arbitrarily close to and greater than $c = -3/4$ (in K_1) that give rise to an attracting fixed point (but no attracting 2-cycle). Verify that as c decreases to and then past $-3/4$ the attracting fixed point becomes indifferent and then splits into an attracting 2-cycle.

Remark 1.83. A different way to look at this is to instead say that as c decreases to $-3/4$ the attracting fixed point merges with a repelling 2-cycle to form an indifferent fixed point exactly at $c = -3/4$. Then as c decreases further, the 2-cycle re-emerges as an attracting 2-cycle and the fixed point becomes repelling. Thus, the fixed point and the 2-cycle exchange "polarity" in this transition. Additional Exercise 1.181 asks you to provide the details.

2. *Julia Set:* We can also see the result of this fundamental change in dynamics by looking at the Julia sets $J(f_c)$ as we vary c. If we start at $c = 0$ and then decrease c, we see the Julia sets $J(f_c)$ change from a circle to a distorted circle. Decreasing c further towards $c = -3/4$, we see that the distorted circle $J(f_c)$ begins to have infinitely many bulbs partially forming as the distorted circle $J(f_c)$ starts pinching in. Exactly at $c = -3/4$ the pinching in for each of the infinitely many distinct bulbs simultaneously becomes complete. Thus we have gone from having one bounded component of $F(f_c)$ for $c \in (-3/4, 1/4)$ consisting of one attracting basin of an attracting fixed point to the situation for $c \in (-5/4, -3/4)$ where there exist infinitely many bounded components of $F(f_c)$ consisting of the attracting basin of an attracting 2-cycle.

3. *Orbit of the origin:* We notice another change in the dynamics as the parameter c moves from K_1 to K_2. The orbit of the origin changes from being attracted to an attracting fixed point to being attracted to an attracting 2-cycle. In each of the examples we have considered where the map $f_c(z) = z^2 + c$ had an attracting cycle, the origin was absorbed into it in the sense that the tail of the orbit (formally $\{f_c^n(0) : n > N\}$ for large N) is nearly identical to the attracting cycle. Another way to say this is that the origin was attracted to the cycle. This is true in general (as we shall soon note in Remark 1.94), and the key fact is that $z = 0$ is a critical point, i.e., $f_c'(0) = 0$.

Exercise 1.84. Verify that $c = 1/4$, the cusp of the cardioid K_1, is an unstable parameter. Use the *Global Complex Iteration Applet for Polynomials* to investigate this with respect to items 1–3 above. ***Try it out!***

Exploration 1.85. What do you think happens to the attracting fixed point and repelling 2-cycle when the parameter c decreases to $-3/4$, but then makes a right turn and starts heading towards $-0.75 + 0.05i$? Investigate this with respect to item 1 (and Remark 1.83) using the *Global Complex Iteration Applet for Polynomials*.

1.4   Critical Points and Critical Orbits

In this section, we address some natural questions you may have been asking as we looked at the dynamics of $f_c(z) = z^2 + c$.

1. Does every such map have an attracting cycle (other than ∞)?

2. How many attracting cycles can f_c have?

3. In the examples we looked at, the orbit of the origin is always attracted to the attracting cycle. Is this true in general?

In this section we see that the key to answering many of these questions involves the notion of critical points, and investigating how the orbit of such points largely determines many key dynamical features. We begin with a definition.

When the power series of an analytic map $f(z)$ at $z_0 \in \mathbb{C}$ has the form $f(z) = f(z_0) + a_k(z - z_0)^k + a_{k+1}(z - z_0)^{k+1} + \ldots$ where $a_k \neq 0$, we say that z_0 maps to $f(z_0)$ with *degree* $v_f(z_0) = k$ (we also call $v_f(z_0)$ the *multiplicity* or *valency*).

Remark 1.86. The condition that z_0 and $f(z_0)$ are in $\mathbb{C}$ will always be met in the examples we look at, and so we leave it to the interested reader to make modifications to this definition when z_0 or $f(z_0)$ is infinity. However, when the term critical point is applied to a point z_0 in the theorems below, we include the possibilities that z_0 or $f(z_0)$ is infinity.

Definition 1.87. We call z_0 a *critical point* of f if $v_f(z_0) > 1$.

If z_0 and $f(z_0)$ are in $\mathbb{C}$, then z_0 is a critical point exactly when $f'(z_0) = 0$ (as in Calculus I). Also, as described in Appendix Section A.6.1 on page 353, since at z_0 the map f is locally a $v_f(z_0)$-to-one mapping, we see that z_0 is a critical point exactly when f is not locally one-to-one.

Definition 1.88. Let f be a rational or entire map. If w is a (super) attracting fixed point of f, then we define the *immediate basin of attraction* $A_f^*(w)$ to be the connected component of $A_f(w)$ that contains w. We point out (but leave it to the interested reader to show) that $A_f^*(w)$ is the component of the Fatou set $F(f)$ that contains w.

Theorem 1.89. *Let w be a (super) attracting fixed point of a non-Möbius rational map f. Then there exists a critical point $z_0 \in A_f^*(w)$ and hence $f^n(z_0) \to w$.*

The proof of Theorem 1.89 is beyond the scope of this text, but it appears as Theorem 7.5.1 in [1]. We note, however, that we have seen this many times now.

Example 1.90. For c in K_1 (see Figure 1.13), we know that $f_c(z) = z^2 + c$ has a finite (super) attracting fixed point p_c. Since the origin is the only critical point (other than ∞, which is fixed) we must then have by Theorem 1.89 that $0 \in A_{f_c}^*(p_c)$ and $f_c^n(0) \to p_c$. This is exactly what we observed (without proof) using the applets in many examples.

To present the corresponding result for attracting cycles we will need the following definition.

Definition 1.91. Let f be a rational or entire map which is non-Möbius. For a (super) attracting p-cycle $w_0, \ldots, w_{p-1}$ of f we define the *immediate basin* of this cycle to be the union of components $\cup_{j=0}^{p-1} F_j$ where F_j is the component of the Fatou set $F(f)$ containing w_j.

Exercise 1.92. The map $f_{-1}(z) = z^2 - 1$ has a super attracting 2-cycle $\{0, -1\}$ and so its immediate basin consists of two components of $F(f_{-1})$, one of which contains 0 and the other which contains -1. Use the *Global Complex Iteration Applet for Polynomials* to investigate the picture shown in Figure 1.14. What are the dynamic properties of the other black bulbs, i.e., components of $F(f_{-1})$? Try to find a pattern to the bulbs so you can experimentally approximate points z_0 such that $f_{-1}^3(z_0) = 0$. How many such points are there? Do the same, but changing 3 to 5. Can you generalize this? **Try it out!**

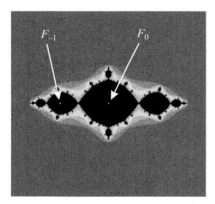

FIGURE 1.14. The immediate attracting basin of the 2-cycle $\{-1, 0\}$ for $f_{-1}(z) = z^2 - 1$ consists of the Fatou components F_{-1} and F_0 containing the cycle points.

Theorem 1.93. *Let f be a rational map which is non-Möbius. Then the immediate basin of each (super) attracting cycle contains a critical point of the map f.*

The proof of this result uses the fixed point version given in Theorem 1.89, the chain rule, and other results that are more than we want to take on at this point. The details can be found in [1] or [3]. Theorem 1.93 leads to the following remark.

Remark 1.94. Since all maps of the form $f_c(z) = z^2 + c$ have only one finite critical point (namely $z = 0$), Theorem 1.93 implies that each such map can have at most one finite (super) attracting cycle and any such cycle must absorb the orbit of the critical point at the origin. Thus we have answered Question 2 and answered Question 3 if f_c has an attracting cycle.

Definition 1.95. Both because 0 is a critical point of the map f_c and because its orbit is important, we call $\{f_c^n(0)\}_{n=1}^\infty$ the *critical orbit* of the map f_c.

We have seen that the critical orbit $\{f_c^n(0)\}_{n=1}^\infty$ plays a special role in understanding the dynamics of the maps $f_c(z) = z^2 + c$. For some c the critical orbit is attracted to an attracting cycle (e.g., $c = 0, -1, 0.278 + 0.534i$) and for other values it is attracted to the super attracting fixed point at ∞ (e.g., $c = 0.3, 4 + i, -2 - 0.3i$). You might wonder then, is it true that for every c value the critical orbit becomes attracted to some (super) attracting cycle? We can use intuition to say that this is probably not true. If we consider the *parameter plane*, also called the c-plane, of all c parameters for the maps f_c, we can think of a tension created by the c values whose critical orbits are attracted to ∞ and the c values whose critical orbits are not. There may be c values where the pull of the critical orbit towards ∞ and the pull of the critical orbit to stay bounded is balanced. Informal reasoning can be the basis for a good guess, but it is far from a proof. We take the easier route and settle the question by looking at the following example.

Example 1.96. Show formally with paper and pencil (or informally with an applet) that for $c = i, 1/4, -5/4, -2$, and $-3/4$, the critical orbit under $f_c(z) = z^2 + c$ is neither attracted to ∞ nor attracted to a finite attracting cycle. Thus we have answered Question 1. *Try it out!*

For $c = i$, you noticed that the critical orbit $i \mapsto i - 1 \mapsto -i \mapsto i - 1 \mapsto -i \mapsto \ldots$ became cyclic, but the cycle did not include 0. Orbits like this have an important role in dynamics and so we give them a special name, as well as the c values which lead to them.

Definition 1.97. We call z_0 a *pre-periodic* point (or *eventually periodic*, but not periodic) under the map f if it is not periodic, but some point on the orbit of z_0 is periodic.

Definition 1.98. We call a parameter c a *Misiurewicz point* if 0 is pre-periodic under f_c.

Thus we see that $c = i$ and $c = -2$ are Misiurewicz points, but $c = 0$ is not since $f_0(z) = z^2$ has a critical orbit $0 \mapsto 0 \mapsto \ldots$ that is periodic, and so not pre-periodic. However, $f_0(z) = z^2$ does have non-critical pre-periodic points, and in Additional Exercise 1.183 you are asked to find them.

If c is a Misiurewicz point, then $J(f_c) = K(f_c)$ (see [3, p. 133]), in which case we call $J(f_c)$ a *dendrite*. Figure 1.15 shows an example.

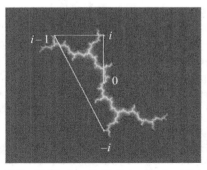

FIGURE 1.15. The dendrite $J(f_i)$ along with the critical orbit $i, i - 1, -i, i - 1, -i, \ldots$

Remark 1.99. To correctly understand pre-periodic points and Misiurewicz points we note the important distinction between orbits that are *pre-periodic* (which we informally described above as those which became cyclic) and those orbits that are attracted to a cycle. The difference is the same as the difference between a sequence approaching a value and a sequence eventually being a value. For example, the sequence $1, 1/2, 1/3, 1/4, \ldots$ approaches 0, but never becomes 0. However, the sequence $2, 1, 1/2, 0, 0, 0, \ldots$ eventually becomes and stays at 0. When using technology such as our applets it can be difficult, if not impossible, to distinguish between the two concepts. For example, set $c = -0.9$ and look at the numerical values of the first 100 points of the orbit of $z_0 = 0$ under f_c. Looking at the 75th and higher orbit values, we see that the points bounce back and forth between what appears to be the same two values. This might lead you to incorrectly conclude that $z_0 = 0$ is pre-periodic. The problem is that the true orbit in this case never exactly bounces back and forth between the exact same values. In fact, it is true (which we leave to the reader to show) that $z_1 < z_3 < z_5 < \ldots$ and $z_0 > z_2 > z_4 > \ldots$ with strict inequalities. However, the odd terms approach $-0.8872983346207418\ldots$ and the even terms approach $-0.1127016653792581\ldots$. Since the applet truncates the data for each z_n, it appears that the odd sequence and even sequence eventually become constant. When trying to see if a point is pre-periodic, technology may mislead you. Careful analysis and proof need to be used.

Let us return to the question, does the critical orbit for every map f_c get attracted to an attracting cycle? We saw that for $c = i, 1/4, -5/4, -2,$ and $-3/4$ this is not so. However, in each case the critical orbit either became cyclic or was attracted to a cycle, though not necessarily an attracting cycle. As mathematicians we naturally wonder, is this always so? Is it possible for a c value to be such that the critical orbit neither becomes cyclic nor is attracted to any cycle? The answer is yes. We do not have the space to explain the deep mathematics to justify this. Nevertheless, asking good questions like this, whether or not we can answer them, is an important part of contributing to mathematics.

So it is not true that every c parameter has a critical orbit that either becomes cyclic or is attracted to a cycle. What is true, however, is that every c parameter has a critical orbit that is either attracted to ∞ or remains bounded. It is this dichotomy that leads us to consider one of the most beautiful objects in all of mathematics, the Mandelbrot set.

Definition 1.100. The *Mandelbrot set* is $M = \{c \in \mathbb{C} : f_c^n(0) \nrightarrow \infty\}$.

We have already encountered some important aspects of M, namely, it contains both K_1 and K_2 in Figure 1.13. We also know that it contains K_n for every n because for f_c to have an attracting n-cycle, the critical orbit must be attracted to it by Theorem 1.93, and thus not be attracted to ∞. The calculation of the sets K_n is an arduous task and a complete description of all such sets has for years stumped mathematicians. Let us therefore use the computer to draw M for us and experimentally investigate the sets K_n and M (keeping in mind that limitations discussed in Remarks 1.55 and 1.99 force us to moderate the confidence that we can place in such pictures). We begin by using the following applet to construct a picture of M.

The *Mandelbrot Set Builder Applet* will color each selected point c in the parameter plane red if $c \notin M$ or black if $c \in M$. Thus, if the the critical orbit under f_c approaches ∞, then c is colored red, otherwise it is colored black. We cannot compute the infinite number of points in the critical orbit, so the applet computes only the number of iterates allowed in the **Maximum Iterations** input box. Setting this to 100 will produce nice results (we encourage you, however, to vary the value to see the effect it has on the picture). The applet will color a selected c value red if and only if one of the calculated critical orbit points lands outside of $\overline{\triangle(0,2)}$. This is justified by the following lemma, which you are asked to prove in Additional Exercise 1.191.

Lemma 1.101. *If the critical orbit $f_c^n(0)$ ever escapes $\overline{\triangle(0,2)}$, then it converges to ∞.*

Exploration 1.102. Experiment with the *Mandelbrot Set Builder Applet* to get a feel for the mathematics that defines the Mandelbrot set. ***Try it out!***

A feature of the Mandelbrot set which stands out is that it is symmetric about the x-axis. Another feature you may have observed is that M has no holes. You are asked to prove these results in Additional Exercises 1.184 and 1.185. The *Mandelbrot Set Builder Applet* is a tool for visualizing M and can lead us to pursue some of its interesting features, but there are some important properties of M it cannot help with, such as whether M is a closed set or not. To do this we need more formal mathematics.

Lemma 1.103. *The Mandelbrot set M is closed.*

Proof. Suppose $c_k \to c^*$ where $c_k \in M$. We will show that $c^* \in M$, thus proving that M is closed. By observing that for f_c the terms of the critical orbit are $0, c, c^2+c, (c^2+c)^2 + c, \ldots$, we see that the nth term can be written $Q_n(c)$ for a polynomial Q_n. Fix $n \in \mathbb{N}$ and note that Q_n is continuous. Since $|Q_n(c_k)| \leq 2$ for each $k \in \mathbb{N}$ by Lemma 1.101, we have $|Q_n(c^*)| \leq 2$ since $|Q_n(c_k)| \to |Q_n(c^*)|$ by the continuity of Q_n. Since this holds for every n, we have shown that the critical orbit of f_{c^*} is contained within $\overline{\triangle(0,2)}$, and thus $c^* \in M$.   □

We close this section by presenting without proof two interesting facts about M. First, the set of Misiurewicz points, each of which is in M (why?), is dense in the boundary ∂M. This means that given an open set U that contains a point in ∂M, the set U must also contain a Misiurewicz point (see [3, p. 133]). Second, ∂M is contained in the closure of the centers of the hyperbolic components defined in Definition 1.80 (see [2, p. 100]). Thus, an open set that contains a point in ∂M also contains the center of a (small) hyperbolic component. Because a center of a hyperbolic component is in the interior of M and is separated away from any other center, we see that ∂M must be quite complicated. We encourage you to take a moment to reflect and fully digest the previous statement. Start by explaining how the set $K_1 \cup K_2$ in Figure 1.13 fails to have this property, but that by attaching many tiny bulbs we can generate a set with this property.

Remark 1.104. Although a lot is known about M, one important question that has stumped mathematicians thus far, is whether or not M contains an open set which does not meet any K_n, that is, an open set of c values for which no f_c has an attracting cycle other than ∞. The conjecture that asserts that this cannot happen is the *density of hyperbolicity conjecture* and remains the focus of research.

1.5   Exploring the Mandelbrot Set

The Mandelbrot set has been called one of the most beautiful objects in mathematics. Our red and black picture created by the *Mandelbrot Set Builder Applet* does not do it justice. The set M has many tiny intricate hairs and bulbs that are hard to see with this bichromatic picture, and the complexity of the picture cries out for a way to zoom in on them.

Initially it was conjectured that the Mandelbrot set is disconnected because low resolution pictures did not show the fine details and thin filaments that connect all the parts of M. However, M is connected (see [1, p. 239]). We cannot prove this here, but a more sophisticated applet will give us pictures that hint at this.

From now on we use the *Parameter Plane and Julia Set Applet*, which colors points in M black, but colors the parameters $c \notin M$ a different shade based on how many iterations it takes for the critical orbit to escape $\overline{\triangle(0, 2)}$. This gives a better feel for the immense detail of the Mandelbrot set M (see Figure 1.16). The applet also shows, for each selected c, a picture of the corresponding Julia set.

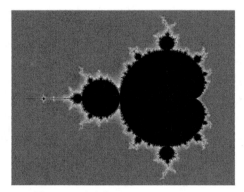

FIGURE 1.16. The Mandelbrot set.

Exploration 1.105. Experiment and play with the *Parameter Plane and Julia Set Applet*, zooming in on parts of M and the Julia sets. Look at not just the geometry of the pictures, but also at the dynamics. What do the bulbs and hairs mean dynamically? Is there a relationship between the geometry and the dynamics? Look for patterns and discover new fine details in the hairs and bulbs. Describe what you find and make a list of observations, questions, and conjectures. You have the tools to explore an infinitely complex world. There are millions of features of M and the related Julia sets to be found by using this applet. In fact, after exploring for a sufficient amount of time, zooming in on the fine details of the sets, you will likely find a picture that no other human has ever seen before! ***Try it out!***

We have quite a menagerie of pictures to see and investigate, including Rabbits, Dragons, and Elephants. We even have Star Clusters, Galaxies, and baby Mandelbrot sets (see Figure 1.17).

In the next section we investigate star cluster sets more carefully so that we can better understand what we are seeing (or not seeing) in such pictures.

FIGURE 1.17. From top left to bottom right we see a Rabbit (Julia set when $c = -0.12 + 0.75i$), a Dragon (Julia set when $c = 0.36 + 0.1i$), Elephants (zoom in on the Mandelbrot set centered at $c = -0.77 + .173i$), a Star Cluster (Julia set when $c = -0.4387 + 0.784i$), a Galaxy (zoom in on the Mandelbrot set centered at $c = -0.75623053 + 0.06418323i$), and a baby Mandelbrot set (zoom in on the Mandelbrot set centered at $c = -1.625$ and adjust the **Parameter plane max iterations** to 150 for a better resolution).

1.5.1 Cantor Dust Sets

We should always be careful when using technology to represent mathematics. What we see is not always an accurate representation of what we are trying to see. For example, using $c = 0.21 + 0.64i$ in the *Parameter Plane and Julia Set Applet*, it might appear that $J(f_c)$ is empty and that all points iterate to ∞ under f_c. If you display the picture in black and white via the **Dynamic plane black/white plot** checkbox on the **Settings tab**, you will see an all white screen, which taken at face value would mean $J(f_c)$ is empty. That, however, is far from the truth and in Additional Exercise 1.186 you are asked to show that there are infinitely many points that do not iterate to ∞. However $J(f_c)$, especially for having so many points, is small in the sense that it does not show up on the computer screen very well. To see the points better we view the color picture and then zoom in far on the non-red parts (e.g., center each zoom in the middle of the largest star cluster) to eventually see regions of black, which represent points which do not iterate to ∞. However, even the regions of black are not what they seem to be. Because we use only a finite number of **Dynamic plane max iterations** some of the black points would ultimately iterate to ∞ if we increased the number of iterations. However, others will not iterate to ∞ and are correctly colored black.

Exercise 1.106. Adjust the **Dynamic plane max iterations** value to make $J(f_c)$, using $c = 0.21 + 0.64i$, harder to see and more accurate, or easier to see and less accurate. Also, use the **Dynamic plane black/white plot** feature to see how it can be used to give you a better picture of $J(f_c)$, keeping in mind that what we see on the computer screen is only an approximation. *Try it out!*

The Star Cluster set $J(f_{0.21+0.64i})$ and other hard-to-see though infinite sets of points are called *Cantor dust* sets. It is hard to see them because each point in the set is disconnected from any other point. That is, for two points z_0 and w_0 in the set, there is a simple closed curve that never meets the Cantor dust set, but winds around z_0 without winding around w_0. Such a set is called *totally disconnected* because the only connected subsets are single points. In Section 1.5.2 we explore one method to prove that certain Julia sets have this property. We first use our applet to illustrate many examples of this phenomenon.

Exploration 1.107. Experiment with the *Parameter Plane and Julia Set Applet* to find examples of Cantor dust Julia sets $J(f_c)$. Can you make a conjecture about which c values correspond to them? *Try it out!*

1.5.2   Connectedness Locus

One of the properties of M that you may have observed in your experimentation is that for each $c \in M$, the Julia set $J(f_c)$ is connected, and for each $c \notin M$, the Julia set $J(f_c)$ is disconnected. (Your informal understanding of connectedness will suffice in this text; however, the following is a formal definition. A compact set E in $\mathbb{C}$ is *disconnected* if there exists a simple closed curve in $\mathbb{C} \setminus E$ which winds around some points of E, but not all points of E.) Move the c value around the parameter plane in the *Parameter Plane and Julia Set Applet* to observe this. This is not a coincidence, and it is because of this that we call M the *connectedness locus* for the family of maps $\{f_c : c \in \mathbb{C}\}$, i.e., $M = \{c \in \mathbb{C} : J(f_c) \text{ is connected}\}$. This is proved by using the following theorem and noting that f_c has its sole finite critical point at the origin.

Theorem 1.108. *Let f be a polynomial of degree greater than or equal to two. Then every finite critical point of f has a bounded orbit if and only if $J(f)$ and $K(f)$ are connected.*

The proof is beyond the scope of this text (see [1, p. 202] for details), but for the maps f_c we will be able to illustrate the reasoning behind the following related dichotomy.

Theorem 1.109. *For the map f_c, either*
(a) $\{f_c^n(0)\}_{n=1}^{\infty}$ is bounded in $\mathbb{C}$ and $J(f_c)$ is connected, or
(b) $f_c^n(0) \to \infty$ and $J(f_c)$ is totally disconnected.

We illustrate the method of proof, though some details requiring advanced methods are left to be found in [3, p. 67]. We first make an important definition.

Definition 1.110. A *compact topological disk (ctd)* is a compact set (i.e., a closed and bounded set) in $\mathbb{C}$ whose boundary is a simple closed smooth path.

Its importance comes from the following lemma and from noting that any ctd is connected.

Lemma 1.111. *If E is a ctd and $c \notin \partial E$, then $f_c^{-1}(E)$ consists of*
(a) one ctd containing 0, if $c \in Int(E)$,
(b) two ctd's, neither containing 0, if $c \notin E$.

We do not prove this, but with the help of our applet we will illustrate it.

Sketch of proof of Theorem 1.109. Choose $r > \max\{2, |c|\}$ such that $|f_c^n(0)| \neq r$ for all $n = 1, 2, \dots$. Define $E_0 = \overline{\Delta(0, r)}$ and $E_n = \{z \in \mathbb{C} : f_c^n(z) \in E_0\}$ for each $n = 1, 2, \dots$. Thus $E_n = f_c^{-1}(E_{n-1})$ for each $n = 1, 2, \dots$ and $c \notin \partial E_n$ for each $n = 0, 1, 2, \dots$. From Additional Exercise 1.191, we know that $|z| > r$ implies $|f_c(z)| > |z| > r$ and $f_c^n(z) \to \infty$. With this information it can be shown that $E_0 \supset E_1 \supset E_2 \supset \dots$ and $K(f_c) = \cap_{n=0}^{\infty} E_n$. We now investigate cases (i) and (ii) of the theorem by analyzing the sets E_n.

Suppose $\{f_c^n(0)\}_{n=1}^{\infty}$ is bounded in $\mathbb{C}$, i.e., $0 \in K(f_c)$. Since $c = f_c(0)$, we have $c \in K(f_c)$, and so $c \in E_n$ for all $n = 0, 1, 2, \dots$. Since $c \notin \partial E_n$, we may then also conclude $0 \in \text{Int}(E_n)$ for all $n = 0, 1, 2, \dots$. Applying Lemma 1.111 inductively we may then show that each E_n is connected as it consists of a single ctd (see Figure 1.18 (left)). Though we omit the details, it follows from a more advanced topological argument that $K(f_c) = \cap_{n=0}^{\infty} E_n$ is connected, which by another advanced topological argument implies $J(f_c) = \partial K(f_c)$ is connected.

Suppose $f_c^n(0) \to \infty$, i.e., $0 \notin K(f_c)$. By the choice of r we know $c \in \text{Int}(E_0)$. However, since 0 does not lie in every E_n (else 0 would lie in $K(f_c)$), there is a positive integer m such that $c \in \text{Int}(E_{m-1})$ but $c \notin E_m$. Thus by Lemma 1.111 we see that E_m consists of one ctd, but $E_{m+1} = f_c^{-1}(E_m)$ consists of two ctd's (see Figure 1.18 (right) where $m = 2$). Applying Lemma 1.111 to each ctd in E_{m+1}, we see that E_{m+2} consists of four ctd's. Proceeding inductively we see that E_{m+k} consists of 2^k ctd's. Although we do not have the tools here to prove it, the size of the ctd's that make up E_{m+k} shrinks to zero as $k \to \infty$, and so we conclude that any two distinct points z and w in $K(f_c)$ must lie in distinct ctd's of E_{m+k} when k is sufficiently large. Hence the boundary of such a ctd containing z, which lies in $A_{f_c}(\infty)$ (why?), is a curve which encloses z but not w.

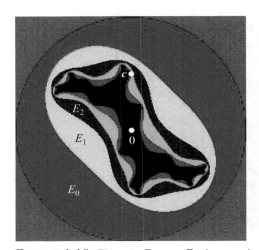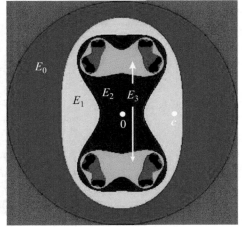

FIGURE 1.18. The sets $E_0, \dots, E_5$ drawn using the *Parameter Plane and Julia Set Applet* with **Dynamic plane escape radius** $r = 2$, **Dynamic plane max iterations** = 5, and **Color sample rate** = 7. In both pictures $E_0 = \overline{\Delta(0, r)}$. The left picture shows $c = i$ with each E_n connected. The right picture shows $c = 0.9$ with each E_0, E_1, and E_2 connected, but subsequent E_n disconnected.

We conclude that $K(f_c)$ is totally disconnected, and thus so is $J(f_c)$ since by utilizing Remark 1.64 we can show that $J(f_c) = \partial K(f_c) = K(f_c)$.   □

So, omitting a few details, we were able to understand how Cantor dust sets naturally arise when considering $J(f_c)$. Another topological property which you may have observed in your explorations is that $K(f)$ has no holes. This we have all the tools to prove, where we define a *hole* as a bounded domain U in $\overline{\mathbb{C}} \setminus K(f)$ such that $\partial U \subseteq K(f)$. (The advanced reader will recognize that the absence of holes in $K(f)$ is equivalent to the set $\overline{\mathbb{C}} \setminus K(f)$ being connected.) We prove this by appealing directly to the definition of $K(f)$ and applying the maximum modulus theorem.

Lemma 1.112. *Let f be a polynomial of degree greater than or equal to two. Then $K(f)$ has no holes.*

Proof. Suppose the set U is a hole in $K(f)$. Let $R > 0$ be such that $|z| > R$ implies $|f(z)| > |z| > R$ (such an R must exist because f is a polynomial of degree greater than or equal to two). From this, we can show $f^n(z) \to \infty$ whenever $|z| > R$. Thus $|f^n(z)| \leq R$ for all $n \in \mathbb{N}$ for any z with a bounded orbit. This means, by the definition of $K(f)$, that we have $K(f) = \{z \in \mathbb{C} : |f^n(z)| \leq R \text{ for all } n \in \mathbb{N}\}$. Since $\partial U \subseteq K(f)$, we see that for any $n \in \mathbb{N}$, the polynomial f^n is bounded by R on ∂U. By a version of the maximum modulus theorem (given as Appendix Corollary A.19 on page 352), we conclude that f^n is bounded by R on U as well. Since this works for all $n \in \mathbb{N}$, we see that $U \subseteq K(f)$, contrary to our assumption that U is a hole in $K(f)$ (that does not meet $K(f)$). This contradiction proves the result.   □

1.5.3   Self-similarity and Symmetry

One of the properties you may have noticed in dynamical plane pictures for the maps f_c is that they are symmetric about the origin. That is, each set $A_{f_c}(\infty)$, $J(f_c)$, and $K(f_c)$ has the property that it contains z if and only if it contains $-z$. You are asked to prove this in Additional Exercise 1.187.

Another property you may have noticed in each set $J(f_c)$ is that small parts of $J(f_c)$ look like other larger parts of $J(f_c)$. We call this property *self-similarity* and note that it is a property of all Julia sets.

Exploration 1.113. Set $c = 0.112 + 0.74667i$ in the *Parameter Plane and Julia Set Applet* and in the dynamic plane (z-plane) window set the x range to be 0.22026 to 0.28746 and set the y range to be 0.4648 to 0.532, and then hit the **Update** button. Notice how similar the picture looks when we zoom in by setting the x range to be 0.2277184 to 0.2475200 and the y range to be 0.50166592 to 0.52146752. Continue zooming in on a point of $J(f_c)$ to see how at all scales (i.e., depth of zoom) the picture, after rotating, looks like the first picture. You can click on the thumbnails at the bottom to see any previously created pictures. Repeat this with other c values to get a sense that this is a general property of Julia sets. *Try it out!*

How about the Mandelbrot set M? Is it self-similar too? The answer is a definite ... sort of. There are places in M that have small pieces that look like larger pieces, but it is not the case, as in the Star Cluster sets above, that the whole set M looks like a bunch of small copies of one piece of itself. The Mandelbrot set is sometimes called quasi-self-similar for this reason.

Exploration 1.114. Set $c = 0.41491386 + 0.60134804i$ in the *Parameter Plane and Julia Set Applet* and zoom in on the Mandelbrot set, centering the zoom on c. What you see are pictures that appear to look the same as you zoom in. If you zoom in on the tips of any of the antennae of the Mandelbrot set you will see a type of self-similarity. Now see what you find when you repeatedly zoom in on a point in the middle of such an antenna. *Try it out!*

Zoom in on the point $c = -1.2418406 - 0.32366967i$ to find what we call a *baby Mandelbrot set*. Set the **Parameter plane max iterations** to a higher value (such as 200 or 300) to see this picture better. You can also change the **Color sample rate** to adjust the color scheme to create a nicer picture. What you have found is not an exact copy of the full Mandelbrot set, as you can tell by the long antennae coming out of it, but it has the unmistakeable look of M. It turns out, due to a deep result, that these baby Mandelbrot sets are dense in the boundary of M, that is, given any neighborhood of a point in ∂M, no matter how small, there exists a baby Mandelbrot set in it. To see it we may have to zoom in pretty far, but it's there. This phenomenon, and more, is discussed in Section 1.7.

Exploration 1.115. Try zooming in on an arbitrary point in the ∂M, and then see if you can find a baby Mandelbrot set. You might need to adjust the **Parameter plane max iterations** to give you a better view. *Try it out!*

1.5.4 Bulbs in M

Let's now try to classify systematically some of what we see in M, in terms of geometry and dynamics. Each point c in the parameter plane (also called the c-plane) corresponds to a function f_c whose Julia set $J(f_c)$, Fatou set $F(f_c)$, and dynamics are viewed in the dynamic plane (also called the z-plane). We will see that M, and the parts that make it up, provide us with a sort of dictionary or index of the types of dynamics we find for $f_c(z) = z^2 + c$.

We have already seen that the cardioid K_1 in Figure 1.16 consists of c values corresponding to maps with attracting fixed points. Off it there are infinitely many bulbs attached that we would like to understand both by their geometrical properties as sets in the c-plane and by the corresponding dynamic properties in the z-plane.

The most prominent bulb off K_1 is the disk $K_2 = \triangle(-1, 1/4)$ representing the parameters corresponding to maps with attracting 2-cycles. The next largest bulb off K_1 is near the top. Select $c = -0.12 + 0.77i$ from it and consider the critical orbit in the dynamic plane to see that there exists an attracting 3-cycle. Use the *Parameter Plane and Julia Set Applet* to observe this by iterating the critical orbit one iterate at a time (by checking the **Show critical orbit** box, then checking the **Iterate orbits** box, and then hitting the + button that appears next to the **Iterate orbits** box). For this c, the Julia set $J(f_c)$ in the dynamic plane is pinched in such a way that three bulbs meet at every pinch point. Let's call the pinch point that corresponds to the immediate basin of this attracting 3-cycle the *main pinch point* (which in this case is near $-0.282 + 0.492i$). Then by iterating the origin one step at a time you can see that each iterate makes roughly a $1/3$ counterclockwise rotation around it. You can use the **Connect orbit points** checkbox to help you track the path of the orbit. Trying other c values from the bulb in the c-plane we find that we always get this same type of behavior. For this reason we call the bulb in M the $1/3$ bulb.

Exercise 1.116. Use the *Parameter Plane and Julia Set Applet* to investigate the dynamics for $c = -0.513 + 0.5693i$. Decide what fraction p/q would best describe the bulb containing this point. *Try it out!*

In the dynamic plane we see an attracting 5-cycle, which cycles around a main pinch point at which five bulbs meet. We see by iterating the origin one step at a time that the iterates make a $2/5$ rotation about the main pinch point in each step. For this reason we call the bulb of M that contains $c = -0.513 + 0.5693i$ the $2/5$ bulb. You should experiment to see this behavior for other c values within this $2/5$ bulb. *Try it out!*

Definition 1.117. Let B be a hyperbolic component of K_q whose boundary meets the boundary of the main cardioid K_1. We call B the *p/q bulb*, and denote it $B_{p/q}$, if for each $c \in B$ the map f_c has an attracting q cycle and each step in the critical orbit in the dynamic plane makes roughly a p/q rotation about the main pinch point.

From what follows it will become clear that for a rational number $p/q, 0 \le p/q \le 1$, there is only one p/q bulb and so we are justified in calling $B_{p/q}$ *the* p/q bulb, as opposed to *a* p/q bulb.

Exercise 1.118. Experiment and prove a relationship between the p/q bulb and its conjugate bulb, that is, the bulb reflected over the x-axis. Find p' and q' if the conjugate of the p/q bulb is the p'/q' bulb. *Try it out!*

Remark 1.119. Theorem 1.79 shows that the multiplier map $\lambda_{p/q}$ maps $B_{p/q}$ conformally onto $\triangle(0, 1)$. The map $\lambda_{p/q}$ extends to be a one-to-one continuous map of $\overline{B_{p/q}}$ onto $\overline{\triangle(0, 1)}$.

It turns out that $\overline{B_{p/q}}$ meets the boundary of the main cardioid ∂K_1 in a single point, which we denote $c_{p/q}$, and call the *root of the p/q bulb*. For example, $c_{1/2} = -3/4$ is the root of the $1/2$ bulb $B_{1/2} = K_2$. In Additional Exercise 1.188 you are asked to investigate the multiplier maps evaluated at the root. Note that since the root $c_{p/q}$ of the p/q bulb lies on the boundary of two hyperbolic components, namely K_1 and $B_{p/q}$, we see that there are two multiplier maps (namely, $\lambda : \overline{K_1} \to \overline{\triangle(0, 1)}$ and $\lambda_{p/q} : \overline{B_{p/q}} \to \overline{\triangle(0, 1)}$) that are defined at each $c_{p/q}$. The root points play a special role and so we wish to understand them further. In particular, we wish to understand the following dynamic property.

Exercise 1.120. The root $c_{p/q}$ of the p/q bulb is the c parameter for which f_c has an indifferent fixed point with multiplier $e^{2\pi i p/q}$, i.e., $c_{p/q} = c(e^{2\pi i p/q})$ where $c(\lambda)$ is the inverse of the multiplier map $\lambda : \overline{K_1} \to \overline{\triangle(0, 1)}$. Since $c_{p/q}$ is a bifurcation parameter, you should describe three dynamical changes (as done in Exercise 1.84) that occur as the parameter c moves from the main cardioid K_1 into $B_{p/q}$ passing through the root $c_{p/q}$. *Try it out!*

Exploration 1.121. Explore and label other bulbs in the same way as in Exercise 1.116. We recommend that you print a large image of M and label each bulb as you go. *Try it out!*

While doing Exploration 1.121 you may have seen a pattern that shows how you can quickly compute p/q for the largest bulb between two bulbs. We explain the pattern, but leave the detailed proofs to be found in [7]. Looking at M we see $B_{1/2}$ and $B_{1/3}$ each attached to the main cardioid, with infinitely many p/q bulbs in between decorating the boundary of the main cardioid. The largest, as we have seen, is $B_{2/5}$. We can correctly "do the math" in this situation by using Farey addition to compute

$$\frac{1}{2} \oplus \frac{1}{3} = \frac{2}{5}.$$

In Farey addition, we add fractions in the unusual way of adding the numerators and adding the denominators. We call the two addends the *Farey parents* (e.g., $\frac{1}{2}$ and $\frac{1}{3}$) and the resulting fraction the *Farey child* (e.g., $\frac{2}{5}$). Correspondingly, we call the bulbs $B_{1/2}$ and $B_{1/3}$ the Farey parents of the Farey child bulb $B_{2/5}$.

Example 1.122. The Farey child of $B_{2/5}$ and $B_{1/3}$, that is, the largest bulb between them, is $B_{3/8}$ since

$$\frac{2}{5} \oplus \frac{1}{3} = \frac{3}{8}.$$

A quick check with the *Parameter Plane and Julia Set Applet* will confirm this. Remember to use the **Connect orbit points** checkbox to help you track the path of the orbit. ***Try it out!***

Remark 1.123. We must be careful with our use of Farey addition of bulbs. The main rule we must be sure to adhere to is: Two bulbs can only be Farey parents if all the bulbs between them are smaller than they are. For example, the largest bulb between the $4/11$ bulb (the one containing $c = -0.292 + 0.633i$) and the $2/5$ bulb is the $3/8$ bulb. Here Farey addition does not work. Because the $3/8$ bulb is larger than the $4/11$ bulb (which you should check on the applet), we know that Farey addition is not applicable.

There is another issue to be dealt with if Farey addition is to help us compute the bulb fraction for all p/q bulbs. Between $B_{1/3}$ and the cusp of the cardioid K_1 there is a largest bulb, but how can we use Farey addition to determine it? The key is to treat the cusp like a Farey parent which is larger than all other bulbs. Additional Exercise 1.190 asks you to determine experimentally what Farey fraction should be used to represent the cusp.

1.5.5   Sub-bulbs of M

Just as the main cardioid K_1 of M has many p/q bulbs attached to it, each p/q bulb has many sub-bulbs attached. Let's investigate them. Use the *Parameter Plane and Julia Set Applet* to view the change in dynamics as we let c vary from $c_1 = -0.16097811 + 0.80545706i$ within the $1/3$ bulb to $c_2 = -0.17462417 + 0.8296561i$ in an attached sub-bulb. We see that the attracting 3-cycle becomes an attracting 15-cycle. The picture of the Julia set with the attracting 15-cycle has features common to Julia sets with an attracting 3-cycle and Julia sets with an attracting 5-cycle. We see that three bulbs of $K(f_{c_1})$ meet at each pinch point. In changing c to c_2, we see that the pinch points from $K(f_{c_1})$ persist, but also, within each bulb of $K(f_{c_1})$, new pinching occurs, such that at each of these newly formed pinch points five bulbs of $K(f_{c_2})$ meet. See Figure 1.19 and use the zooming

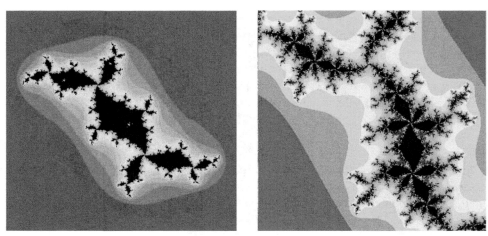

FIGURE 1.19. Left shows $J(f_{c_1})$ and right shows $J(f_{c_2})$, slightly magnified.

features in the *Parameter Plane and Julia Set Applet* to see the pattern repeat at all scales (remembering to adjust the **Dynamic plane max iterations** value as needed).

Small Project 1.124. Investigate the behavior seen above by entering into a variety of sub-bulbs attached to a variety of p/q bulbs. Can you discern a pattern? If shown an example of a Julia set such as the one in Figure 1.20, can you identify which sub-bulb the c value came from?

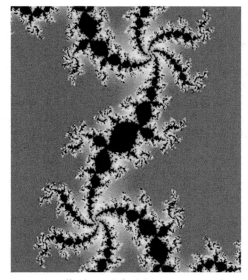

FIGURE 1.20. How can you tell where to find the c parameter for this picture of $J(f_c)$?

1.5.6   Limbs and Antennae in M

The Mandelbrot set M is a connected set. We can formally describe this by saying that a simple closed curve in the complement of M must either wind around no point of M or must wind around all of M. That M is connected is not easy to prove and we shall not attempt to do it, but we will use it to describe some other important aspects of M. In particular, we make use of the fact that we can naturally disconnect M into two connected pieces by removing a root $c_{p/q}$ of any p/q bulb (which we also will not prove here).

Definition 1.125. The set $M \setminus \{c_{p/q}\}$ consists of two connected sets, one that contains the cardioid K_1 and the other containing $B_{p/q}$, which we call the p/q *limb*.

On each p/q limb there is the main p/q bulb with infinitely many sub-bulbs attached. Each of these sub-bulbs has an infinite number of tinier sub-sub-bulbs, and so on. However, the p/q limb contains more than just these bulbs. It also contains what we informally call *antennae* made out of thin filaments that reach out from the p/q bulb. Looking straight above the 1/3 bulb there is a junction where three equally spaced filaments, which we call *spokes*, meet (see Figure 1.21). The *main* or *principal spoke* is the one that attaches to the 1/3 bulb, and the shortest spoke is a 1/3 rotation about the junction point from the main spoke. This relationship between the short spoke and the p/q value occurs frequently, but not always (e.g., examine $B_{1/5}$, which contains the point $c = 0.39 + 0.33i$, using the *Parameter Plane and Julia Set Applet*). This generalization has been made into a formal theorem, but to do so, the notion of shortest had to be changed slightly (see [7]). We also note that if you zoom in on other junctions found on the antennae on the 1/3 limb, you will often, but not always, see three spokes meeting at the junction. This seeming coincidence cries out for further exploration. First we make a definition.

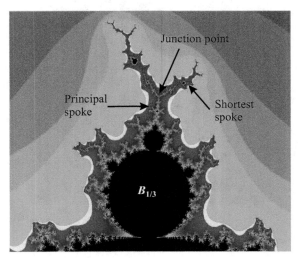

FIGURE 1.21. Illustration of main junction point of 1/3 limb of M.

Definition 1.126. When q spokes meet at a junction point, we will call it a junction point of *order q*.

Exploration 1.127. Investigate other p/q limbs and their principal spokes to get an idea of how often the spoke that is a p/q rotation from the main spoke is the shortest. Is there a pattern for the longest spoke? *Try it out!*

Exploration 1.128. Investigate other p/q limbs and the junction points within them to get an idea of how often the junctions there are of order q. What is the order of the other junctions? Is there a pattern to the types of junction orders on a p/q limb? *Try it out!*

In the course of exploring the p/q limbs you may have seen many baby Mandelbrot sets like the one in Figure 1.22. To see them clearly you will want to adjust the **Parameter plane max iterations** value to 300, 400, 500, or even 1,000 or more depending on the size of the baby Mandelbrot set that you want to see. Also, adjusting the **Color sample rate** can help. We mentioned before that baby Mandelbrot sets are everywhere (they are dense in ∂M); however, even though they are all seeming copies of the original M set, they are not all the same. They have different antennae sprouting from them. If we pay attention to these antennae and to how many meet at the junction points, can we get an idea of how the baby Mandelbrot set is related to the p/q limb it is in? For example, in Figure 1.22 we see that the junction points near the tips of the antennae are all of order three. Furthermore, we see that along these filaments, closer to the baby Mandelbrot set, we have junction points of order five. Can this information be a clue to help you find where this baby Mandelbrot set is? Imagine playing a game where your friend shows you a picture of a baby Mandelbrot set. Can you win the game by locating it? These are hard questions, but maybe by investigating them you will find some partial answers, connections, or maybe pose some interesting questions of your own.

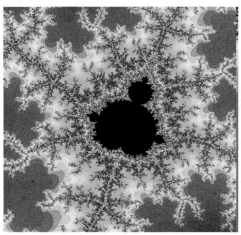

FIGURE 1.22. Which p/q limb contains this baby Mandelbrot set? Off of which sub-bulb of $B_{p/q}$ is it?

In your investigation of the antennae, bulbs, and limbs of M you may have also noticed that some of the parameter plane pictures, specifically enlarged areas of the tips of the antennae of M, look very much like some of the dynamic plane pictures, specifically enlarged areas of the tips of certain Julia sets. The pictures can look so much alike that it can be confusing which is which (see Figure 1.23). In [20] it was shown that zooming in

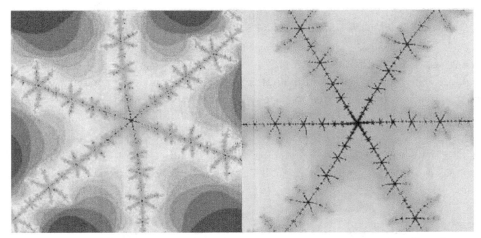

FIGURE 1.23. Which is part of the Mandelbrot set and which is part of a Julia set?

the parameter plane near a Misiurewicz point c (see Definition 1.98) shows a portion of the Mandelbrot set that is a rotation of an enlargement of the Julia set of f_c in the dynamic plane near c. This is curious given that the parameter plane and the dynamic plane are, on the face of things, different animals. (Most mathematical discoveries are not accompanied by a shout of 'Eureka!', but rather a quietly spoken 'Huh, now that's curious.') Why do parts of the Mandelbrot set seem identical to parts of certain Julia sets? The proof is too complicated for these pages, but we can see examples. Use the *Parameter Plane and Julia Set Applet* to zoom in very far on the point $c = 0.4711819 + 0.3541484i$ in both the parameter plane and the dynamic plane (in the dynamic plane, c is the the first point in the critical orbit). Then compare the portion of M near c to the portion of $J(f_c)$ near c. Up to a rotation they seem identical.

1.5.7 Fixed points for $f_c(z) = z^2 + c$

In this section we look at the dynamics near the fixed points of $f_c(z) = z^2 + c$.

Example 1.129. Using the *Parameter Plane and Julia Set Applet* set $c = -0.513 + 0.5693i$ in the parameter plane and choose seed $z_0 = -0.4034 + 0.2896i$ (near the main pinch point) in the dynamic plane. Zoom in close to the fixed point (use the **Plot fixed points** box to color the fixed points purple) and then iterate the seed value one step at a time and observe the behavior. Iterate several other seeds near the pinch point to see the same behavior. We see that the main pinch point is a repelling fixed point whose multiplier λ has argument very close to $(2/5)2\pi$.

Exercise 1.130. Using paper and pencil (along with your calculator or computer) solve for the fixed points of f_c where $c = -0.513 + 0.5693i$ as in Example 1.129. Compute their multipliers and verify that the pinch point's multiplier λ has argument very close to $(2/5)2\pi$. Use the *Complex Function Iterator Applet* or *Global Complex Iteration Applet for Polynomials* to iterate the map $z \mapsto \lambda z$ for seed values near the repelling fixed point at the origin. Compare the dynamics of $z \mapsto \lambda z$ near the origin and $z \mapsto f_c(z)$ near the pinch point. ***Try it out!***

Exploration 1.131. Investigate other c values in the 2/5 limb to see if the repelling fixed point has a multiplier with argument very close to $(2/5)2\pi$. You can do this experimentally by checking the behavior of orbits that start near the repelling fixed point (use the **Plot fixed points** feature in the *Parameter Plane and Julia Set Applet* to locate the fixed points). Relate the argument of the multiplier to the dynamics near the repelling fixed point and note how the argument corresponds to the spiralling we see in $K(f_c)$ near the repelling fixed point. How would this behavior change in other p/q limbs? ***Try it out!***

We saw the role of the fixed point of f_c that also serves as a main pinch point in the set $K(f_c)$. What can you observe about the other fixed point? How does it differ from the main pinch point? Does it have a special dynamic role? Can you prove anything or make a conjecture about it?

1.5.8   Concluding Remarks about the Family $f_c(z) = z^2 + c$

Though we have seen and discussed many aspects of the Mandelbrot set there are many more things to know. We encourage you to delve deeper into the topics we discussed and pursue your own line of questions. There are many fascinating things left to discover in this infinite playground of mathematics. ***Go enjoy it!***

We end this section with a remark about one of the goals in the introduction to this chapter, to introduce, investigate, and understand what we can about chaotic dynamical systems. It is easy to believe that systems involving the weather, the stock market, or the motion of the bodies in the Milky Way galaxy are chaotic. There are many uncontrollable variables in these systems, and changing one, even just a little, may affect the other variables. What we see with this family of maps $f_c(z) = z^2 + c$ is also chaos though it exists without unknown variables or random processes. This simple system of one complex variable and one complex parameter produces chaotic behavior of an unimaginably wide and rich sort. This gives us a small taste of what is truly a global phenomenon; chaos is all around us, even in what seems like the simplest of systems. For a non-technical look at some history of chaos, chaos theory, and the chaoticians who investigate it, we recommend *Chaos* by James Gleick [15].

1.5.9   Other Uni-critical Families of Polynomials

In this section we investigate the dynamics of polynomials $P_c(z) = z^d + c$ where $d = 2, 3, 4, \ldots$ (we have already studied $d = 2$). Since these maps have only one finite critical point (at the origin), we can analyze the parameter space of these maps in the same way as for the maps $f_c(z) = z^2 + c$. We define the *dth degree Mandelbrot set* to be $M_d = \{c \in \mathbb{C} : P_c^n(0) \nrightarrow \infty\}$. As we were told for the Mandelbrot set, the condition that the critical orbit $\{P_c^n(0)\}$ tends to ∞ is equivalent to the condition that some point in this orbit escapes $\overline{\triangle(0, 2)}$. Steps to prove this, and more, are given in Additional Exercise 1.191.

You are encouraged to explore the dynamics of such maps using the *Parameter Plane and Julia Set Applet*. Where $z^2 + c$ appears, use the drop down menu to select $z^d + c$, and then enter an integer $d > 2$. Additional Exercise 1.192 asks you to prove some symmetries exhibited in the parameter plane and the dynamic plane. Additional Exercise 1.193 asks you to show that M_d is the connectedness locus for this new family.

Large Project 1.132. Investigate these families of maps using the *Parameter Plane and Julia Set Applet* in the same way as as we did for $f_c(z) = z^2 + c$. Try to find patterns and relationships in the bulbs and antennae. Develop your own questions, make conjectures, and describe M_d.

1.6 Transcendental Dynamics

In this section we investigate the dynamics of three different families of maps: $E_c(z) = ce^z$, $S_c(z) = c \sin z$, and $C_c(z) = c \cos z$, where $c \in \mathbb{C} \setminus \{0\}$ is a parameter. We first consider their dynamic properties for fixed c, then we study the parameter plane, that is, we investigate what changes occur in the dynamics when c varies. The functions have one striking difference from the maps $P_c(z) = z^d + c$ studied above. They are *transcendental entire* maps, i.e., are analytic on all of $\mathbb{C}$ (entire), but are not polynomials. Hence, they not only fail to have an attracting fixed point at ∞, they fail to even be defined at ∞. These maps have an essential singularity at ∞ which means, among other things, that they cannot be defined at ∞ in any continuous way (Appendix Example B.14 on page 362 illustrates this). However, it turns out that ∞ and its basin play a central role. Instead of the basin of ∞ being in the Fatou set as we have for polynomials, we have the opposite.

Proposition 1.133. *For the maps $E_c(z) = ce^z$, $S_c(z) = c \sin z$, and $C_c(z) = c \cos z$, where $c \in \mathbb{C} \setminus \{0\}$, the Julia set is the closure of the attracting basin of ∞, e.g., $J(E_c) = \overline{A_{E_c}(\infty)}$.*

This proposition asserts a striking difference between polynomial dynamics and the dynamics of these transcendental maps. Though the proof of this is beyond the scope of this text (see [6, 9] as general references on the dynamics of E_c), we use it, in particular, to program the computer to illustrate their Julia sets. To do so, we must first understand how, or rather in what direction, a point z_0 can iterate to ∞ under the maps. We first investigate $E_c(z) = ce^z$.

Remark 1.134. For those who are familiar with *Picard's theorem* (see [1, p. 242]), we give some idea of why the Julia set of certain transcendental entire functions is related to the closure of the points which iterate to ∞. Picard's Theorem says that given an entire map g with an essential singularity at ∞ and given any neighborhood U of ∞ (no matter how small), g will map $U \setminus \{\infty\}$ infinitely often onto the entire complex plane $\mathbb{C}$ except for at most one point. This means that, with the exception of at most one point, for each $w \in \mathbb{C}$ there are infinitely many points $z \in \mathbb{C}$ such that $g(z) = w$ (the map e^z illustrates this with an exceptional value at $w = 0$). Hence, there is a high degree of sensitive dependence as some points near ∞ must have very different orbits even in the first application of the map (let alone after repeated iteration of the map).

1.6.1 Exponential Dynamics

In this section we investigate the dynamics of $E_c(z) = ce^z$. Because $E_c(z) = ce^z = ce^x e^{iy}$ where $x = \operatorname{Re} z$ and $y = \operatorname{Im} z$, we have $|E_c(z)| = |c|e^x$. Hence $E_c(z)$ is large, and thus close to ∞, when $x = \operatorname{Re} z$ is large. If $\operatorname{Re} z > 50$, then $|E_c(z)| > |c|e^{50}$ is extremely large for any c that is not so small that a computer would recognize it as 0. We use this to justify the algorithm implemented in the *Parameter Plane and Julia Set Applet* for drawing $J(E_c)$, which you can access by selecting ce^z from the function menu. A point in the dynamic plane is colored based on how many iterates it takes to escape, by which we mean have its real part become greater than 50. Thus the colored points represent $A_{E_c}(\infty)$, which by Proposition 1.133 must look the same as the Julia set. Points colored black do not escape, at least not after iterating the number of times set in the **Dynamic plane max iterations** box, and so these points represent the Fatou set. Since, as you are asked to prove in Additional Exercise 1.194, $E_c^n(z_0) \to \infty$ if and only if Re $E_c^n(z_0) \to +\infty$, we say that points that iterate to ∞, do so in the direction of the positive real axis.

Remark 1.135. A point z may be identified by the algorithm as being in the Julia set when it is not. However, we can show that if that occurs, there would have to be a point very close to z that does lie in the Julia set. Hence, for a visual approximation of the Julia set, this technical issue does not pose a serious concern. Large Project 1.195, however, asks you to examine this matter in depth.

Example 1.136. We show that for $c = 0.2$ the Julia set $J(E_{0.2})$ is what we will call a *Cantor bouquet*. Our proof will not use the applet, nor its algorithm, but will employ only the notion of sensitive dependence on initial conditions. However, we find it useful to first view the Julia set in Figure 1.24 as drawn by the *Parameter Plane and Julia Set Applet*.

Because points in black do not escape and other points iterate to ∞ (in the direction of the positive real axis), we see that the colored set depicts $J(E_{0.2})$. It appears from the picture that $J(E_{0.2})$ contains large open sets, but this is an artifact of our algorithm's inability to iterate infinitely many times. Were we able to set the **Dynamic plane max iterations** equal to infinity we would not find any open sets in $J(E_{0.2})$. We explain why in a moment, but first we use the zooming feature to see that what appears to be tiny fingers in $J(E_{0.2})$ are actually made of tinier fingers, which are themselves made of even tinier fingers and so on. ***Try it out!*** We see self similarity again. We now explain why the picture of $J(E_{0.2})$ shows fingers inside of fingers.

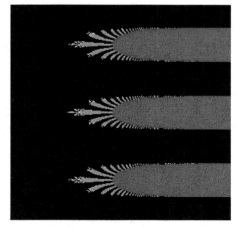

FIGURE 1.24. A portion of the Cantor bouquet Julia set of $E_{0.2}$ shown for $0 \le x \le 10$ and $-10 \le y \le 10$.

We begin by showing that $E_{0.2}$ has a real attracting fixed point. By the intermediate value theorem, $E_{0.2}$ has a fixed point p for some real value $0 < p < 1$ since $E_{0.2}(x) - x$ is positive for $x = 0$ and negative for $x = 1$. Also, p is attracting since $|E'_{0.2}(p)| < 1$.

We now show that the half plane $H = \{\operatorname{Re} z < 1\}$ is contained in the attracting basin $A_{E_{0.2}}(p)$. Set $\eta = |E'_{0.2}(1)| = 0.2e$ and note that $|E'_{0.2}(z)| < \eta < 1$ for $z \in H$. Thus, for $z \in H$, we have $|E_{0.2}(z) - p| = |E_{0.2}(z) - E_{0.2}(p)| = |\int_p^z E'_{0.2}(s)\,ds| \le \eta|z - p|$. Hence the action of $E_{0.2}$ is to move points in H closer to p by a factor of at least η. Using induction, along with the fact that $E_{0.2}(H) \subseteq H$ (verify), it can be shown that for $z \in H$ we have $E_{0.2}^n(z) \to p$, i.e., $H \subseteq A_{E_{0.2}}(p)$.

Exercise 1.137. Use the *Parameter Plane and Julia Set Applet* with $E_{0.2}(z) = 0.2e^z$ to see this contraction. In the dynamic plane window with $-2 \le \operatorname{Re} z \le 2$ and $-2 \le \operatorname{Im} z \le 2$ (which will be all black), pick two points and iterate the map one step at a time to see that after each step the points are closer together, as they iterate toward p. *Try it out!*

The technique that shows how a bound on the derivative can be used to show contraction of a map is general, and very useful. We interrupt our proof of the Cantor bouquet result to state a lemma for future use, which you are asked to prove in Additional Exercise 1.196. First, recall that a set D is called *convex* if for any points $z, w \in D$, the line segment connecting z and w is a subset of D.

Lemma 1.138 (Contraction Lemma). *Let f be analytic and map the convex domain D into itself. Suppose that $|f'(z)| < \eta < 1$ for $z \in D$. Then f contracts distances by a factor of at least η on D, i.e., $|f(z) - f(w)| < \eta|z - w|$ for $z, w \in D$. Furthermore, $|f^n(z) - f^n(w)| < \eta^n|z - w|$ for $z, w \in D$. Hence, if $a \in D$ is a fixed point of f, then $D \subseteq A_f(a)$.*

Returning to the proof that $J(E_{0.2})$ is a Cantor bouquet we now consider $E_{0.2}^{-1}(H) = \{z \in \mathbb{C} : E_{0.2}(z) \in H\}$. Use the transformation properties of the exponential map to justify the picture in Figure 1.25 showing that $\mathbb{C} \setminus E_{0.2}^{-1}(H)$ consists of infinitely many components, which we call *fingers*. Label them C_k for $k \in \mathbb{Z}$, each C_k being a translate by $2\pi i k$ of C_0. We also note that $H \subseteq E_{0.2}^{-1}(H)$.

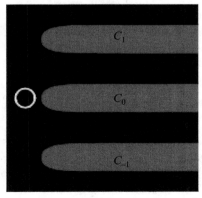

FIGURE 1.25. In this window of the complex plane, the set $E_{0.2}^{-1}(H)$ is colored in black with components of its complement in red for $-2 \le x \le 18$ and $-10 \le y \le 10$. For reference, the unit circle is shown in white.

Each C_k is mapped *conformally* (a conformal map is one-to-one and analytic) onto the half plane $\{\operatorname{Re} z \geq 1\}$, and thus each C_k contains a preimage under $E_{0.2}$ of each C_j for all $j \in \mathbb{Z}$. The preimages, shown in red in Figure 1.26, are sub-fingers of C_k with what we call *gaps* (in black) in between. The gaps, as in Figure 1.25, extend all the way to ∞ in the direction of the positive x-axis.

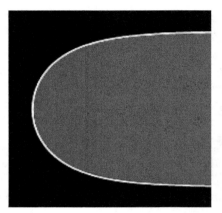 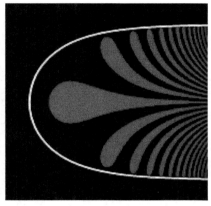

FIGURE 1.26. The left image shows the tip of the finger C_0 and the right image shows a portion of $E_{0.2}^{-2}(H) = E_{0.2}^{-1}(E_{0.2}^{-1}(H))$ for $1 \leq x \leq 6$ and $-2 \leq y \leq 2$. The red sub-fingers do not meet the (white) boundary of C_0.

We continue to take inverse images of H to see that, for $n \in \mathbb{N}$, the set $E_{0.2}^{-n}(H)$ has a complement $\mathbb{C} \setminus E_{0.2}^{-n}(H)$ consisting of fingers inside of fingers inside of fingers, and so on. More precisely, we make the following definition.

Definition 1.139. We call each component of $\mathbb{C} \setminus E_{0.2}^{-n}(H)$ a *stage n finger*.

Figure 1.25 depicts stage 1 fingers in red and the right picture in Figure 1.26 depicts stage 2 fingers in red. The gaps (in black) are then portions of $E_{0.2}^{-2}(H)$ that separate the fingers from each other. We encourage you to investigate stage n fingers more closely using the *Parameter Plane and Julia Set Applet* by setting both the **Dynamic plane max iterations** and the **Dynamic plane min iterations** to n, while setting the **Escape criterion** to $\operatorname{Re} z > 1$. Each stage n finger is contained in a stage $n - 1$ finger for $n \geq 2$. The boundary of a stage n finger for $n \geq 2$ is mapped by $E_{0.2}$ onto the boundary of a stage $n - 1$ finger and thus is mapped by $E_{0.2}^n$ onto the vertical line $\{\operatorname{Re} z = 1\}$.

It will be important to understand how thick fingers can be and so we make the following definition.

Definition 1.140. Let F be a stage n finger. We define the *thickness* of F to be the value of t such that F cannot contain an open Euclidean disk with diameter strictly larger than t, but for any value $s < t$, the set F contains an open Euclidean disk with diameter s, i.e., $t = \sup\{2r > 0 : \Delta(z, r) \subseteq F \text{ for some } z \in F\}$.

Exercise 1.141. Prove that the thickness of a stage 1 finger C_k is π by using the mapping properties of the exponential map. ***Try it out!***

The set $F_n = \mathbb{C} \setminus E_{0.2}^{-n}(H)$ is the union of all stage n fingers and because a stage n finger is contained in a stage $n-1$ finger we see that $F_1 \supset F_2 \supset \cdots$ and $F_1 \cap F_2 \cap \cdots \cap F_n = F_n$. Thus we regard $\cap_{n=1}^{\infty} F_n$ as the union of the stage infinity fingers. They, together with the point at ∞, comprise the Julia set, a proposition we state as follows.

Proposition 1.142. *The Julia set* $J(E_{0.2}) = \cap_{n=1}^{\infty} F_n \cup \{\infty\} = \overline{\mathbb{C}} \setminus A_{E_{0.2}}(p)$.

Proof. Let $J = \cap_{n=1}^{\infty} F_n \cup \{\infty\}$. We then have, whose proofs we leave to the exercises: (a) any two fingers are separated by an infinitely long gap (of black) points in $A_{E_{0.2}}(p)$ (Additional Exercise 1.197), and (b) the thickness of the stage n fingers shrinks to zero as n goes to infinity (Additional Exercise 1.198). Facts (a) and (b) show that J cannot contain any open set (Additional Exercise 1.199). Hence for $z \in J$ there are points arbitrarily close to it that lie in $\cup_{n=1}^{\infty} E_{0.2}^{-n}(H)$, and thus iterate to p. Since the orbit of every $z \in J$ lies forever in $\{\text{Re } z \geq 1\}$ by the definition of J (and thus differs from an orbit which converges to p), we get that z is sensitive to initial conditions. So $z \in J(E_{0.2})$. Likewise, for a point not in J, it and nearby points which also lie in the open set $A_{E_{0.2}}(p)$ have the same dynamic behavior (they all iterate towards p), placing the point in the Fatou set $F(E_{0.2})$. Hence $J = J(E_{0.2})$. $\square$

Exploration 1.143. Use the *Parameter Plane and Julia Set Applet* to track the orbit of nearby points in $J(E_{0.2})$ to see the sensitivity to initial conditions. It is a bit difficult to pick points that stay in the viewing window for more than a few steps, but by looking at the data in the **Orbits 1 and 2** tab at the bottom, you can see the orbits separate from each other. You can also zoom in on the picture so that you can choose your initial seed values close to each other, and then zoom out to see their orbits. ***Try it out!***

1.6.2 Hairs and Endpoints

Thinking again about this construction of $J(E_{0.2})$ we now describe this Cantor bouquet by describing the hairs (previously called the stage infinity fingers) that make up the set. This section is not crucial to the rest of the text so we will not provide all of the details to prove everything claimed, but we will discuss some fascinating aspects of what we saw in Example 1.136.

Pick a stage 1 finger C_{k_1}, and from the infinite number of sub-fingers within C_{k_1} pick a sub-finger $C_{k_1 k_2}$. From the infinite number of sub-sub-fingers within that, pick a sub-sub-finger $C_{k_1 k_2 k_3}$. Continuing so that each $C_{k_1 k_2 \ldots k_n}$ is a stage n finger in $C_{k_1 k_2 \ldots k_{n-1}}$, we see that $\gamma = \cap_{n=1}^{\infty} C_{k_1 k_2 \ldots k_n}$ must, by (b) be infinitely thin (i.e., contain no open disk of any positive radius), but also stretch to ∞ in the positive x-direction. We call γ a *hair*, and note that there are infinitely many of them in $J(E_{0.2})$ corresponding to the different ways one can choose $k_1, k_2, \ldots$. Each hair γ is a curve, an image of a continuous map $h_\gamma : [0, \infty) \to \mathbb{C}$ such that $h_\gamma(t) \to \infty$ as $t \to \infty$. We call $h_\gamma(0)$ the *endpoint* of the hair γ. There are a few important details necessary to prove all this. One is showing that there is something left in the set $\gamma = \cap_{n=1}^{\infty} C_{k_1 k_2 \ldots k_n}$, or even in the larger set in $\cap_{n=1}^{\infty} F_n$. Perhaps, the left tips of the stage n fingers move farther and farther to the right so that the set $\cap_{n=1}^{\infty} F_n$ is empty. These matters can be pursued in [6]. Additional Exercise 1.200, however, gives another way to see that there are infinitely many hairs (shown to have finite endpoints) in $J(E_{0.2})$.

As in the construction, each hair is separated from another by an infinite (though possibly thin) gap that stretches out to ∞ in the direction of the positive x-axis. Thus in $\mathbb{C}$ we would say these hairs are all separated from each other. However, by including ∞, where all these hairs meet, we produce a set $J(E_{0.2})$ that is a connected subset of the sphere $\overline{\mathbb{C}}$.

We conclude this section with remarks about two fascinating properties of the Cantor bouquet $J(E_{0.2})$. To discuss the first properly we need a more general definition of what it means for a subset of $\overline{\mathbb{C}}$ to be connected.

Definition 1.144. A set C is *connected* in $\overline{\mathbb{C}}$ if there do not exist open subsets U and V of $\overline{\mathbb{C}}$ such that (i) $U \cap C \neq \emptyset$, (ii) $V \cap C \neq \emptyset$, (iii) $C \subseteq U \cup V$, and (iv) $U \cap V \cap C = \emptyset$. A more intuitive definition is: A set $C \subseteq \overline{\mathbb{C}}$ is connected in $\overline{\mathbb{C}}$ if we cannot break up (disconnect) C into a disjoint union of non-empty sets A and B such that no sequence from A converges to a point in B, and vice versa.

We can now present (without proof) two startling properties of the Cantor bouquet $J(E_{0.2})$.

Property 1. Let $\mathcal{E}$ denote the set of endpoints of all the hairs in $J(E_{0.2})$. Then $\mathcal{E}^* = \mathcal{E} \cup \{\infty\}$ is connected in $\overline{\mathbb{C}}$. Informally, this says that the endpoints together with ∞ are somehow bunched so tightly that it is impossible to disconnect the set. However, and this is the strange part, removing ∞ from $\mathcal{E}^*$ creates a *totally disconnected* set $\mathcal{E}$ (i.e., the only connected subsets of $\mathcal{E}$ are those that contain only a single point). We can show this easily by noting that between any two hairs there is a gap in $F(E_{0.2})$, thus showing that the endpoints cannot lie in the same connected subset of $\mathcal{E}$. The gap extends to ∞ and so cannot be used to create a disconnection of $\mathcal{E}^*$. It remains to be shown that no other method of disconnecting $\mathcal{E}^*$ works either. This is too complicated for the present text, but the details are in [21].

Property 2. The Hausdorff dimension of a set in $\overline{\mathbb{C}}$ is a number that relates to the size of the set. It is defined (see [11] for further details) so that it shares many of the properties of our usual notion of dimension and so we refer to it as a dimension though it need not be an integer. The Hausdorff dimension of a set measures, in a sense, how much space the set fills up. The Hausdorff dimension of any smooth curve, in particular any hair in $J(E_{0.2})$ (which is shown to be smooth in [5]), is 1, whether the endpoint is included or not. Let $\mathcal{H}$ denote the union of all *open* hairs, that is, of hairs with endpoints removed. Thus $\mathcal{H} = J(E_{0.2}) \setminus \mathcal{E}^*$. The hairs are separated by gaps that prevent them from accumulating too much and filling up a lot of space. So it seems plausible that $\mathcal{H}$ has Hausdorff dimension 1, which it does. What is astounding is that the set of endpoints $\mathcal{E}$ has dimension 2. Thus the endpoints seem to bunch up so that the set of all of them fills up two dimensions of space. Since for each infinitely long hair there is only one endpoint, it seems impossible that the dimension of the set of endpoints could be larger than the dimension of the set of open hairs. But this is what happens. Details are in [18, 19].

1.6.3   Critical Orbits, Exploding Julia Sets, Parameter Space for E_c

As we have seen, critical points dictate much about the dynamics of a rational map, especially for the polynomial $f_c(z) = z^2 + c$. The exponential map $E_c(z) = ce^z$, however, has no critical points (verify), but has 0 as what we call an asymptotic value and this will play a similar role.

Definition 1.145. An *asymptotic value* for an entire function f is a value $a \in \overline{\mathbb{C}}$ such that there exists a curve $\gamma : [0, +\infty) \to \mathbb{C}$ tending to ∞ (i.e., $\gamma(t) \to \infty$ as $t \to +\infty$) such that $f(\gamma(t)) \to a$ as $t \to +\infty$.

For E_c we may choose the curve to be the negative real axis, along which the values $E_c(z) \to 0$ since $\operatorname{Re} z \to -\infty$. The following theorem, whose proof can be found in [6], parallels Theorems 1.89 and 1.93 for rational maps.

Theorem 1.146. *Let f be a transcendental entire map. Then the immediate basin of each (super) attracting cycle of f contains a critical point of f or an asymptotic value of f.*

From this we see that the orbit $\{E_c^n(0)\}_{n=1}^{\infty}$ plays a special role and so we call it the *critical orbit*.

Fact 1. An attracting cycle of E_c attracts the critical orbit $\{E_c^n(0)\}_{n=1}^{\infty}$.

Fact 1 follows from Theorem 1.146. The next fact has a proof beyond the scope of this text, but we shall nonetheless make frequent use of it.

Fact 2. If $E_c^n(0) \to \infty$, then $J(E_c) = \overline{\mathbb{C}}$.

Fact 2 is different from anything we have seen. The super attracting fixed point at ∞ for any polynomial implies that we could never have an empty Fatou set.

Fact 2 also leads us to discover the fascinating exploding Julia sets we can see when we vary c in the map E_c. Example 1.136 showed that $J(E_{0.2})$ is a Cantor bouquet of hairs and $F(E_{0.2})$ is the attracting basin of a finite attracting fixed point $p \approx 0.26$. If we increase c from $c = 0.2$ to $c = 0.5$ we see that $E_{0.5}^n(0) \to +\infty$ and so, by Fact 2, we have $J(E_{0.5}) = \overline{\mathbb{C}}$. Hence some real value c^* between $c = 0.2$ and $c = 0.5$ is a bifurcation point where the dynamics drastically change. What we see is that as c grows from being smaller than c^* to larger than c^* the Julia set goes from being confined to $\{\operatorname{Re} z \geq 1\}$ to exploding to be all of $\overline{\mathbb{C}}$. You are asked to determine c^* and show this exploding behavior in Additional Exercise 1.201.

So we get different dynamic behavior for E_c depending on whether the critical orbit escapes to ∞. We use this dichotomy to color the parameter plane based on how many iterations it takes for the critical orbit to escape (that is, have real part become greater than 50), leaving c colored black if the critical orbit does not escape. Investigate the parameter space using the *Parameter Plane and Julia Set Applet*. Set the window in the parameter plane to show $[-3.2, 3.2] \times [-3, 3]$ to find a cardioid shaped region, with many bulbs attached. You may want to set the **Parameter plane max iterations** value to 40, 60, 80, or 100 to see the cardioid in more detail. Because of the computing power needed to compute the complex exponential iterates, each picture may take a while for the applet to complete.

Exploration 1.147. Experiment with the applet to see if you can guess the dynamical significance of the cardioid and some of the bulbs. ***Try it out!***

Let's investigate the exponential parameter plane as we did for the family $f_c(z) = z^2 + c$ by calculating the set L_1 of c values for which E_c has an attracting fixed point w.

For w to be an attracting fixed point we require $E_c(w) = w$ and $|E_c'(w)| < 1$. The first condition yields $ce^w = w$ which used with the second gives $|w| = |ce^w| < 1$. Notice w is both the fixed point and the multiplier. Thus we are looking for c such that $c = we^{-w}$ for $|w| < 1$. By moving w around the unit circle, the c values trace out the cardioid-like set in the parameter plane pictured in Figure 1.27. Use the *ComplexTool* applet to draw this.

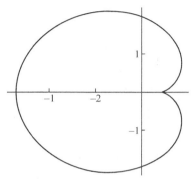

FIGURE 1.27. This cardioid-like shape is the boundary of L_1, the set of c parameters that give rise to an attracting fixed point for the map $E_c(z) = ce^z$.

This cardioid-like L_1 has similarities with the cardioid K_1 in the Mandelbrot set. For example, we can find p/q bulbs attached to its boundary. The p/q bulbs might not be located where you first expect. The $1/3$ bulb in the Mandelbrot set M is at the top of the cardioid K_1 whereas the top of L_1 has the $2/5$ bulb attached.

Exploration 1.148. Use the *Parameter Plane and Julia Set Applet* to investigate the new locations of the p/q bulbs. One way to determine their locations is to experiment long enough to find a pattern (can you use Farey addition here too?). Another way is to use the multiplier map, or its inverse, which you are asked to determine in Additional Exercise 1.202. Be sure the settings for the applet have values that produce good pictures. You may want to check the values for the escape criterion and max and min iterations which you adjusted earlier, or reload the applet to reset them. ***Try it out!***

Exploration 1.149. Where is the $1/2$ bulb located? Does it have a sub-bulb that corresponds to attracting 4-cycles? ***Try it out!***

Exercise 1.150. Whether or not you think you can solve it, try to find an interesting mathematical question to ask about the dynamics of exponentials. Try to draw parallels with the family $f_c(z) = z^2 + c$, or show where parallels do not hold. ***Try it out!***

1.6.4   Trigonometric Dynamics

Recall that $\sin z = \frac{-i}{2}(e^{iz} - e^{-iz})$ and $\cos z = \frac{1}{2}(e^{iz} + e^{-iz})$, and set $S_c(z) = c \sin z$ and $C_c(z) = c \cos z$. Additional Exercise 1.203 asks you to show that orbits under the maps $S_c(z)$ and $C_c(z)$ escape to ∞ in the direction of the positive or negative imaginary axis. Thus, using Proposition 1.133 to note $J(S_c) = \overline{A_{S_c}(\infty)}$ and $J(C_c) = \overline{A_{C_c}(\infty)}$, the algorithm in the *Parameter Plane and Julia Set Applet* for creating the Julia sets for the exponential maps $E_c(z) = ce^z$ can be modified. An orbit under either S_c or C_c is deemed to escape to ∞ if any point in the orbit has imaginary part larger than 50 in absolute value. Additional Exercise 1.204 shows that neither S_c nor C_c has a finite asymptotic value, but both have infinitely many critical points. This may appear to present a difficulty since a map having more than one critical orbit would pose a challenge to coloring and analyzing the parameter space as we did for $z^d + c$ and ce^z. However, as we ask you to verify in Additional Exercise 1.205, the maps S_c and C_c have only one critical orbit. The critical orbit for S_c is $\{S_c^n(c)\}_{n=1}^\infty$ and the critical orbit for C_c is $\{C_c^n(c)\}_{n=1}^\infty$.

As in the exponential case we have key facts that will aid our understanding of the dynamics of these trigonometric functions.

Fact 1. Any attracting cycle of S_c attracts its critical orbit $\{S_c^n(c)\}_{n=1}^\infty$.

Fact 2. Any attracting cycle of C_c attracts its critical orbit $\{C_c^n(c)\}_{n=1}^\infty$.

Fact 3. If $S_c^n(c) \to \infty$, then $J(S_c) = \overline{\mathbb{C}}$.

Fact 4. If $C_c^n(c) \to \infty$, then $J(C_c) = \overline{\mathbb{C}}$.

Because the orbit $\{S_c^n(c)\}_{n=1}^\infty$ is critical to the dynamics of the map S_c, we can color the parameter plane as we did for the family of maps E_c. The *Parameter Plane and Julia Set Applet* will color c based on how many iterations it takes for the critical orbit to escape, i.e., have imaginary part larger than 50 in absolute value. It is left black if the critical orbit does not escape. The applet for C_c works similarly. Use it to experiment and investigate the dynamics of S_c and C_c. If you look closely in the parameter planes, you will find a familiar friend. ***Try it out!***

1.7 The Mandelbrot Set is Universal

You may have noticed that the parameter planes for the maps $f_c(z) = z^2 + c$, $C_c(z) = c \cos z$, and $S_c(z) = c \sin z$ drawn by the *Parameter Plane and Julia Set Applet* show black regions that contain sets resembling the Mandelbrot set M. See Figure 1.28.

The appearance of copies of the Mandelbrot set in these diverse settings is unexpected. Why should the Mandelbrot set, a picture of the fundamental dichotomy (whether the critical orbit escapes or not) in the parameter space of quadratic maps $f_c(z) = z^2 + c$, have anything to do with the parameter space of the transcendental maps $C_c(z) = c \cos z$ and $S_c(z) = c \sin z$? As it turns out copies of the Mandelbrot set appear in so many parameter planes that it is a fundamental mathematical object, like π or e, that arises in many unexpected places. As baby Mandelbrot sets are dense in the boundary of the full Mandelbrot set, so are they dense in many parameter spaces. The main reason, which we state only loosely, is that often iterates of maps have dynamic behavior like the dynamic behavior of iterates of a quadratic polynomial when the maps are considered on small domains. Thus, as a parameter is changed, we see the same changing behavior in the parameter plane as we do in the family $f_c(z) = z^2 + c$ corresponding to the original Mandelbrot set. In [22] you can find the details showing that small Mandelbrot sets are dense in the bifurcation locus for what are called *holomorphic families* of rational maps.

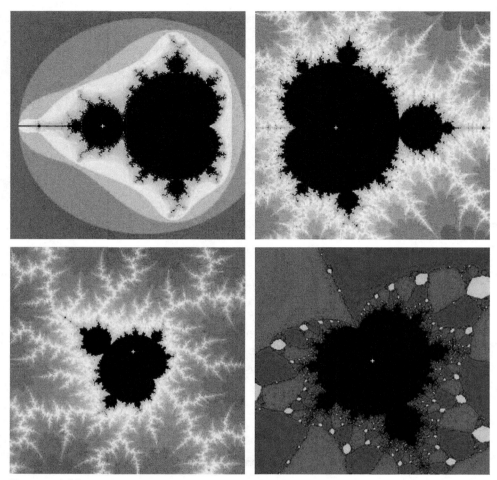

FIGURE 1.28. The original Mandelbrot set M (upper left), a copy of M containing $c = 3.2$ in the parameter space for $c \cos z$ (upper right), a copy of M containing $c = 2.17 + 1.3i$ in the parameter space for $c \sin z$ (lower left), and a copy of M containing $\rho = 0.906 + 0.423i$ in the parameter space for Newton's method applied to the map $p_\rho(z) = z(z-1)(z-\rho)$.

1.8   Concluding Remarks and New Directions

We investigated the chaos that arises in a number of complex dynamical systems. We saw how Newton's method can (and often must) produce chaotic behavior, giving beautiful fractal images. To investigate chaos more generally, we looked at iteration of analytic maps that were not generated as the Newton map of a polynomial. We found that even in the seemingly simple class $f_c(z) = z^2 + c$ there was a richness and complexity we could not have imagined. Similar features were present in the dynamics of the entire maps $E_c(z) = ce^z$, $S_c(z) = c \sin z$, and $C_c(z) = c \cos z$, though some new phenomena also appeared. Throughout, an important part of our investigations was to track, if we could, how the behavior of the systems changed when the maps were changed. We did this by varying the parameter c. In some sense, however, these were really mild perturbations of the maps since many of the salient features of the maps (such as the super attracting fixed point at ∞ for $f_c(z) = z^2 + c$ or the essential singularity at ∞ for the transcendental maps) persisted no matter how the c values were changed.

1.8.1 Perturbation with a Pole

There are many interesting ways to perturb a map that produce beautiful mathematics. The maps $F_c(z) = z^d + c/z^m$ with $d, m \in \mathbb{N}$ are, for $c \in \mathbb{C} \setminus \{0\}$, perturbations of the dynamically well understood map $z \mapsto z^d$. Unlike the $f_c(z) = z^2 + c$ perturbations of $z \mapsto z^2$, they add a new dimension to the analysis due to the pole at the origin. For c small, F_c behaves like $z \mapsto z^d$, but only as long as z is sufficiently far from the origin. Near the origin the pole changes the dynamics considerably from $z \mapsto z^d$. These systems lead to fascinating mathematics that can sometimes be represented by pictures such as Figure 1.29 showing features not seen before in this chapter.

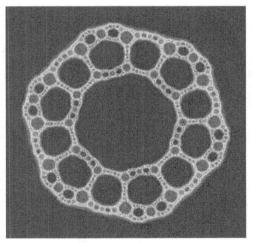

FIGURE 1.29. A Sierpinski curve Julia set for $F(z) = z^3 + 0.13i/z^3$. The regions in shades of orange are in $A_F(\infty)$ and the points in yellow represent $J(F)$.

Large Project 1.151. Investigate the dynamics of the maps $F_c(z) = z^d + c/z^m$. You can use the *Parameter Plane and Julia Set Applet* and your analytic paper and pencil techniques. We have not described how the parameter plane is being drawn. It is up to you to determine this and to figure out what you can on your own. Of all the maps mentioned in this chapter, these are the newest and least studied.

1.8.2 Random Dynamics

In the examples of perturbed maps throughout this chapter, though the perturbation could be mild by changing a simple parameter or more severe by adding a pole as in F_c, there was one fundamental assumption made regarding the perturbed system – once the map was perturbed, it was fixed and iterated again and again to create the dynamics. Another way to perturb a system is to allow the map to change at every step in the orbit.

Suppose we start with two maps f and g. The usual iteration dynamics considers the orbit generated by $f^n(z_0)$ or by $g^n(z_0)$. It is natural to ask what happens if at each stage of the orbit either map f or g can be applied. In some cases the dynamics can be much more fun. We study the behavior of orbits $h_{i_1}(z_0), h_{i_2}(h_{i_1}(z_0)), h_{i_3}(h_{i_2}(h_{i_1}(z_0))), \ldots$ where the maps h_i are chosen to be either f or g. We use the term *random dynamics* when the choices of maps are made at random. Such systems are connected to iterated function systems and their attractor sets (see [8]), such as the van Koch curve and Sierpinski triangle.

Instead of investigating such attractor sets, we look in another direction. We investigate what we call a *random basin of attraction* as follows. Fix a point $z_0 \in \mathbb{C}$ and randomly select h_1 to be either f or g, each with probability $1/2$. Then set $z_1 = h_1(z_0)$. Randomly select h_2 to be either f or g and set $z_2 = h_2(z_1) = h_2(h_1(z_0))$. Continue to produce a *random orbit* $z_0, z_1, z_2, \ldots$, and consider the probability that it converges to infinity.

Example 1.152 (Devil's Staircase). Let $f(x) = 3x$ and $g(x) = 3x - 2$ be defined on the real line $\mathbb{R}$. For $x_0 \in \mathbb{R}$ generate a random orbit $x_0, x_1, x_2, \ldots$. Let $P(x_0)$ denote the probability that it converges to $+\infty$, noting that we distinguish between convergence to $+\infty$ and $-\infty$. It is plausible that if $x_0 > 1$, then $P(x_0) = 1$ since no matter how we choose f and g at each step, we have the orbit points grow larger and positive. Similarly, if $x_0 < 0$, then $P(x_0) = 0$ since both f and g force the orbit points to grow larger and negative. What happens for $0 \leq x_0 \leq 1$, however, is more interesting. We show the graph of P, called the Devil's Staircase, in Figure 1.30.

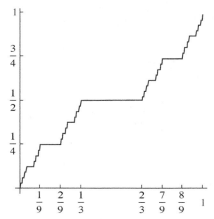

FIGURE 1.30. The Devil's Staircase is the graph of the probability $P(x)$ that a random orbit generated by the maps $f(x) = 3x$ and $g(x) = 3x - 2$ tends to $+\infty$.

An interesting property of this graph is that between any two steps (i.e., intervals where $P'(x) = 0$), we have infinitely many steps. Also, the sum of the lengths of the steps between 0 and 1 is exactly 1, so we say that this graph is almost always flat, i.e., almost always has $P'(x) = 0$. However, $P(x)$ is both continuous and increasing from $y = 0$ to $y = 1$ as x goes from 0 to 1.

Exercise 1.153. Properties of the Devil's Staircase.
(a) Show that $P(x) = 0$ for $x < 0$ and $P(x) = 1$ for $x > 1$.
(b) Show that $P(0) = 0$ and $P(1) = 1$.

(c) Show that $P(x)$ is increasing (not strictly), i.e., $x < y$ implies $P(x) \leq P(y)$.

(d) Show that $P(x) = 1/2$ and $P'(x) = 0$ on the interval $(1/3, 2/3)$.

(e) Show that $P(x) = 1/4$ and $P'(x) = 0$ on the interval $(1/9, 2/9)$.

(f) Find where $P(x) = 3/4$.

(g) In parts (d)-(f), you located the largest step (corresponding to $(1/3, 2/3)$), and the two next largest steps. Describe the four next largest steps and the pattern relating all steps.

(h) Show that the sum of the lengths of the steps is 1.

(i) Show that $P(x)$ is continuous.

Example 1.154 (Devil's Colosseum). The Devil's Colosseum is a higher dimensional analog of the Devil's Staircase, and is also understood through random dynamics. Instead of functions defined on the real line, we let $g_1(z) = z^2/4$ and $g_2(z) = z^2 - 1$ be complex-valued maps, and then use their second iterates to define $f(z) = g_1^2(z)$ and $g(z) = g_2^2(z)$.

We now investigate the random basin of attraction for ∞ as follows. Fix a point $z_0 \in \mathbb{C}$. For the random orbit $z_0, z_1, z_2, \ldots$, let $P(z_0)$ be the probability it tends to ∞. For z_0 large $(|z_0| > 10$ will do), we have $P(z_0) = 1$ since no matter what choices of f or g we make at each step, the orbit of z_0 will go to ∞. Though it is not as simple, we can also show that $P(z_0) = 0$ for z_0 near zero. In between, however, we get interesting behavior. The function $P(z)$ has a graph illustrated in Figure 1.31 called the Devil's Colosseum and has properties similar to those of the Devil's Staircase. To climb out from the bottom, you would walk a path like the Devil's Staircase.

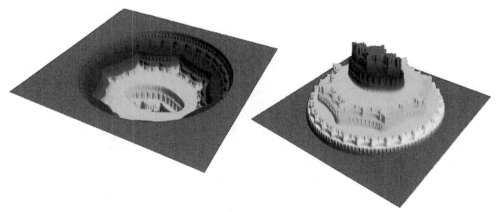

FIGURE 1.31. The left picture shows the Devil's Colosseum, the graph of the probability $P(z)$ that a random orbit tends to ∞. The picture on the right is the inverted Devil's Colosseum, sometimes called a fractal wedding cake. The first pictures of these objects were created and studied by Hiroki Sumi (see [25]).

We can describe the dynamics that underlie the Devil's Colosseum in terms of stability of orbits, and in doing so introduce the following extension of the notions of the Fatou and Julia sets. Let $F(\langle f, g \rangle)$ denote the Fatou set of the random system, defined to be the set of points $z_0 \in \mathbb{C}$ such that every randomly generated sequence of maps h_{i_n} produces a stable orbit in the sense that nearby seed values have similar orbits when the sequence h_{i_n} is used. We define the Julia set of the random system to be $J(\langle f, g \rangle) = \overline{\mathbb{C}} \setminus F(\langle f, g \rangle)$. We can then describe the oddity of the Devil's Colosseum. The function P is continuous on $\mathbb{C}$, yet it is constant on each component of the Fatou set even though the Julia set contains no open set (see [25]). It is these components of the Fatou set that make up the various horizontal levels, or steps, in Figure 1.31. The Julia set is the set on which $P(z)$ varies (since $P(z)$ does not vary, i.e., it is flat, on $F(\langle f, g \rangle)$). The Julia set is pictured in Figure 1.32.

FIGURE 1.32. The Julia set $J(\langle f, g \rangle)$ in $\mathbb{C}$ is the set on which $P(z)$ varies. This picture was drawn via the computer application Julia 2.0 (see [23]).

Remark 1.155. The random dynamics in the Devil's Colosseum produces something that cannot happen with usual iteration dynamics. Given a polynomial $f(z)$, define $P(z_0)$ to be the probability of an orbit with seed z_0 tending to ∞. Since there is only one map to choose, there is no randomness involved, and $P(z_0)$ must be either 0 or 1. We have $P(z) = 1$ for $z \in A_f(\infty)$ and $P(z) = 0$ for $z \in K(f)$. The function P is then discontinuous at every point $z \in \partial A_f(\infty) = \partial K(f) = J(f)$. Thus we see that the random dynamics exhibited by the Devil's Colosseum, where $P(z)$ is continuous on all of $\mathbb{C}$, is a change from iteration dynamics.

Exercise 1.156. You are encouraged to investigate random dynamical systems visually by generating your own pictures similar to the Devil's Colosseum. Sample code is provided in Appendix 1.C.

There is a large world of chaotic phenomena to explore. Whether you stay within the confines of the topics of this book or venture into new areas, there is an infinite amount of mathematics to discover. We encourage you to explore and make your own contributions to mathematics.

1.9 Additional Exercises

Orbits, Examples, and Fixed Points

Exercise 1.157. Prove $f^n(x) \to +\infty$ for all $x \in \mathbb{R}$ when $f(x) = e^x$. The advanced reader can show that the convergence is uniform, i.e., given $M > 0$, there exists an integer $N > 0$ such that $n > N$ implies $f^n(x) > M$ for all $x \in \mathbb{R}$.

Exercise 1.158. For $f(x) = \sin x$ where $x \in \mathbb{R}$ is given in radians prove that $f^n(x) \to 0$ for all $x \in \mathbb{R}$. Hint: Apply the mean value theorem to show that $|\sin x| < |x|$ for all $x \in \mathbb{R} \setminus \{0\}$. The advanced reader can show that the convergence is uniform, i.e., given $\varepsilon > 0$, there exists an integer $N > 0$ such that $n > N$ implies $|f^n(x) - 0| < \varepsilon$ for all $x \in \mathbb{R}$.

Exercise 1.159. For $f(x) = \cos x$ where $x \in \mathbb{R}$ is given in radians prove that $f^n(x) \to 0.739085...$ for all $x \in \mathbb{R}$ where $x^* = 0.739085...$ is the number such that $\cos x^* = x^*$. Hint: Apply the mean value theorem to show that $|\cos x - x^*| < |x - x^*|$ for all $x \in \mathbb{R} \setminus \{x^*\}$. The advanced reader can show that the convergence is uniform, i.e., given $\varepsilon > 0$, there exists an integer $N > 0$ such that $n > N$ implies $|f^n(x) - x^*| < \varepsilon$ for all $x \in \mathbb{R}$.

Exercise 1.160. For $g(z) = \sin z$, show that $g^n(\pm i\varepsilon) \to \infty$ for any $\varepsilon > 0$. Hint: Show that $g(iy)$ is purely imaginary and $|g(iy)| > |y|$ for each real $y \neq 0$.

Complex Newton's Method

Exercise 1.161. Let F be the Newton map of a polynomial f of degree two or more.
(a) Suppose $F^n(z_0) \to a$ for some $z_0 \in \mathbb{C}$ and some finite point $a \in \mathbb{C}$. Prove $f(a) = 0$.
(b) Show that $F(\infty) = \infty$ and so it is necessary that part (a) consider only finite points.

Exercise 1.162. Prove that the Newton map of $f(z)/f'(z)$ has super attracting fixed points at the roots of f regardless of the orders of the roots of f. Hint: Given a root α of f of order k, determine the order of the root of $f(z)/f'(z)$.

Exercise 1.163. Construct examples of analytic maps f to justify the statement that there is no universal $r^* > 0$ such that $\triangle(\alpha, r^*) \subseteq A_F(\alpha)$ for all analytic maps f with a root at α, where F denotes the Newton map of f.

Exercise 1.164. Let $p(z) = (z - \alpha)^k (z - \alpha_1) \ldots (z - \alpha_s)$ be a polynomial of degree $n = k + s$, where the α_j need not be distinct. Follow the steps below to show that $r = \frac{d(2k-1)}{2n-1}$ is a radius of convergence for the Newton map F at α, where $d = \min_{j=1,\ldots,s} |\alpha - \alpha_j|$.
(a) For the Newton map $F(z) = z - \frac{p(z)}{p'(z)}$, we wish to show that $|z - \alpha| < r$ implies $|F(z) - \alpha| < |z - \alpha|$. This says that the action of F is to move points in the disk $\triangle(\alpha, r)$ closer to α. Show that it implies that $F^n(z) \to \alpha$ as $n \to \infty$ when $|z - \alpha| < r$.
(b) Show that $|F(z) - \alpha| < |z - \alpha|$ is equivalent to $\frac{|F(z)-\alpha|}{|z-\alpha|} = \left|1 - \frac{p(z)}{(z-\alpha)p'(z)}\right| < 1$.
(c) Show that $\left|1 - \frac{1}{w}\right| = \left|\frac{w-1}{w}\right| = \left|\frac{w-1}{w-0}\right| < 1$ holds if and only if $\operatorname{Re} w > 1/2$.
(d) Justify the following. The term $w = \frac{(z-\alpha)p'(z)}{p(z)} = (z - \alpha)\left[\frac{k}{z-\alpha} + \sum_{j=1}^{s} \frac{1}{z-\alpha_j}\right] = k + \sum_{j=1}^{s} \frac{z-\alpha}{z-\alpha_j}$. Thus $\operatorname{Re} w > \frac{1}{2}$ exactly when $\operatorname{Re} \sum_{j=1}^{s} \frac{z-\alpha}{z-\alpha_j} > \frac{1}{2} - k$, which holds, in particular, when $\sum_{j=1}^{s} \left|\frac{z-\alpha}{z-\alpha_j}\right| < k - \frac{1}{2}$.

(e) Show that if $|z - \alpha| < r$, then $|z - \alpha_j| > d - r$, where $d = \min_j |\alpha - \alpha_j|$. Hence
$$\sum_{j=1}^{s} \left| \frac{z-\alpha}{z-\alpha_j} \right| < s \frac{r}{d-r} = (n-k) \frac{r}{d-r}.$$

(f) Combine the above to reach the desired conclusion. Also, note that for $z \in \Delta(\alpha, r)$ we have $|F(z) - \alpha| < |z - \alpha|$, a stronger statement than saying that $\Delta(\alpha, r) \subseteq A_f(\alpha)$. For this reason, the value $r = \frac{d(2k-1)}{2n-1}$ may be called a *radius of contraction* for Newton's method at α.

Small Project 1.165. Fix an integer $n \geq 2$ and let $f(z) = z^n - 1$. Express the value for r found in Exercise 1.164 in terms of only n. See if you can improve on it for the radius of contraction by investigating the proof applied to the maps $f(z) = z^n - 1$. Hint: You may wish to calculate an $r^* > 0$ such that $|F'(z)| < 1$ on $\Delta(\alpha, r^*)$ for each root α of f, and then apply the Contraction Lemma 1.138.

Small Project 1.166. This is an open-ended project to investigate whether the value for r found in Exercise 1.164 can be improved if we know something about the geometry of the roots of p. See if you can improve on the value for the radius of contraction by investigating the proof in Exercise 1.164 for polynomials f of degree 3 with distinct roots $\alpha_p, \beta_p, \gamma_p$. If the roots all lie on a line can you get more from the proof? What if the roots form an equilateral triangle? Are there geometric configurations of the roots that allow for better results than what is given in Exercise 1.164? Can you generalize to higher degree polynomials? What if you allow for multiple roots of f? If a paper and pencil result is too hard, provide a conjecture based on examples you consider with the *Complex Newton Method Applet*. Whether or not you can answer them, give some related questions that one might consider. Remember, asking questions is important, even if you can't answer them. Hint: You may wish to calculate an $r^* > 0$ such that $|F'(z)| < 1$ on $\Delta(\alpha, r^*)$ for each root α of f, and then apply the Contraction Lemma 1.138.

Global Conjugation

Exercise 1.167. Prove that if $ab \neq 0$, then the maps $z \mapsto az$ and $z \mapsto bz$ are globally conjugate if and only if $a = b$ or $a = 1/b$.

Exercise 1.168. Prove that if $ab \neq 0$, then the maps $z \mapsto z + a$ and $z \mapsto z + b$ are globally conjugate.

Exercise 1.169. Prove that a quadratic map $f(z) = az^2 + bz + d$ is globally conjugate to one and only one map of the form $z \mapsto z^2 + c$.

Exercise 1.170. Prove that a rational map R is globally conjugate to a polynomial if and only if there exists $w \in \overline{\mathbb{C}}$ with $R^{-1}(\{w\}) = \{w\}$.

Newton Map of a Cubic Polynomial

Exercise 1.171. Prove that the attracting basin of an attracting fixed point of a rational or entire map is an open set in $\overline{\mathbb{C}}$, and so does not contain any of its boundary points. Hint: Use Theorem 1.17 together with the fact that if f is a continuous map, then $f^{-1}(U)$ is open whenever U is open.

Exercise 1.172. Use the *Complex Newton Method Applet* to investigate the behavior of Newton's method applied to the maps $f_1(z) = z(z-1)(z-0.908-0.423i)$ and $f_2(z) = z(z-1)(z-0.913-0.424i)$. Describe the behavior you see, especially for the black seed values where Newton's method fails.

Iteration of an Analytic Function

Exercise 1.173.

(a) Show that for the map $f(z) = z^2$ there is a point $z_0 \in C(0, 1)$ whose orbit is dense on $C(0, 1)$, that is, $\overline{\{f^n(z_0) : n \in \mathbb{N}\}} = C(0, 1)$. Hint: On $C(0, 1)$, we have $f(e^{i2\pi\theta}) = e^{i4\pi\theta}$, so the angle $\theta \in [0, 1)$ is doubled (mod 1). Consider $\theta_0 \in [0, 1)$ given in binary form as $\theta_0 = 0.0\,1\,00\,01\,10\,11\,000\,001\,010\,011\,100\,101\,110\,111\,0000\ldots$ where spaces have been included in the binary expansion to illustrate the pattern.

(b) Use (a) to justify the second claim in Remark 1.46.

(c) Because of computer limitations, any point you choose on $C(0, 1)$ will have a computed orbit that eventually becomes periodic, although it may take a large number of orbit points to see this. Justify this statement.

Exercise 1.174. Complete the proof of Theorem 1.48.

(a) Use Schwarz's Lemma (Appendix Theorem A.21 on page 353) to show that $|f'(a)| < 1$ for an attracting fixed point $a \in \mathbb{C}$ of an analytic map f. Hint: Consider a small disk $\triangle(a, r)$ where f is attracting and investigate the map $g : \triangle(0, 1) \to \triangle(0, 1)$ defined by $g(z) = \frac{f(rz+a)-a}{r}$.

(b) Suppose $a \in \mathbb{C}$ is a repelling fixed point of an analytic map f. Use the fact that f is locally one-to-one with a locally defined inverse to show $|f'(a)| > 1$. Hint: Since $|f'(a)| \geq 1$ (else a would be attracting), we have that f is locally one-to-one (see Appendix Section A.6.1 on page 354 regarding the open mapping theorem and the Inverse Function Theorem A.24). Letting h be f^{-1} defined on a small disk $\triangle(a, r)$, show that a is an attracting fixed point for h (since it was repelling for f) and then apply part (a) to h. Use this to argue for the conclusion.

Exercise 1.175. Let f be a map that is analytic at ∞ such that $f(\infty) = \infty$. Let $\lambda = 1/f'(\infty)$. Prove that if $|\lambda| < 1$, then ∞ is an attracting fixed point according to Definition 1.16. Also, prove that if $|\lambda| > 1$, then ∞ is a repelling fixed point according to Definition 1.44. Hint: Noting that the map $k(z) = 1/f(1/z)$ has a fixed point at the origin with multiplier λ (see Appendix Lemma B.20 on page 363), study how the map $\phi(z) = 1/z$ transfers information about k at the origin to information about f at ∞.

Exercise 1.176. Let $g(z)$ be a polynomial of degree greater than or equal to two. Prove that $\partial A_g(\infty) = \partial K(g)$, and by Theorem 1.59 we have $J(g) = \partial A_g(\infty) = \partial K(g)$.

Exercise 1.177.

(a) For $c = -0.467 + 0.513i$ determine the attracting fixed point p_c of f_c.

(b) Write the multiplier λ for p_c in polar form.

(c) Use the *Global Complex Iteration Applet for Polynomials* to iterate a seed z_0 near p_c one step at a time (use the zoom feature to get a better look) to see how the orbit approaches p_c. The way this orbit approaches p_c is related to the multiplier λ. Describe the connection.

(d) Use the *Complex Function Iterator Applet* to iterate the map $z \mapsto \lambda z$ for seed values near the attracting fixed point at the origin. Describe the convergence.

(e) Compare (c) and (d), explaining in as much mathematical detail as you can what you find. Hint: Consider the Taylor series of f_c expanded around the point p_c.

(f) Choose various c in K_1 and study the convergence of orbits to the attracting fixed point using the applet. Use the picture from the applet to approximate the argument of the multiplier. Note: Some c values will be easy to work with, while others are not as easy.

Exercise 1.178.

(a) For $c = -0.92568 + 0.22512i$ determine the attracting 2-cycle $\{u_c, v_c\}$ of f_c.

(b) Write the multiplier λ for $\{u_c, v_c\}$ in polar form.

(c) Use the *Global Complex Iteration Applet for Polynomials* to iterate a seed z_0 near one of the cycle points u_c one step at a time (use the zoom feature to get a better look) to see how every other orbit point approaches u_c. The way this orbit approaches u_c is related to the multiplier λ. Describe the connection.

(d) Use the *Complex Function Iterator Applet* to iterate the map $z \mapsto \lambda z$ for various seed values near the attracting fixed point at the origin. Describe the convergence.

(e) Compare (c) and (d), explaining in as much mathematical detail as you can what you find. Hint: Consider the Taylor series of the iterate f_c^2 expanded around the point u_c.

(f) Choose any c in K_2 and study the convergence of orbits to the attracting 2-cycle using the applet. Use the picture from the applet to approximate the argument of the multiplier. Note: Some c values will be easy to work with, while others are not as easy.

Exercise 1.179. Determine the form of the multiplier map on K_2. The function $\lambda : K_2 \to \triangle(0, 1)$ maps $c \in K_2$ to the multiplier of the attracting 2-cycle of f_c. Show that it can be extended to a map $\lambda : \overline{K_2} \to \overline{\triangle(0, 1)}$ that is continuous, one-to-one, and onto.

Exercise 1.180. Without appealing to Theorem 1.79, prove that if f_{c_0} has an attracting n-cycle, then for all c close to c_0, the map f_c has an attracting n-cycle. Conclude that K_n is an open set. Hint: You may use Rouché's Theorem (see [1, p. 294]) to prove the general result that if the coefficients of a polynomial P are sufficiently close to the corresponding coefficients of a polynomial Q of the same degree, then the roots of P and the roots of Q are close. Apply this to the polynomials $P_1(z) = f_{c_0}^n(z) - z$ and $Q_1(z) = f_c^n(z) - z$.

Exercise 1.181. Mathematically justify Remark 1.83 on page 34.

Critical Points and Critical Orbits

Exercise 1.182. Let $f(z) = z^3 - 1$ and let F denote the corresponding Newton map. Show that F cannot have an attracting cycle of any order other than the fixed points at the roots of f. Hint: Check the behavior of each critical point of F and consider Theorem 1.89.

Note: More can be said. It is true that Newton's method fails to find a root of f only for seed values from $\partial A_F(1) = \partial A_F(e^{2\pi i/3}) = \partial A_F(e^{-2\pi i/3})$. In fact, $\overline{\mathbb{C}} = A_F(1) \cup A_F(e^{2\pi i/3}) \cup A_F(e^{-2\pi i/3}) \cup \partial A_F(1)$. The method of proof is along the same lines but uses a more powerful set of results. The results imply that if Newton's method fails for all seed values in an open set of points, then there is a critical point of F whose orbit does not converge to any of the roots of f. The key ingredients to this proof are Sullivan's No Wandering Domains Theorem, the classification of forward invariant components of Fatou sets, and the role of critical points in parabolic domains, Siegel disks, and Herman rings. The results can be found in [1].

Exercise 1.183. Show that the periodic points for the map $f(z) = z^2$ are of the form $e^{i2\pi p/q}$ where $p, q \in \mathbb{N}$ with p/q in lowest terms and q odd. Show that the pre-periodic points have the same form, but with q even. Hint: Show that if $p/q = \frac{n}{2^k - 1}$ for $k \geq 1, n \geq 1$, then $e^{i2\pi p/q}$ has a period of k or some divisor of k. Then use the fact that every rational number p/q with q odd can be expressed in this form. This can be proved by using the fact that the element 2 in the multiplicative group consisting of those elements of $\mathbb{Z}_q$ that are relatively prime to q has a finite multiplicative order which we call k.

Exercise 1.184. One feature of the Mandelbrot set that stands out is that it is symmetric about the x-axis. Prove this by showing that $c \in M$ if and only if $\bar{c} \in M$. Hint: Compare the critical orbits $\{f_c^n(0)\}$ and $\{f_{\bar{c}}^n(0)\}$.

Exercise 1.185. Prove that the Mandelbrot set M has no holes (i.e., its complement $\overline{\mathbb{C}} \setminus M$ is connected) by appealing directly to its definition. Hint: Suppose that U is a bounded domain of $\overline{\mathbb{C}} \setminus M$ such that $\partial U \subseteq M$. Apply Appendix Corollary A.19 on page 352 (a version of the maximum modulus theorem) to the maps Q_n on $\overline{U}$, where Q_n are as in the proof of Lemma 1.103.

Exploring the Mandelbrot Set

Exercise 1.186. Let P be a polynomial of degree $n \geq 2$. Show that the filled-in Julia set $K(P)$ has infinitely many points as follows.

(a) Show that $P(z)$ has a fixed point z_0 by applying the fundamental theorem of algebra.

(b) Show that if z_0 is such that $P^{-1}(\{z_0\}) = \{z_0\}$, then $P(z) = z_0 + a(z - z_0)^n$ for some $a \neq 0$ from which it can be shown that $K(P) = \overline{\Delta(z_0, r)}$ where $r = |a|^{1/(1-n)}$.

(c) Show that if $P^{-1}(\{z_0\}) \neq \{z_0\}$, then there exists an infinite sequence of distinct points z_{-n} such that $\cdots \mapsto z_{-n} \mapsto z_{-n+1} \mapsto \cdots \mapsto z_{-2} \mapsto z_{-1} \mapsto z_0$. Hence, $\{z_{-n}\} \subseteq K(P)$.

It is true that $J(P)$ contains uncountably many points, but to show this requires more effort (see [1, p. 95]).

Exercise 1.187. Prove that for $c \in \mathbb{C}$, we have $z \in A_{f_c}(\infty)$ if and only if $-z \in A_{f_c}(\infty)$. Do the same for $J(f_c)$ and $K(f_c)$.

Exercise 1.188. Since the root $c_{p/q}$ of the p/q bulb in the Mandelbrot set lies on the boundary of two hyperbolic components, namely K_1 and $B_{p/q}$, Theorem 1.79 implies that there are two multiplier maps that are defined at $c_{p/q}$. The map $\lambda : K_1 \to \Delta(0, 1)$

can be extended to be defined at $c_{p/q}$, as can $\lambda_{p/q} : B_{p/q} \rightarrow \triangle(0, 1)$. Does $\lambda(c_{p/q}) = \lambda_{p/q}(c_{p/q})$? What does it mean if they are the same? different?

Exercise 1.189. Find a c value in $B_{5/32}$. Formal proof is not required.

Exercise 1.190.

(a) Experiment with the *Parameter Plane and Julia Set Applet* to determine what Farey fraction should be used to represent the cusp of the cardioid K_1 in the Farey procedure for finding child bulbs. We treat the cusp as a Farey parent which is larger than all other bulbs.

(b) It is true (though not easy to prove) that starting with Farey parents $B_{1/2}$ and the cusp one can compute through Farey addition all of the p/q bulbs attached to the upper half of K_1. Illustrate this by producing a Farey family tree that contains the lineage of $B_{5/32}$ up to the ancestors $B_{1/2}$ and the cusp of K_1.

(c) Use Exercise 1.118 and part (b) to show that all of the p/q bulbs attached to K_1 can be identified through Farey addition.

Other Uni-critical Families of Polynomials

Exercise 1.191. Let $P_c(z) = z^d + c$ for $d = 2, 3, 4, \ldots$.

(a) Show that if $|z| \geq |c|$ and $|z| > 2$, then there exists $\varepsilon > 0$ such that $|P_c(z)| > |z|(1 + \varepsilon)$.

(b) Use induction to show that if $|z| \geq |c|$ and $|z| > 2$, then $P_c^n(z) \rightarrow \infty$.

(c) Apply (b) to prove $P_c^n(0) \rightarrow \infty$ as $n \rightarrow \infty$ if and only if $|P_c^k(0)| > 2$ for some $k \in \mathbb{N}$. In particular, $M_d \subseteq \overline{\triangle(0, 2)}$.

(d) Prove that the filled-in Julia set of P_c is contained in $\overline{\triangle(0, 2)}$ when $|c| \leq 2$. Show by example that this does not hold when $|c| > 2$, i.e., it is not the case that for all c we have $P_c^n(z) \rightarrow \infty$ whenever $|z| > 2$.

Exercise 1.192.

(a) Use the *Parameter Plane and Julia Set Applet* to explore the parameter plane for the family $P_c(z) = z^d + c$, and then identify and explain the symmetry you see there.

(b) Use the *Parameter Plane and Julia Set Applet* to explore the Julia sets of maps in the family $P_c(z) = z^d + c$, and then identify and explain the symmetry you see there.

Exercise 1.193. For an integer $d > 2$, prove M_d is the connectedness locus for the family of maps $P_c(z) = z^d + c$, i.e., show that $M_d = \{c \in \mathbb{C} : J(P_c) \text{ is connected}\}$.

Transcendental Dynamics

Exercise 1.194. Prove that $E_c^n(z_0) \rightarrow \infty$ if and only if $\mathrm{Re}\, E_c^n(z_0) \rightarrow +\infty$, where $E_c(z) = ce^z$ and $c \in \mathbb{C} \setminus \{0\}$.

Large Project 1.195. This large project, to determine the accuracy of the exponential Julia set applet pictures, is open-ended and should be attempted only after Section 1.6.1 has been covered.

As mentioned in Remark 1.135 on page 54, the algorithm used to illustrate the exponential Julia set $J(E_{0.2})$ can color some seed values z_0 non-black (which it does when at least one of the orbit points $E_{0.2}(z_0), \ldots, E_{0.2}^{20}(z_0)$ lands in the set $\{\operatorname{Re} z > 50\}$), that are actually not in $J(E_{0.2})$. If z_0 maps to $E_{0.2}(z_0) = 51 + \pi i$, it will not be left black. However, we see that $E_{0.2}^2(z_0) = -0.2e^{51} \in H = \{z : \operatorname{Re} z < 1\} \subseteq A_{E_{0.2}}(p)$, which implies $z_0 \in A_{E_{0.2}}(p) = F(E_{0.2})$. The goal of the project is to estimate how close points like z_0 are to points in $J(E_{0.2})$. This is, in a sense, a measure of the error in the algorithm.

One method to use is to consider an open rectangle R centered at z_0 and its expanded image $E_{0.2}(R)$. If it meets a point that has an orbit that truly tends to ∞, then there is a corresponding point in R that lies in $J(E_{0.2})$. So you must consider how much R gets expanded, and how big the expansion needs to be before you know that the expanded image $E_{0.2}(R)$ contains a point with an orbit that truly tends to ∞. Hint: Consider the open interval $(q, +\infty)$ and its $2\pi i k$-translates where q is the repelling fixed point for $E_{0.2}$.

Also, you should consider what estimates you can get if w_0 is such that $\operatorname{Re} E_{0.2}(w_0) < 50$ but $E_{0.2}^2(w_0) = 51 + \pi i$. Thus w_0 takes two steps before the algorithm gives it a non-black color, as opposed to z_0 that used only one step. The point w_0 is not in $J(E_{0.2})$, but there are points nearby which are in $J(E_{0.2})$. Can you estimate how near these points are?

Some related questions are: Is there a relationship between the size of the viewing window and the accuracy of the pictures? Can you quantify this, in general or in specific cases? How does the relationship depend on the value 50 that was chosen to determine the escape criterion or on the value 20 that determined how many orbit points might be checked? If you set the **Dynamic plane min iterations** to 2 (which means that a seed value is iterated twice before the escape condition is tested, i.e., a seed z_0 is given a non-black color only if at least one of the points $z_2, \ldots, z_{20}$ has real part greater than 50), do we get a more accurate or less accurate picture?

This is a technical project. Some of these questions are addressed in [10].

Exercise 1.196.

(a) Prove the Contraction Lemma 1.138 stated on page 55. Hint: Consider the argument preceding the statement of the lemma.

(b) Show that if f is a one-to-one analytic function such that $|f'(z)| > \gamma > 1$ for all z in some domain D and $f(D)$ is convex, then f is expanding on D, i.e., we have $|f(z) - f(w)| \geq \gamma |z - w|$ for all $z, w \in D$. Hint: Consider f^{-1}, which by the Inverse Function Theorem A.24 on page 354, is analytic.

Exercise 1.197. Prove that two fingers (of any stage) constructed in Example 1.136 are separated by an infinitely long gap of black points in $A_{E_{0.2}}(p)$. Hint: The horizontal lines $K_n = \{z \in \mathbb{C} : \operatorname{Im} z = (2n + 1)\pi\}, n \in \mathbb{N}$ lie in $A_{E_{0.2}}(p)$ and separate the stage 1 fingers. Show that the inverse image of these lines under $E_{0.2}$ separates the stage 2 fingers, and then proceed inductively.

Exercise 1.198. Follow the steps (a)-(d) to prove that the thickness of a stage n finger is no greater than π/γ^{n-1}, where $\gamma = \ln 5 > 1$.

(a) Prove that a stage n finger is contained in the right half plane $\{\operatorname{Re} z \geq \gamma\}$ by showing $\operatorname{Re} E_{0.2}(z) \leq |E_{0.2}(z)| < 1$ whenever $\operatorname{Re} z < \gamma$.

(b) Suppose $E_{0.2}(\triangle(z_0, r)) \subseteq \{\text{Re } z > \gamma\}$ for some $r < \pi$. Show that $E_{0.2}(\triangle(z_0, r)) \supseteq \triangle(E_{0.2}(z_0), \gamma r)$ by filling in the details of the following argument. Define a branch L of the inverse of $E_{0.2}$ on the convex set $\{\text{Re } z > \gamma\}$ such that $L(E_{0.2}(z_0)) = z_0$ (given by $L(z) = \text{Log } z + \ln 5 + 2\pi i n$ for some $n \in \mathbb{Z}$). For $|z - z_0| = r$, we then have $r = |z - z_0| = |L(E_{0.2}(z)) - L(E_{0.2}(z_0))| \leq (1/\gamma)|E_{0.2}(z) - E_{0.2}(z_0)|$ since $|L'(z)| = |1/z| < 1/\gamma$ when $\text{Re } z > \gamma$. Because $E_{0.2}$ is univalent on $\triangle(z_0, r)$, it follows that $E_{0.2}$ maps the circle $|z - z_0| = r$ to a simple closed curve C that encloses $\triangle(E_{0.2}(z_0), \gamma r)$. (At this point one could complete the proof by applying the Argument Principle A.20 on page 352 to see that the number of zeros of $E_{0.2}(z) - E_{0.2}(a)$ is equal to one for all values a in the interior of C.) Applying the Maximum Modulus Theorem A.18 on page 352 to $L(z) - z_0$ on $E_{0.2}(\triangle(z_0, r))$, we see that $L(\triangle(E_{0.2}(z_0), \gamma r)) \subseteq \triangle(z_0, r)$, and the result follows.

(c) Use parts (a) and (b) to prove that if $\triangle(z_0, r)$ is contained in a stage n finger for $n \geq 2$, then $\triangle(E_{0.2}(z_0), \gamma r)$ is contained in a stage $(n - 1)$ finger.

(d) Use Exercise 1.141 and induction to complete the proof.

Exercise 1.199. Prove that the set J in Proposition 1.142 cannot contain an open set.

Exercise 1.200. Follow the steps below to construct infinitely many (though not all) of the hairs in $J(E_{0.2})$. Each hair is the image of a continuous map $h : [0, \infty) \to \mathbb{C}$ such that $h(t) \to \infty$ as $t \to \infty$.

(a) Prove that there exists a repelling fixed point $q \in \mathbb{R}$ such that for any $x > q$ we have $E_{0.2}(x) > x$ and therefore $E_{0.2}^n(x) \to +\infty$. Use this to conclude that the interval $h_0 = [q, +\infty)$ is a hair in $J(E_{0.2})$. We call it a straight hair since it extends to ∞ in a straight line.

(b) Use the $2\pi i$ periodicity of $E_{0.2}$ to show that for each $k \in \mathbb{Z}$ the set h_k, defined to be the $k2\pi i$ translate of h_0, is also a hair in $J(E_{0.2})$. We call each h_k a stage 1 hair, which could be called the main hair in the corresponding stage 1 finger.

(c) Argue that inside of the stage 1 finger C_0, there exist infinitely many hairs $h_{0,k}$ in $J(E_{0.2})$ such that $E_{0.2}(h_{0,k}) = h_k$. No $h_{0,k}$ is a straight hair, except for $h_{0,0}$ which equals $h_0 = [q, +\infty)$.

(d) Use the $2\pi i$ periodicity of $E_{0.2}$ to show that for each $j, k \in \mathbb{Z}$ the set $h_{j,k}$, defined to be the $j2\pi i$ translate of $h_{0,k}$, is also a hair in $J(E_{0.2})$. We call each $h_{j,k}$ a stage 2 hair.

(e) Repeat the arguments to show that there exist infinitely many stages of hairs in $J(E_{0.2})$.

(f) If we let H_n denote all the stage n hairs, then $\cup_{n=1}^{\infty} H_n \cup \{\infty\}$ is not quite all of $J(E_{0.2})$. However, we can show that $J(E_{0.2}) = \overline{\cup_{n=1}^{\infty} H_n} \cup \{\infty\}$ by showing that each point in the hairs described in Section 1.6.2 is a limit of points from the collection of stage n hairs.

Exercise 1.201. Determine the exact real bifurcation value c^* between $c = 0.2$ and $c = 0.5$ for the family of maps E_c. Show that for $c < c^*$ we have $\{\text{Re } z < 1\} \subseteq F(E_c)$, but for $c > c^*$ we have $F(E_c) = \emptyset$.

Exercise 1.202. For each $c \in L_1$, the map $E_c(z) = ce^z$ has an attracting fixed point with multiplier $\lambda(c)$ (see Figure 1.27 on page 60). Thus we have a multiplier map $\lambda : L_1 \to \Delta(0, 1)$.

(a) Find the inverse of the multiplier map.

(b) Use the inverse of the multiplier map to find c where the 7/13 bulb attaches to the cardioid L_1. Hint: What kind of fixed point will there be at this c value?

Exercise 1.203. Show that orbits under the maps $S_c(z) = c \sin z$ and $C_c(z) = c \cos z$ escape to ∞ in the direction of the positive or negative imaginary axis:

(a) Show that $S_c^n(z) \to \infty$ if and only if $|\operatorname{Im} S_c^n(z)| \to +\infty$.

(b) Show that $C_c^n(z) \to \infty$ if and only if $|\operatorname{Im} C_c^n(z)| \to +\infty$.

Exercise 1.204. Show that neither $S_c(z) = c \sin z$ nor $C_c(z) = c \cos z$ has a finite asymptotic value:

(a) Examine $|\sin(x + iy)|$ to show that if S_c had a finite asymptotic value, then the curve γ (along which S_c has a finite asymptotic value) would be vertically bounded, that is, lie in a horizontal strip $\{|\operatorname{Im} z| \le M\}$.

(b) Show that $\sin z$ maps the imaginary axis into itself.

(c) Show that $\sin z$ maps the vertical line $\{\pi/2 + iy : y \in \mathbb{R}\}$ into $[1, +\infty)$.

(d) Use the fact that $\sin z$ is 2π periodic along with parts (a), (b), and (c) to show that $\sin z$ cannot have a finite asymptotic value.

(e) Show that S_c cannot have a finite asymptotic value since $\sin z$ does not.

(f) Show that C_c cannot have a finite asymptotic value since $C_c(z) = S_c(z + \pi/2)$.

Exercise 1.205.

(a) Explain in what sense $\{S_c^n(c)\}$ is the only critical orbit of S_c.

(b) Explain in what sense $\{C_c^n(c)\}$ is the only critical orbit of C_c.

Definitions and Properties of the Julia and Fatou Sets

Exercise 1.206. Let a_n and b_n for $n \ge 0$ be sequences of positive real numbers such that $1 = a_0 > b_0 > a_1 > b_1 > \dots$ and $a_n \to 0$. Note $b_n \to 0$ also.

(a) Construct a strictly increasing function $g : [0, +\infty) \to [0, +\infty)$ such that g is differentiable on $(0, +\infty)$ and $g(a_n) = a_n$ and $g(b_n) = b_n$ with $g'(a_n) = 0$ and $g'(b_n) = 2$. Construct g so that it does not fix any points other than the a_n, the b_n, and 0. Hence each a_n is a super attracting fixed point of g and each b_n is a repelling fixed point of g. Sketch a graph of g to convince yourself that this can be done. It is not necessary that you produce a formula for g.

(b) Define a function $f : \mathbb{C} \to \mathbb{C}$ by $f(re^{i\theta}) = g(r)e^{i\theta}$. Consider the dynamics of points both on and near the circles $|z| = a_n$ and $|z| = b_n$ to see that each "repelling circle" $|z| = b_n$, which is fixed by f, must lie in $J(f)$.

(c) Show that $0 \in F(f)$. Hint: Use the fact that $f(\Delta(0, a_n)) \subseteq \Delta(0, a_n)$ for each $n \ge 0$.

(d) Show that (b) and (c), together with the fact that $b_n \to 0$, implies that 0 is not in the interior of $F(f)$.

Exercise 1.207. Use the definitions of Fatou and Julia set to prove the statements in Proposition 1.213 parts (a)-(d). Hint: For the complete invariance statements, use the Open Mapping Theorem A.23 on page 354 to show that when it is non-constant, f is an open map, i.e., if U is an open set in the domain of f, then the image $f(U)$ is also an open set in $\overline{\mathbb{C}}$. For (d), use Montel's Theorem 1.210 and part (c).

Exercise 1.208. Show that the set of repelling cycles for $f(z) = z^2$ is dense in $J(f) = C(0, 1)$ by explicitly solving for the set of p-periodic points for each $p \in \mathbb{N}$.

1.10   Bibliography

[1] Alan F. Beardon, *Iterations of Rational Functions*. Springer-Verlag, New York, 1991.

[2] Bodil Branner. The Mandelbrot set. In *Chaos and Fractals (Providence, RI, 1988)*, volume 39 of *Proc. Sympos. Appl. Math.*, pages 75–105. Amer. Math. Soc., Providence, RI, 1989.

[3] Lennart Carleson and Theodore W. Gamelin, *Complex Dynamics*. Springer-Verlag, New York, 1993.

[4] Ruel V. Churchill and James Ward Brown, *Complex Variables and Applications*. McGraw-Hill Book Co., New York, eighth edition, 2009.

[5] M. Viana da Silva, The differentiability of the hairs of $exp(Z)$. *Proc. Amer. Math. Soc.*, 103(4): 1179–1184, 1988.

[6] Robert L. Devaney. Chapter 4—Complex Exponential Dynamics. In F. Takens, Henk Broer, and B. Hasselblatt, editors, *Handbook of Dynamical Systems*, volume 3 of *Handbook of Dynamical Systems*, pp. 125–223. Elsevier Science, 2010.

[7] Robert L. Devaney and Mónica Moreno Rocha, Geometry of the antennas in the Mandelbrot set. *Fractals*, 10(1): 39–46, 2002.

[8] Robert L. Devaney, *A First Course in Chaotic Dynamical Systems*. Addison-Wesley Studies in Nonlinearity. Addison-Wesley Publishing Company Advanced Book Program, Reading, MA, 1992.

[9] Robert L. Devaney, A survey of exponential dynamics. In *Proceedings of the Sixth International Conference on Difference Equations*, pp. 105–122, CRC. Boca Raton, FL, 2004.

[10] Marilyn B. Durkin, The accuracy of computer algorithms in dynamical systems. *Internat. J. Bifur. Chaos Appl. Sci. Engrg.*, 1(3): 625–639, 1991.

[11] Kenneth Falconer, *Fractal Geometry—Mathematical Foundations and Applications*. John Wiley & Sons, Chichester, 1990.

[12] Pierre Fatou, Sur les équations fonctionnelles. *Bull. Soc. Math. France*, 47: 161–271, 1919.

[13] Pierre Fatou, Sur les équations fonctionnelles. *Bull. Soc. Math. France*, 48: 208–314, 1920.

[14] Pierre Fatou, Sur les équations fonctionnelles. *Bull. Soc. Math. France*, 48: 33–94, 1920.

[15] James Gleick, *Chaos*. Penguin Books, New York, 1987.

[16] John G. Hocking and Gail S. Young, *Topology*. Dover Publications Inc., New York, second edition, 1988.

[17] Gaston Julia, Mémoire sur l'itération des fonctions rationnelles. *Journal de Math. Pure et Appl.*, pp. 47–245, 1918.

[18] Bogusława Karpińska, Area and Hausdorff dimension of the set of accessible points of the Julia sets of λe^z and $\lambda \sin z$. *Fund. Math.*, 159(3): 269–287, 1999.

[19] Bogusława Karpińska, Hausdorff dimension of the hairs without endpoints for $\lambda \exp z$. *C. R. Acad. Sci. Paris Sér. I Math.*, 328(11): 1039–1044, 1999.

[20] Tan Lei, Ressemblance eintre l'ensemble de Mandelbrot set l'ensemble de Julia au voisinage d'un point de Misiurewicz. In A. Douady and J. H. Hubbard, editors, *Étude dynamique des polynômes complexes. Partie II*, volume 85 of *Publications Mathématiques d'Orsay [Mathematical Publications of Orsay]*, pp. 139–152. Université de Paris-Sud, Département de Mathématiques, Orsay, 1985.

[21] John C. Mayer, An explosion point for the set of endpoints of the Julia set of $\lambda \exp(z)$. *Ergodic Theory Dynam. Systems*, 10(1): 177–183, 1990.

[22] Curtis T. McMullen, The Mandelbrot set is universal. In *The Mandelbrot set, theme and variations*, volume 274 of *London Math. Soc. Lecture Note Ser.*, pp. 1–17. Cambridge Univ. Press, Cambridge, 2000.

[23] Rich Stankewitz, Wendy Conatser, Trey Butz, Ben Dean, Yun Li, and Kristopher Hart, *JULIA 2.0 Fractal Drawing Program*.
http://rstankewitz.iweb.bsu.edu/JuliaHelp2.0/Julia.html.

[24] Norbert Steinmetz, *Rational Iteration*, volume 16 of *de Gruyter Studies in Mathematics*. Walter de Gruyter & Co., Berlin, 1993.

[25] Hiroki Sumi, Random complex dynamics and semigroups of holomorphic maps. Proc. London. Math. Soc. 102(1): 50–112, 2011.

1.A   Appendix: Definitions and Properties of the Julia and Fatou Sets

Here we present the formal definitions of the Fatou set and Julia set of a rational or entire map. We also provide the statement of Montel's theorem, a major tool in complex dynamics, and a proposition stating some of the properties of the Fatou and Julia sets.

The spherical distance between the points $z, w \in \overline{\mathbb{C}}$ is $\sigma(z, w)$.

Definition 1.209 (Fatou Set). Let f be a rational or entire map. The Fatou set is the set $F(f) = \{z \in \overline{\mathbb{C}} :$ for every $\varepsilon > 0$ there exists $\delta > 0$ such that $\sigma(z, w) < \delta$ implies $\sigma(f^n(z), f^n(w)) < \varepsilon$ for all $n \in \mathbb{N}\}$.

Thus, if $z \in F(f)$ and ε is small, then for w to have an orbit ε-*similar* to the orbit of z (by which we mean that corresponding orbit points are never more than ε apart) we need only to choose w close enough to z, i.e., within a distance δ. Another way to interpret this definition is to say that $F(f)$ is the set of points z such that for any $\varepsilon > 0$ there exists $\delta > 0$ such that $f^n(\Delta_\sigma(z, \delta)) \subseteq \Delta_\sigma(f^n(z), \varepsilon))$ for all $n \in \mathbb{N}$, i.e., a tiny neighborhood of z (of size δ measured with the spherical metric) will have an orbit that stays tiny (of size no greater than ε measured with the spherical metric) along the entire orbit (see Figure 1.33). This fails to happen with the map $f(z) = z^2$ for any neighborhood of a point on the unit circle $C(0, 1)$.

A major tool used in dynamics is Montel's theorem, which we state without proof in a setting most easily applied to the dynamics of interest to us here. (See [3, p. 9] to find it stated in terms of normal families.)

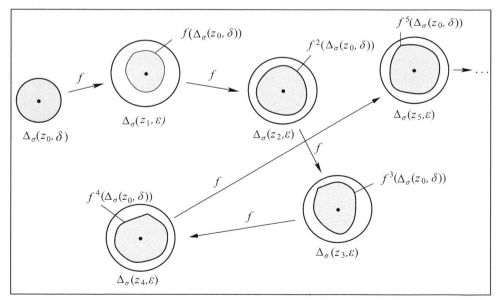

FIGURE 1.33. This picture illustrates how, for a point $z_0 \in F(f)$, a tiny neighborhood $\Delta_\sigma(z_0, \delta)$ has a forward image that stays tiny (within an ε neighborhood of each orbit point z_n) for the entire orbit. We illustrate this up to the fifth iterate, but the last arrow with the dots indicates that this happens for all iterates f^n.

Theorem 1.210 (Montel's Theorem). *Let $U \subseteq \overline{\mathbb{C}}$ be an open set in the domain set of a rational or entire map f. If the family of maps $\{f^n : n \in \mathbb{N}\}$ omits three points $z, w, v \in \overline{\mathbb{C}}$, i.e., $f^n(U) \cap \{z, w, v\} = \emptyset$ for all $n \in \mathbb{N}$, then $U \subseteq F(f)$. In particular, if $f(U) \subseteq U$ and $\overline{\mathbb{C}} \setminus U$ contains three or more points, then $U \subseteq F(f)$.*

Montel's theorem can make it quick to show that $f(z) = z^2$ has $F(f) = \overline{\mathbb{C}} \setminus C(0, 1) = \Delta(0, 1) \cup (\overline{\mathbb{C}} \setminus \overline{\Delta(0, 1)})$. Since $f(\Delta(0, 1)) \subseteq \Delta(0, 1)$, the theorem gives $\Delta(0, 1) \subseteq F(f)$. Similarly, we can show $\overline{\mathbb{C}} \setminus \overline{\Delta(0, 1)} \subseteq F(f)$. Also, from Section 1.3.2 or by considering the definition of the Fatou set, we know that $C(0, 1)$ does not contain any points in the Fatou set $F(f)$. Hence, we conclude $F(f) = \overline{\mathbb{C}} \setminus C(0, 1)$.

Remark 1.211. Though its proof is too advanced for this text, it is true that if f is a rational or entire map and $z_0 \in F(f)$, then there exists $r > 0$ such that $\Delta_\sigma(z_0, r) \subseteq F(f)$, i.e., $F(f)$ is an open set. However, if f is not rational or entire, then $F(f)$ need not be open. To see this, consider the map $f(z)$ defined to be zero when $z = x + iy$ with $x, y \in \mathbb{Q}$ and $f(z) = z$ otherwise. In this case $f^n = f$ for all $n \in \mathbb{N}$ and $F(f) = \{0\}$. See Additional Exercise 1.206 for such an example where f is continuous but not analytic.

Definition 1.212 (Julia Set). The Julia set of a rational or entire map f is defined to be

$$J(f) = \overline{\mathbb{C}} \setminus F(f).$$

For a point z to be in $J(f)$ there must be points w that are arbitrarily close to z, but which fail to have orbits similar to the orbit of z. More precisely, for a point z to be in $J(f)$ there must be $\varepsilon > 0$ such that for every $\delta > 0$, there exists w within a distance δ of z such that the orbit of w is not ε-similar to the orbit of z.

In Additional Exercise 1.207 you are asked to show the following properties (a)-(d). Properties (e) and (f), however, require some tools from complex analysis which are a bit beyond the level of this text (see [1, p. 148] and [24, p. 38] for details).

Proposition 1.213 (Properties of the Julia and Fatou Sets). *Let f be a rational or entire map. Then:*

(a) $F(f)$ is an open set in $\overline{\mathbb{C}}$ and thus $J(f)$ is a closed set in $\overline{\mathbb{C}}$.

(b) $F(f)$ is completely invariant, i.e., $f(F(f)) \subseteq F(f)$ and $f^{-1}(F(f)) \subseteq F(f)$.

(c) $J(f)$ is completely invariant, i.e., $f(J(f)) \subseteq J(f)$ and $f^{-1}(J(f)) \subseteq J(f)$.

(d) $J(f)$ contains an open set if and only if $J(f) = \overline{\mathbb{C}}$.

(e) The set of repelling periodic cycles of f is dense in $J(f)$. That is, each repelling periodic cycle of f is in $J(f)$ and for every open set U that intersects $J(f)$, there is a repelling periodic point z that lies in U.

(f) Let A be the set of z in $J(f)$ such that the orbit of z is dense in $J(f)$ (i.e., for every open set U that intersects $J(f)$, there is an orbit point z_n that lies in U). Then A is dense in $J(f)$ (i.e., for every open set V that intersects $J(f)$, there is a point $z \in A$ that lies in V).

Remark 1.214. Additional Exercises 1.208 and 1.173 illustrate properties (e) and (f) for $f(z) = z^2$. We see that, both in this example and in general, the Julia set contains a dense set of points that in some sense are the ultimate in predictable behavior. These are the periodic points. Nothing could be more predictable than to have the orbit follow a finite set of points over and over again since any orbit point can be only one of a few values. However, the Julia set also contains points with an orbit that is dense in the Julia set. Such orbits are unpredictable in the sense that an orbit point could be close to any of the infinitely many points in the Julia set. Thus we see that inside the sensitive dependence that defines chaos (and thus defines the Julia set $J(f)$) there lies an intriguing mix of predictability and unpredictability.

1.B Appendix: Global Conjugation and Möbius Map Dynamics

In this appendix we show that the global conjugation defined in Section 1.2.6 can be used to quickly classify and understand the dynamics of a *Möbius map*, i.e., a map of the form $f(z) = \frac{az+b}{cz+d}$ where $ad - bc \neq 0$. The classification depends on the number and type of fixed points of f. We begin with the following proposition.

Proposition 1.215. *Non-identity Möbius maps can have only exactly one or exactly two fixed points in $\overline{\mathbb{C}}$.*

Proof. Let $f(z) = \frac{az+b}{cz+d}$ be a non-identity Möbius map. Suppose that $f(\infty) \neq \infty$, i.e, $c \neq 0$. Solving for fixed points of f, i.e., solving the equation $f(z) = z$, yields the equation $cz^2 + (d - a)z - b = 0$, which has two distinct roots or one double root in $\mathbb{C}$. If $c = 0$, then f is a linear map and so has one fixed point at ∞ and possibly a second fixed point in $\mathbb{C}$.   □

We now describe the dynamics of a non-identity Möbius map f based on how many fixed points it has.

Case 1: Suppose $f(z)$ fixes only ∞. We can quickly show that f has the form $f(z) = z + \beta$ for some $\beta \in \mathbb{C} \setminus \{0\}$. Thus $f^n(z) = z + n\beta$, and hence $f^n(z) \to \infty$ as $n \to \infty$ for all $z \in \overline{\mathbb{C}}$ as in Example 1.49. We also see that $J(f) = \{\infty\}$.

Case 2: Suppose $f(z)$ fixes only $w \in \mathbb{C}$. Let $\psi(z) = \frac{1}{z-w}$ and define $g(z) = \psi f \psi^{-1}(z)$, noting that g is also Möbius. By Exercise 1.29 the map $g(z)$ has only one fixed point at ∞. From Case 1 we see that $g(z) = z + \alpha$ for some $\alpha \in \mathbb{C} \setminus \{0\}$ and $g^n(z) \to \infty$ for all $z \in \overline{\mathbb{C}}$. Hence $f^n(z) = \psi^{-1} g^n \psi(z) \to \psi^{-1}(\infty) = w$ for all $z \in \overline{\mathbb{C}}$. So, if f is Möbius with unique fixed point w, then $f^n(z) \to w$ for all $z \in \overline{\mathbb{C}}$ and $J(f) = \{w\}$.

Case 3: Suppose that f fixes 0 and ∞. Then $f(z) = kz$ for some $k \in \mathbb{C} \setminus \{1\}$ (verify) and thus $f^n(z) = k^n z$. We have three categories based on $|k|$:

(a) If $|k| < 1$, then $f^n(z) \to 0$ for all $z \in \overline{\mathbb{C}} \setminus \{\infty\}$ and $J(f) = \{\infty\}$.

(b) If $|k| > 1$, then $f^n(z) \to \infty$ for all $z \in \overline{\mathbb{C}} \setminus \{0\}$ and $J(f) = \{0\}$.

(c) If $|k| = 1$, then f is a rotation $z \mapsto e^{i\theta} z$ for some $\theta \in (0, 2\pi)$ whose dynamics are easy to understand, and $F(f) = \overline{\mathbb{C}}$.

Case 4: Suppose that f fixes w_1 and w_2 where $w_1 \neq w_2$. Defining $\psi(z) = \frac{z-w_1}{z-w_2}$ and setting $g(z) = \psi f \psi^{-1}(z)$, we see that g falls into Case 3. As in Case 2, we can now understand the dynamics of f as a change of coordinates of the dynamics of g by noting that $f^n(z) = \psi^{-1} g^n \psi(z)$. We have one of:

(a) $f^n(z) \to w_1$ for all $z \in \overline{\mathbb{C}} \setminus \{w_2\}$ and $J(f) = \{w_2\}$.

(b) $f^n(z) \to w_2$ for all $z \in \overline{\mathbb{C}} \setminus \{w_1\}$ and $J(f) = \{w_1\}$.

(c) f is conjugate to a rotation $z \mapsto e^{i\theta} z$ for some $\theta \in (0, 2\pi)$, and thus $F(f) = \overline{\mathbb{C}}$.

Remark 1.216. In Cases 2 and 4, the Möbius map $\psi(z)$ was used to move the fixed points of f to more convenient locations so that the simple dynamics of Cases 1 and 3 could be related to the dynamics of f. This is an advantage to using global conjugation: it allows us to reposition special points in more convenient places before we do our analysis. The technique can be used to simplify the analysis of Möbius map dynamics and we can use it with higher degree maps as well. See Section 1.2.7 where a conjugation of the Newton map of a quadratic function simplifies the analysis. Additional Exercises 1.167–1.170 explore global conjugation further.

1.C Appendix: Code for Drawing Random Dynamics Pictures

Because of its complexity, the following algorithm is not well suited for an applet. However, we provide Matlab code that can be used to generate pictures like the Devil's Colosseum in Figure 1.31. The four files main.m, f.m, g.m, and proced2.m need to be created separately, but stored in the same folder. When main.m is compiled using Matlab, a picture will be generated.

```
% ********* This begins the file main.m ********************
% This file requires the files (functions) g.m, f.m, and
% proced2.m to be in the same folder when this file is
% compiled. Upon compilation this file will generate a graph
% of the probability that a random orbit generated by the
% maps f and g will escape to infinity.

h = 0.1; % Determines step size in mesh
         % of points to be plotted
Maxd = 10; % Determines how many random steps can be taken

cntx = 0; % Counter for x coordinate
clear x1;
for x = -5:h:5
 cntx = cntx + 1;
 cnty = 0; % Counter for y coordinate
 clear y1;
  for y = -5:h:5
    cnty = cnty + 1;
```

```
        n = 0;
        x1(cntx) = x;
        y1(cnty) = y;
        [z(cntx,cnty), n] = proced2(x, y, 1, n, Maxd);
    end
end

% % Uncomment the following lines to export/save data sets
% % containing the x(i) coordinates, y(j) coordinates, and
% % z(i,j) values which when plotted make up the graph.
% % These data sets can then be imported into another
% % application (such as Maple or Mathematica) and plotted
% % and otherwise manipulated there.
%
% save DevilDataX3.dat x1 -ASCII;
% save DevilDataY3.dat y1 -ASCII;
% save DevilDataZ3.dat z -ASCII;

figure
surfc(x1, y1, z)
lighting phong
shading interp %flat %interp

clear;
% ********* This ends the file main.m ********************

% ********* This begins the file f.m ********************
function z = f(x,y)
z1 = [x*x-y*y-1,2*x*y];
z = [z1(1)*z1(1)-z1(2)*z1(2)-1,2*z1(1)*z1(2)];
% ********* This ends the file f.m ********************

% ********* This begins the file g.m ********************
function z = g(x,y)
z1 = [(x*x-y*y)/4,x*y/2];
z = [(z1(1)*z1(1)-z1(2)*z1(2))/4,z1(1)*z1(2)/2];
% ********* This ends the file g.m ********************

% ********* This begins the file proced2.m ****************
% This recursively defined procedure/function will compute
% the value s which for proced2(x, y, 1, n, Maxd) (here I
% used q=1) represents the probability that the point (x,y)
% will have a random orbit escape (have modulus > K) in the
% first M steps of the orbit.
```

```
function [s, n] = proced2(x, y, q, n0, M)
n = n0 + 1;
p = 0.5; % Probability that map f is chosen at each step
K = 10;  % Escape radius for random orbit
s = 0.0;
z=f(x,y); % file f.m must be in folder next to this file
w=g(x,y); % file g.m must be in folder next to this file

if n < M  % M is the max number of recursion steps allowed
    if z(1)*z(1)+z(2)*z(2) > K*K
        s = s + p*q;
    else
        [d, n] = proced2(z(1), z(2), p*q, n, M);
        s = s + d;
    end

    if w(1)*w(1)+w(2)*w(2) > K*K
        s = s + (1-p)*q;
    else
        [d, n] = proced2(w(1), w(2), (1-p)*q, n, M);
        s = s + d;
    end
end
% ********* This ends the file proced2.m ******************
```

2

Soap Films, Differential Geometry, and Minimal Surfaces

Michael J. Dorff (text),
James S. Rolf (software)

2.1 Introduction

Minimal surfaces are beautiful geometric objects with interesting properties that can be studied with the help of computers. Some standard examples of minimal surfaces in $\mathbb{R}^3$ are the plane, Enneper's surface, the catenoid, the helicoid, and Scherk's doubly periodic surface (see Figure 2.1; the images shown are part of the surfaces, which continue on).

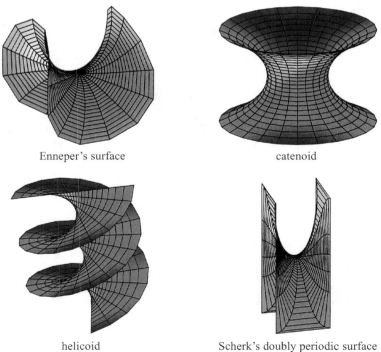

Enneper's surface   catenoid

helicoid   Scherk's doubly periodic surface
FIGURE 2.1. Examples of some minimal surfaces.

Minimal surfaces are related to soap films that result when a wire frame is dipped in soap solution. To get a sense of this connection, consider the following problem.

Steiner Problem: Four houses are located so that they form the vertices of a square that has sides of length one mile. Neighbors want to connect their houses with a road of least length. What should the shape of the road be?

FIGURE 2.2. What is the shortest path connecting the 4 vertices?

Possible solutions include:

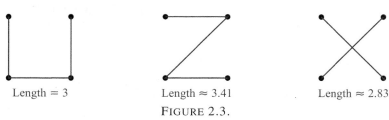

Length = 3   Length ≈ 3.41   Length ≈ 2.83

FIGURE 2.3.

However, none is the solution. The correct solution in Figure 2.4 has length of $1 + \sqrt{3} \approx 2.7$ miles. For more information about Steiner problems see [5] or [13].

FIGURE 2.4. The shortest path connecting these 4 vertices.

How can we generalize this problem? One way is to have n vertices. So the problem becomes, given n cities, find a connected system of straight line segments of shortest length such that any pair of cities is connected by a path of line segments.

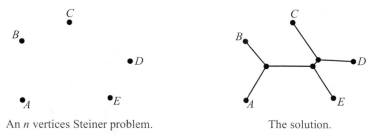

An n vertices Steiner problem.   The solution.

FIGURE 2.5.

Another way to generalize is to move up a dimension. What is the analogue of the Steiner problem in one dimension higher? The Steiner problem minimizes distance (a 1-dimensional object) in a plane (a 2-dimensional object). Soap film minimizes area (a 2-dimensional object) in space (a 3-dimensional object).

Is there a connection between the two- and three-dimensional optimization problems? Consider the soap film created by dipping a cube frame into soap solution shown in Figure 2.6. The soap film creates the minimum surface area for a surface with a cube as its boundary. If we projected it onto the plane, the resulting shape would look like the solution to the Steiner problem with four vertices shown in Figure 2.4.

FIGURE 2.6. Soap film formed by a cube.

Water molecules exert a force on each other. Near the surface of the water there is a greater force pulling the molecules toward the center of the water, creating surface tension that tends to minimize the surface area of the shape. Soap solution has a lower surface tension than water and this permits the formation of soap films that also tend to minimize geometric properties such as length and area. For more information see [22].

Minimal surfaces can be created by dipping wire frames into soap solution.

Example 2.1. By dipping into soap solution a wire frame of a slinky (or helix) with a straw connecting the ends of the slinky, we can create part of the minimal surface known as the *helicoid*.

Example 2.2. By dipping a 3-dimensional version of the wire frame in Figure 2.8 (a box frame missing two parallel edges on the top and two parallel edges on the bottom) into

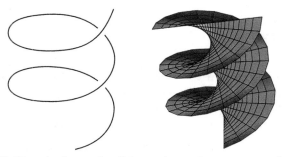

FIGURE 2.7. The wire frame of a slinky can be used to create part of the helicoid.

FIGURE 2.8. The wire frame of a box missing four edges can be used to create part of Scherk's doubly periodic surface.

soap solution we can create part of the minimal surface known as *Scherk's doubly periodic surface*.

Exploration 2.3. The minimal surfaces shown in Figure 2.1 can be formed by dipping a wire frame in soap solution. Determine the shape of the wire frame that creates (a) Enneper's surface and (b) the catenoid. **Try it out!**

Remark 2.4. To get a soap film of the part of Enneper's surface shown in Figure 2.1, we can dip a wire frame that matches the seams along a baseball. Dipping it in soap solution produces two minimal surfaces. The first is half of the sphere of a baseball and the other is the complementary half. If you start with one, you can deform it into the other by carefully blowing air into the soap film. There is a third, mysterious and unseen minimal surface one passes through while doing this, one that is unstable. It cannot remain in existence–disturbances cause it to deform into another surface.

One active area of minimal surface theory is the study of complete embedded minimal surfaces. These are minimal surfaces that are boundaryless (complete) and have no self-intersections (embedded). The plane, the catenoid, the helicoid, and Scherk's doubly periodic surface are examples. Enneper's surface is not embedded, because it has self-intersections as its domain increases (see Exploration 2.11).

To begin to understand minimal surfaces, we need some tools from differential geometry discussed in Section 2. Section 3 defines a minimal surface and discusses some examples and properties. Section 4 uses complex analysis to study minimal surfaces and introduces the Weierstrass representation formula to describe and study their properties efficiently. These sections are fundamental and should be read first. In Sections 5–6, we explore ideas that lead to beginning research problems. Sections 5 and 6 are independent. In Section 5 we present the Weierstrass representation in the form of the Gauss map and height differential, which is the basis for much current research about minimal surfaces in $\mathbb{R}^3$. Section 6 connects ideas about minimal surfaces with planar harmonic mappings in geometric function theory (i.e., the study of complex analysis from a geometric viewpoint). In this chapter, there are three applets that can be accessed online at www.maa.org/ebooks/EXCA/applets.html

1. *DiffGeomTool* is used to visualize and explore basic differential geometry concepts in $\mathbb{R}^3$ such as the graph of a parametrization of a surface, curves on a surface, tangent planes on a surface, and unit normals on a surface.

2. *MinSurfTool* is used to visualize and explore minimal surfaces in $\mathbb{R}^3$ by using various forms of the Weierstrass representation.

3. *ComplexTool* is used to plot the image of domains in $\mathbb{C}$ under complex-valued functions.

Each section contains examples, exercises, and explorations that involve the applets. You should do all the exercises and explorations, many of which present surfaces and concepts that will be used later in the chapter (there are additional exercises at its end). There are also short and long projects that are suitable as research problems. The goal of the chapter is not to give a comprehensive treatment of the topic, but to engage the reader with the general notions, questions, and techniques of the area, and to encourage the reader to pose and pursue the reader's own questions. To understand the nature and purpose of this text, the reader should read the Introduction.

2.2 Differential geometry

Our goal is to develop the mathematics necessary to investigate minimal surfaces in $\mathbb{R}^3$. They minimize area locally and can be thought of as saddle surfaces. At each point, the bending upward in one direction is matched with the bending downward in the orthogonal direction. The bending is known mathematically as curvature. So, to understand and investigate minimal surfaces, we should first understand curvature, a quantity that is measured using the tools of differential geometry. Differential geometry is a field of mathematics in which calculus is applied to geometric shapes.

We begin our discussion of differential geometry by looking at a surface in $\mathbb{R}^3$. Every point on a surface $M \subset \mathbb{R}^3$ can be designated by a point, $(x, y, z) \in \mathbb{R}^3$. It can also be represented by two parameters. Let D be an open set in $\mathbb{R}^2$. Then M can be represented by a function $\mathbf{x} : D \to \mathbb{R}^3$, where $\mathbf{x}(u, v) = (x_1(u, v), x_2(u, v), x_3(u, v))$, so M is the image of $\mathbf{x}(D)$. We will require that $\mathbf{x}$ be differentiable, so each coordinate function $x_k(u, v)$ has continuous partial derivatives of all orders in D. Such a function or mapping is called a *parametrization*.

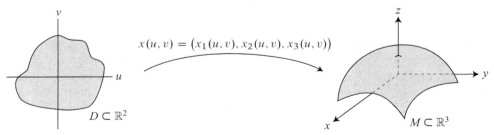

FIGURE 2.9. The parameterization of a surface.

Let's consider two examples.

Example 2.5. Enneper's surface is a minimal surface formed by bending a disk into a saddle surface. It can be parametrized by

$$\mathbf{x}(u, v) = \left(u - \frac{1}{3}u^3 + uv^2, v - \frac{1}{3}v^3 + u^2v, u^2 - v^2 \right),$$

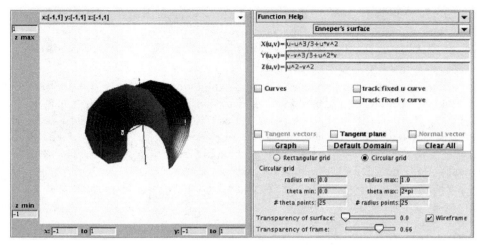

FIGURE 2.10. Enneper's surface.

where u, v are in a disk of radius r. We can use the applet, *DiffGeomTool*, to graph the parametrization of Enneper's surface (Figure 2.10). Open *DiffGeomTool* and enter the coordinate functions of the parametrization as

$$X(u, v) = u - 1/3 * u \wedge 3 + u * v \wedge 2$$
$$Y(u, v) = v - 1/3 * v \wedge 3 + u \wedge 2 * v$$
$$Z(u, v) = u \wedge 2 - v \wedge 2$$

into the appropriate boxes. In the gray part on the right side of the applet, click on **Circular grid** with radius min: 0.0, radius max: 1.0, theta min: 0.0, and theta max: 2π (in the applet enter pi for π). This is because we want our u, v values to be the unit disk. Then click the **Graph** button. To rotate the graph, place the cursor on the image of the surface, and then click on and hold the left button on the mouse as you move the cursor. To increase the size of the image of the surface click on the left button on the mouse; to decrease it, click on the right mouse button.

Example 2.6. If a heavy flexible cable is suspended between two points at the same height, then it takes the shape of a curve whose equation is $\alpha(t) = a \cosh(t/a)$ and is called a catenary from the Latin word that means "chain". In calculus, we discuss rotating a curve about a line to get a surface of revolution. We can apply this idea to the catenary to get a

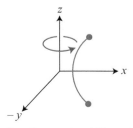

FIGURE 2.11. Creating a catenoid by rotating a catenary.

surface known as the catenoid. A *catenoid* is a surface that can be generated by rotating a catenary on its side about the z-axis (see Figure 2.11). A catenoid is also a minimal surface as we will verify in Section 2.3. How do we parametrize this catenoid? If we let $x_1 = a \cosh v$ $(-\infty < v < \infty)$ and $x_3 = av$, then $r(v) = (a \cosh v, av)$ is a parametrization of the catenary curve on its side in the xz-plane. Rotating a line about an axis is a circular motion, and a circle can be parametrized by $(\cos u, \sin u)$. So, we can parametrize the rotation of the catenary curve about the z-axis by multiplying $a \cosh v$ by $\cos u$ for the x_1-coordinate function, and multiplying $a \cosh v$ by $\sin u$ for the x_2-coordinate function. Hence, we get a parametrization for the catenoid

$$\mathbf{x}(u, v) = \left(a \cosh v \cos u, a \cosh v \sin u, av \right).$$

Using *DiffGeomTool*, we can graph it with $a = 1$ by clicking on **Rectangular grid**, and setting the boxes to $0 \leq u \leq 2\pi$ and $-\frac{2\pi}{3} \leq v \leq \frac{2\pi}{3}$ (see Figure 2.12). Enter

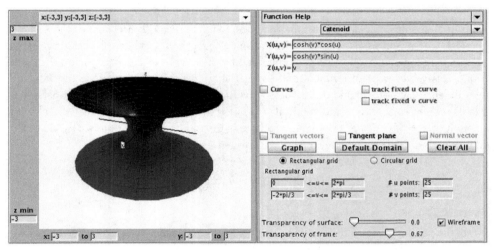

FIGURE 2.12. The catenoid.

$\cosh v, \cos u$, and $\sin u$ as $\cosh(v)$, $\cos(u)$, and $\sin(u)$. Check what happens if you change the u, v values. For example, try

(a) $\pi \leq u \leq 2\pi, \quad -\frac{2\pi}{3} \leq v \leq \frac{2\pi}{3}$

(b) $0 \leq u \leq 2\pi, \quad 0 \leq v \leq \frac{2\pi}{3}$

(c) $0 \leq u \leq 2\pi, \quad -\frac{\pi}{4} \leq v \leq \frac{\pi}{4}$

(d) $0 \leq u \leq 2\pi, \quad -\pi \leq v \leq \pi$.

Exercise 2.7. A torus is a surface (but not a minimal surface) formed by rotating a circle in the xz-plane about the z-axis. If the circle has radius b and is centered at a distance of a from the origin, then its parametrization is

$$\mathbf{x}(u, v) = \left((a + b \cos v) \cos u, (a + b \cos v) \sin u, b \sin v \right),$$

where a, b are fixed, $0 < u < 2\pi$, and $0 < v < 2\pi$.

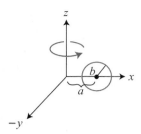

FIGURE 2.13. Creating a torus by rotating a circle.

(a) Show how to derive the parametrization.

(b) Use *DiffGeomTool* to sketch the graph of a torus with $a = 3$ and $b = 2$. Use **Rectangular grid** with $0 < u < 2\pi$, and $0 < v < 2\pi$.

Note: It will be helpful to keep a well organized notebook with answers to the exercises, because the same calculations will be used several times. ***Try it out!***

In pre-calculus, we say that a function of one variable $y = F(x)$ satisfies the vertical line test. The graph of $F(x)$ is a 1-dimensional object in $\mathbb{R}^2$, created by plotting points $(u, F(u))$ in the plane. Thus, it is parametrized by the map $u \to (u, F(u))$ from $\mathbb{R}$ to $\mathbb{R}^2$. Analogously, we speak of a function of two variables $z = f(x, y)$, where (x, y) lies in a two-dimensional domain and f satisfies the vertical line test (here a line is vertical when it is parallel to the z-axis). The graph of $f(x, y)$ is a two-dimensional surface in $\mathbb{R}^3$ with a height of $z = f(x, y)$ at (x, y). An example is the minimal surface known as Scherk's doubly periodic surface which can be parametrized by

$$\mathbf{x}(u, v) = \left(u, v, \ln\left(\frac{\cos u}{\cos v}\right)\right).$$

Exercise 2.8.

(a) In the parametrization of Scherk's doubly periodic surface, what are the restrictions on the u and v values in the domain?

(b) Use *DiffGeomTool* and your answer from part (a) to sketch a graph of Scherk's doubly periodic surface with $-0.48\pi \le u, v \le 0.48\pi$.

(c) Scherk's doubly periodic surface is an example of the graph of a function. For any function $f(x, y)$, find a parametrization of its graph.

Try it out!

Exercise 2.9. Let r be a differentiable curve whose derivative does not vanish (i.e., $r'(v) \neq 0$ for all v in the domain) and let r lie in a plane in $\mathbb{R}^3$. A *surface of revolution* is a surface that is formed by rotating r about an axis in the plane such that the curve does not intersect the axis. The catenoid and torus are examples. Let $r(v) = (f(v), 0, g(v))$ be such a curve in the xz-plane.

(a) Find a parametrization for the surface of revolution generated by rotating the curve about the z-axis.

(b) Check that your answer in (a) matches the parametrizations of the catenoid and the torus.

Try it out!

Exploration 2.10. For the torus $T_{a,b}$ whose parametrization is given in Exercise 2.7, use *DiffGeomTool* to plot $T_{3,2}$. Describe what happens to the shape of $T_{a,b}$ as a gets smaller and b gets larger. Explain this in terms of how we derived the parametrization of the torus. Hint: In *DiffGeomTool*, plot the tori

$$T_{2.7,2} \qquad T_{2.4,2} \qquad T_{2,2} \qquad T_{3,2.4} \qquad T_{3,2.7} \qquad T_{3,3}.$$

What happens when $a < b$? ***Try it out!***

Exploration 2.11. As mentioned earlier, Enneper's surface is not embedded, because it has self-intersections. Use *DiffGeomTool* and the parametrization in Example 2.5 to graph Enneper's surface with the domain being a disk of radius 1.

(a) What happens to Enneper's surface as the radius r of the disk increases?

(b) Estimate the largest value of r for which Enneper's surface has no self-intersections.

(c) Assuming that the intersection occurs on the z-axis, prove your result from part (b).

Try it out!

So far we have discussed how a function (i.e., a parametrization) models a surface. Our goal is to determine the bending or curvature of curves on a surface. To do this, we will use the parametrization of a surface to discuss the tangent plane and normal vector at a point on the surface. Suppose $\mathbf{x}(u, v)$ is a parametrization of a surface $M \subset \mathbb{R}^3$. If we fix $v = v_0$ and let u vary, then $\mathbf{x}(u, v_0)$ depends on one parameter and is known as a *u-parameter curve*. Likewise, we can fix $u = u_0$ and let v vary to get a *v-parameter curve* $\mathbf{x}(u_0, v)$ (see Figure 2.14). Tangent vectors for the u-parameter and v-parameter curves are computed by differentiating the component functions of $\mathbf{x}$ with respect to u and v, respectively. That is, $\mathbf{x}_u$ and $\mathbf{x}_v$ are the tangent vectors defined by

$$\mathbf{x}_u = \left(\frac{\partial x_1}{\partial u}, \frac{\partial x_2}{\partial u}, \frac{\partial x_3}{\partial u} \right), \qquad \mathbf{x}_v = \left(\frac{\partial x_1}{\partial v}, \frac{\partial x_2}{\partial v}, \frac{\partial x_3}{\partial v} \right).$$

For each point $p = x(u_0, v_0)$ on the surface, we get two vectors by substituting u_0 and v_0 in $\mathbf{x}_u$ and $\mathbf{x}_v$. When we have a parametrization of a surface, we will require that $\mathbf{x}_u$ and $\mathbf{x}_v$ be linearly independent (i.e., not be constant multiples of each other). The span of $\mathbf{x}_u$ and $\mathbf{x}_v$ (i.e., the set of all vectors that can be written as a linear combination of $\mathbf{x}_u, \mathbf{x}_v$) then forms a plane called the *tangent plane*.

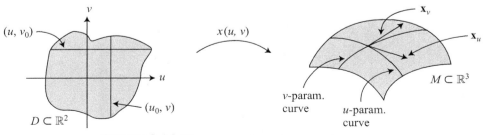

FIGURE 2.14. The u-parameter and v-parameter curves.

Definition 2.12. The *tangent plane* of a surface M at a point p is

$$T_p M = \{\mathbf{v} \,|\, \mathbf{v} \text{ is tangent to } M \text{ at } p\}.$$

Definition 2.13. The *unit normal* to a surface M at a point $p = \mathbf{x}(a, b)$ is

$$\mathbf{n}(a, b) = \left. \frac{\mathbf{x}_u \times \mathbf{x}_v}{|\mathbf{x}_u \times \mathbf{x}_v|} \right|_{(a,b)}.$$

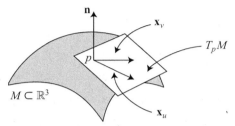

FIGURE 2.15. A tangent plane, $T_p M$, and unit normal vector, $\mathbf{n}$.

Not every surface has a well-defined choice of a unit normal $\mathbf{n}$. Those that do not are called *non-orientable*. An example is the Möbius strip. An orientable strip can be modeled by taking a strip of paper and taping the ends together. In this model, the strip has an outside and an inside–you cannot get from one side to the other without crossing an edge. A Möbius strip is modeled by taking the same strip of paper and twisting one end before taping. This forms a strip in which you can get from one side to the other without crossing over an edge. The relationship between a Möbius strip and its unit normal $\mathbf{n}$ is further explored in Exercise 2.137 in the Additional Exercises at the end of this chapter. The unit normal, $\mathbf{n}$, is orthogonal to the tangent plane at p (see Figure 2.15). If the surface M is oriented, then there are two unit normals at $p \in M$, one pointing outward and one inward. The definition of $\mathbf{n}$ chooses one of them.

Example 2.14. A torus is parametrized by

$$\mathbf{x}(u, v) = \Big((3 + 2 \cos v) \cos u, (3 + 2 \cos v) \sin u, 2 \sin v \Big)$$

where $0 < u, v < 2\pi$. For $v_0 = \frac{\pi}{3}$, the u-parameter curve is

$$\mathbf{x}\left(u, \frac{\pi}{3}\right) = (4 \cos u, 4 \sin u, \sqrt{3}).$$

For $u_0 = \frac{\pi}{2}$, the v-parameter curve is

$$\mathbf{x}\left(\frac{\pi}{2}, v\right) = (0, 3 + 2 \cos v, 2 \sin v).$$

We have

$$\mathbf{x}_u(u, v) = (-(3 + 2 \cos v) \sin u, (3 + 2 \cos v) \cos u, 0)$$
$$\mathbf{x}_v(u, v) = (-2 \sin v \cos u, -2 \sin v \sin u, 2 \cos v).$$

The u-parameter curve, $\mathbf{x}\left(u, \frac{\pi}{3}\right)$ and the v-parameter curve, $\mathbf{x}\left(\frac{\pi}{2}, v\right)$ intersect on the torus at $p = \mathbf{x}\left(\frac{\pi}{2}, \frac{\pi}{3}\right)$. The tangent vectors to the u- and v-parameter curves at p are

$$\mathbf{x}_u\left(\frac{\pi}{2}, \frac{\pi}{3}\right) = (-4, 0, 0)$$

$$\mathbf{x}_v\left(\frac{\pi}{2}, \frac{\pi}{3}\right) = (0, -\sqrt{3}, 1).$$

They span the tangent plane, $T_p M$. We compute that

$$\mathbf{x}_u\left(\frac{\pi}{2}, \frac{\pi}{3}\right) \times \mathbf{x}_v\left(\frac{\pi}{2}, \frac{\pi}{3}\right) = (-4, 0, 0) \times (0, -\sqrt{3}, 1) = (0, 4, 4\sqrt{3}).$$

Hence,

$$\mathbf{n}\left(\frac{\pi}{2}, \frac{\pi}{3}\right) = \left(0, \frac{1}{2}, \frac{\sqrt{3}}{2}\right).$$

We can use *DiffGeomTool* to display the u-parameter curve, the v-parameter curve, $\mathbf{x}_u$, $\mathbf{x}_v$, and $\mathbf{n}$. Enter the parametrization in this example for the torus. Then click **Curves**. A **Point location** box, a **fixed u** box, and a **fixed v** box will appear. In the **Point location** box, enter $\frac{\pi}{2}$ into the first box (i.e., the fixed u value) and $\frac{\pi}{3}$ into the second box (i.e., the fixed v value). Make sure that the **fixed u** box is clicked but not the **fixed v** box. Then click the **Graph** button. The v-parameter curve will appear. If you now click the **track fixed u curve** box, a slider will appear. Moving it with the cursor will move the point along the v-parameter curve on the torus. Now, click on the **fixed u** box and then click the **Graph** button again. The u-parameter curve will appear. By clicking on the **track fixed v curve** box and moving the slider, the point along the v-parameter curve will move. Next, click on the following boxes separately followed by the **Graph** button: **Tangent vectors** box, **Tangent plane** box, and **Normal vector** box. This will cause these geometric objects to appear. You should convince

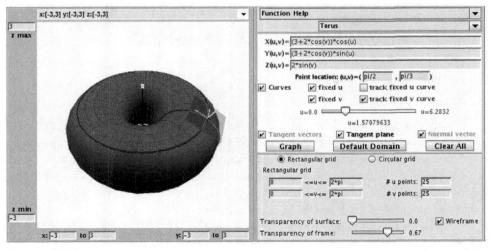

FIGURE 2.16. The torus with u, v-parameter curves, the tangent vectors, the tangent plane, and the normal vector.

yourself that the images of the vectors at $(u, v) = (\frac{\pi}{2}, \frac{\pi}{3})$ match the computed values in Example 2.14.

Exercise 2.15. For a surface of revolution (see Exercise 2.9) parametrized by

$$\mathbf{x}(u, v) = \big(f(v) \cos u, \, f(v) \sin u, \, g(v) \big)$$

the u-parameter curves are called *parallels* and are the curves formed by horizontal slices, while the v-parameter curves are called *meridians* and are the curves formed by vertical slices. Describe the parallels and meridians for the catenoid in Example 2.6 and the torus in Exercise 2.7. ***Try it out!***

Exercise 2.16. A catenoid has parametrization

$$\mathbf{x}(u, v) = \big(\cosh v \cos u, \cosh v \sin u, v \big)$$

with $0 < u < 2\pi$ and $-\frac{2\pi}{3} < v < \frac{2\pi}{3}$.

(a) Use *DiffGeomTool* to sketch the u-parameter curve, $\mathbf{x}(u, 0)$, and the v-parameter curve, $\mathbf{x}(0, v)$, on the catenoid. Also, sketch the vectors $\mathbf{x}_u(0, 0)$, $\mathbf{x}_v(0, 0)$, and $\mathbf{n}(0, 0)$.

(b) Compute the vectors $\mathbf{x}_u(0, 0)$, $\mathbf{x}_v(0, 0)$, and $\mathbf{n}(0, 0)$.

Try it out!

Exercise 2.17. Scherk's doubly periodic surface has parametrization

$$\mathbf{x}(u, v) = \left(u, v, \ln \left(\frac{\cos u}{\cos v} \right) \right)$$

with $-0.48\pi < u, v < 0.48\pi$.

(a) Use *DiffGeomTool* to sketch the u-parameter curve, $\mathbf{x}\big(u, \frac{\pi}{4}\big)$ and the v-parameter curve, $\mathbf{x}\big(\frac{\pi}{4}, v\big)$ on Scherk's doubly periodic surface (make sure you use $-0.48\pi < u, v < 0.48\pi$). Sketch the vectors $\mathbf{x}_u\big(\frac{\pi}{4}, \frac{\pi}{4}\big)$, $\mathbf{x}_v\big(\frac{\pi}{4}, \frac{\pi}{4}\big)$, and $\mathbf{n}\big(\frac{\pi}{4}, \frac{\pi}{4}\big)$. You can slide them by clicking the **track fixed u curve** box. The collection of vectors, $(\mathbf{x}_u, \mathbf{x}_v, \mathbf{n})$, is known as a *moving frame* or *Frenet frame* of the curve. The way they vary in $\mathbb{R}^3$ as the frame moves along the curve describes how the curve twists and turns in $\mathbb{R}^3$. For more details, see [18] or [21].

(b) Compute the vectors $\mathbf{x}_u\big(\frac{\pi}{4}, \frac{\pi}{4}\big)$, $\mathbf{x}_v\big(\frac{\pi}{4}, \frac{\pi}{4}\big)$, and $\mathbf{n}\big(\frac{\pi}{4}, \frac{\pi}{4}\big)$.

Try it out!

We will use the normal vector $\mathbf{n}$ to define the curvature of a curve on a surface. As we mentioned in the introduction, at each point on a minimal surface, the bending upward in one direction is matched with the bending downward in the orthogonal direction. We will use this definition of curvature in our definition of a minimal surface. A plane containing the normal $\mathbf{n}$ will intersect the surface M in a curve, α. For each α we can compute its curvature, which measures how fast the curve pulls away from its tangent line at p. A curve in $\mathbb{R}^3$ can be parametrized by a function of one variable, say $\alpha(t)$, where $\alpha : [a, b] \to \mathbb{R}^3$. The parametrization is not unique.

Exercise 2.18. Find two parametrizations of the unit circle in the xy-plane. ***Try it out!***

The lack of uniqueness can cause difficulties. To eliminate them we will require our parametrization to be a unit speed curve.

Definition 2.19. A curve α is a *unit speed curve* if $\left|\alpha'(t)\right| = 1$.

If a parametrization of a regular curve $\alpha(t)$ is not of unit speed, we can reparametrize it by arclength to get a unit speed curve $\alpha(s)$ (you probably saw this in third-semester calculus, and you may want to review how to reparametrize a curve by arclength). We will assume that curves are unit speed curves $\alpha(s)$.

Given a curve α, we want to discuss its curvature (or bending). We quantify the amount that the curve bends at each point p by measuring how rapidly it pulls away from its tangent line at p. The amount of bending is the rate of change of the angle θ that tangents make with the curve at p which is the second derivative.

Definition 2.20. The *curvature* of a unit speed curve α at s is $\left|\alpha''(s)\right|$.

Example 2.21. A torus is parametrized by

$$\mathbf{x}(u, v) = \Big((3 + 2\cos v)\cos u, (3 + 2\cos v)\sin u, 2\sin v\Big),$$

where $0 < u, v < 2\pi$. Let's compute the curvature for the u-parameter curves and the v-parameter curves. The v-parameter curves (or meridians) are the curves formed by vertical slices of the torus, and hence are circles of radius $b = 2$. To compute their curvature, we take their parametrization

$$\mathbf{x}(v) = \mathbf{x}(u_0, v) = \Big((3 + 2\cos v)\cos u_0, (3 + 2\cos v)\sin u_0, 2\sin v\Big),$$

where u_0 is a fixed value. Since $|\mathbf{x}'(v)| = 2$, we reparametrize $\mathbf{x}(v)$ as a unit speed curve by replacing v with $\frac{s}{2}$ to get

$$\mathbf{x}(s) = \left(\Big(3 + 2\cos\left(\frac{s}{2}\right)\Big)\cos u_0, \Big(3 + 2\cos\left(\frac{s}{2}\right)\Big)\sin u_0, 2\sin\left(\frac{s}{2}\right)\right).$$

We then compute the curvature of the v-parameter curves to be

$$\left|\mathbf{x}''(s)\right| = \frac{1}{2}.$$

The u-parameter curves (or parallels) are the curves formed by horizontal slices of the torus, and so are circles of radius $3 + 2\cos v_0$, where $v_0 \in (0, 2\pi)$ is fixed. The radii vary between 1 and 5. The curves are parametrized by

$$\mathbf{x}(u) = \mathbf{x}(u, v_0) = \Big((3 + 2\cos v_0)\cos u, (3 + 2\cos v_0)\sin u, 2\sin v_0\Big),$$

which can be reparametrized as the unit speed curve

$$\mathbf{x}(s) = \left((3 + 2\cos v_0)\cos\left(\frac{s}{3 + 2\cos v_0}\right), (3 + 2\cos v_0)\sin\left(\frac{s}{3 + 2\cos v_0}\right), 2\sin v_0\right).$$

Computing the curvature yields

$$\left|\mathbf{x}''(s)\right| = \frac{1}{3 + 2\cos v_0},$$

which varies between $\frac{1}{5}$ and 1.

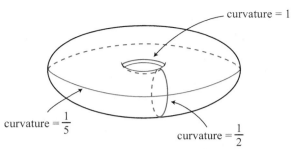

FIGURE 2.17. The curvature of the meridians and parallels on a torus.

Exercise 2.22. Verify that replacing v with $\frac{s}{2}$ in $\mathbf{x}(u, v)$ in the previous example gives a unit speed curve. ***Try it out!***

Exercise 2.23. Compute the curvatures of the meridians and parallels of the catenoid parametrized by

$$\mathbf{x}(u, v) = \left(a \cosh v \cos u, a \cosh v \sin u, av \right).$$

Try it out!

Exercise 2.24. The curve parametrized by $\alpha(t) = (a \cos t, a \sin t, bt)$ is a *helix*, which is a spiral that rises with a pitch of $2\pi b$ on the cylinder $x^2 + y^2 = a^2$.

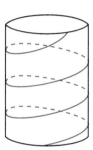

FIGURE 2.18. A helix in a cylinder.

We can create a surface by connecting a line from the axis $(0, 0, bt)$ through the helix $(a \cos t, a \sin t, bt)$ at height bt. This ruled surface is a minimal surface known as a *helicoid*. All minimal surfaces, including the helicoid, can be parametrized in several ways. We will use the parametrization of the helicoid

$$\mathbf{x}(u, v) = \left(a \sinh v \cos u, a \sinh v \sin u, au \right).$$

(a) Compute the curvatures of the u-parameter curves and v-parameter curves of the helicoid (making the v-parameter curve into a unit speed curve is not easy).

(b) Use *DiffGeomTool* to graph the helicoid with $a = 1$ (see Figure 2.19).

Try it out!

Exercise 2.25. From Example 2.21 you may have conjectured that the curvature of a circle of radius r is $\frac{1}{r}$. This is correct. Prove it. ***Try it out!***

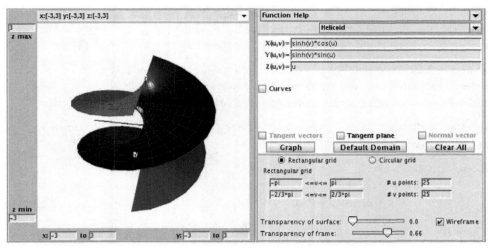

FIGURE 2.19. Helicoid.

Let's return to surfaces. Suppose we have a curve $\sigma(s)$ on a surface M. We can determine the unit tangent vector, $\mathbf{w}$, of σ at $p \in M$ and the unit normal, $\mathbf{n}$, of M at $p \in M$. Note that $\mathbf{w}$ and $\mathbf{n}$ span a plane $\mathcal{P}$ that intersects M creating a curve $\alpha(s)$.

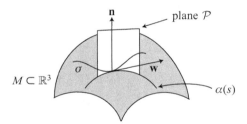

FIGURE 2.20. The normal curvature.

Definition 2.26. The *normal curvature* in the $\mathbf{w}$ direction is

$$k(\mathbf{w}) = \alpha'' \cdot \mathbf{n}.$$

We know $\alpha'' \cdot \mathbf{n} = |\alpha''||\mathbf{n}| \cos \theta$, where θ is the angle between $\mathbf{n}$ and α''. Hence $\alpha'' \cdot \mathbf{n}$ is the projection of α'' onto the unit normal (whence, the name *normal* curvature). The normal curvature measures how much the surface bends towards $\mathbf{n}$ as you travel in the direction of the tangent vector $\mathbf{w}$ starting p. As we rotate the plane about $\mathbf{n}$, we get a set of curves on the surface each of which has a value for its curvature. Let k_1 and k_2 be the maximum and minimum curvature values at p. The directions in which the normal curvature attains its absolute maximum and absolute minimum values are known as the *principal directions*.

Definition 2.27. The *mean curvature* (i.e., average curvature) of a surface M at p is

$$H = \frac{k_1 + k_2}{2}.$$

It turns out that k_1 and k_2 come from orthogonal tangent vectors. The mean curvature depends on p. It can be shown that H does not change if we choose any two orthogonal vectors and use their curvature values to compute H at p. Note that from the definition we see that if the principal curve with normal curvature k_j is bending toward the unit normal $\mathbf{n}$, then $k_j > 0$ and if it is bending away from $\mathbf{n}$, then $k_j < 0$.

Example 2.28. At a point on a sphere of radius a, all the curves α are circles of radius a and hence have the same curvature, $1/a$. Since the curves are bending away from $\mathbf{n}$, $k_1 = -1/a = k_2$, and the mean curvature is $-1/a$.

Exercise 2.29. Determine the mean curvature at points on the cylinder parametrized by $\mathbf{x}(u, v) = (a\cos u, a\sin u, bv)$. ***Try it out!***

Exercise 2.30. Determine the points on the torus

$$\mathbf{x}(u, v) = \big((a + b\cos v)\cos u, (a + b\cos v)\sin u, b\sin v\big)$$

where $H > 0$, $H = 0$, or $H < 0$. ***Try it out!***

In the next section we will define a minimal surface in terms of mean curvature. Using our current definition it is tedious to compute the mean curvature of a surface, because we have to find the principle curvatures for every point on the surface. We would like to have an explicit expression to compute mean curvature at a point p on the surface simply by substituting the coordinates of p into the formula. There is a formula for mean curvature using the coefficients of the first and second fundamental forms for a surface. Since $\alpha(s) = (\mathbf{x}(u(s), v(s)))$ the chain rule gives us that $\alpha'(s) = \mathbf{x}_u \dfrac{du}{ds} + \mathbf{x}_v \dfrac{dv}{ds}$. From here on we will denote $\dfrac{du}{ds}$ by du and $\dfrac{dv}{ds}$ by dv. Because α is a unit speed curve,

$$
\begin{aligned}
1 =& |\alpha'|^2 = \alpha' \cdot \alpha' \\
=& (\mathbf{x}_u \, du + \mathbf{x}_v \, dv) \cdot (\mathbf{x}_u \, du + \mathbf{x}_v \, dv) \\
=& \mathbf{x}_u \cdot \mathbf{x}_u \, du^2 + 2\mathbf{x}_u \cdot \mathbf{x}_v \, dudv + \mathbf{x}_v \cdot \mathbf{x}_v \, dv^2 \\
=& E \, du^2 + 2F \, dudv + G \, dv^2.
\end{aligned}
\tag{2.1}
$$

The terms $E = \mathbf{x}_u \cdot \mathbf{x}_u$, $F = \mathbf{x}_u \cdot \mathbf{x}_v$, and $G = \mathbf{x}_v \cdot \mathbf{x}_v$ are known as the *coefficients of the first fundamental form*. They describe how lengths on a surface are distorted as compared to their measurements in $\mathbb{R}^3$.

Next, recall $k(\mathbf{w}) = \alpha'' \cdot \mathbf{n}$. Note that $\alpha' \cdot \mathbf{n} = 0$, and so $(\alpha' \cdot \mathbf{n})' = 0$, which implies $\alpha'' \cdot \mathbf{n} + \alpha' \cdot \mathbf{n}' = 0$, and thus $\alpha'' \cdot \mathbf{n} = -\alpha' \cdot \mathbf{n}'$. Similarly, $-\mathbf{x}_u \cdot \mathbf{n}_u = \mathbf{x}_{uu} \cdot \mathbf{n}$. So

$$
\begin{aligned}
k(\mathbf{w}) =& -\alpha' \cdot \mathbf{n}' \\
=& -(\mathbf{x}_u \, du + \mathbf{x}_v \, dv) \cdot (\mathbf{n}_u \, du + \mathbf{n}_v \, dv) \\
=& -\mathbf{x}_u \cdot \mathbf{n}_u \, du^2 - (\mathbf{x}_u \cdot \mathbf{n}_v + \mathbf{x}_v \cdot \mathbf{n}_u) \, dudv - \mathbf{x}_v \cdot \mathbf{n}_v \, dv^2 \\
=& \mathbf{x}_{uu} \cdot \mathbf{n} \, du^2 + 2\mathbf{x}_{uv} \cdot \mathbf{n} \, dudv + \mathbf{x}_{vv} \cdot \mathbf{n} \, dv^2 \\
=& e \, du^2 + 2f \, dudv + g \, dv^2.
\end{aligned}
$$

The terms $e = \mathbf{x}_{uu} \cdot \mathbf{n}$, $f = \mathbf{x}_{uv} \cdot \mathbf{n}$, and $g = \mathbf{x}_{vv} \cdot \mathbf{n}$ are called the *coefficients of the second fundamental form*. They describe how much the surface bends away from the tangent plane.

Example 2.31. A catenoid can be parametrized by

$$\mathbf{x}(u, v) = \Big(a \cosh v \cos u, a \cosh v \sin u, av \Big).$$

Using this parametrization, we compute that

$$\mathbf{x}_u = (-a \cosh v \sin u, a \cosh v \cos u, 0)$$
$$\mathbf{x}_v = (a \sinh v \cos u, a \sinh v \sin u, a).$$

The coefficients of the first fundamental form are

$$E = \mathbf{x}_u \cdot \mathbf{x}_u = a^2 \cosh^2 v;$$
$$F = \mathbf{x}_u \cdot \mathbf{x}_v = 0;$$
$$G = \mathbf{x}_v \cdot \mathbf{x}_v = a^2 \cosh^2 v.$$

What do E, F, and G tell us? Let $(u_0, v_0) \in D$ be a point in the domain and let's take a small square with a vertex there. Because $\mathbf{x}_u \cdot \mathbf{x}_v = F = 0$, we know that the orthogonal lines from the u-parameter curve and the v-parameter curve remain orthogonal on the catenoid. That is, small squares will be mapped to small rectangles (technically, they are not rectangles but the term serves our purpose). Because $E = G$ we know that $|\mathbf{x}_u| = |\mathbf{x}_v|$ and adjacent sides of the image rectangle will have the same length. So small squares in the domain D will be mapped to small squares on the catenoid. For simplicity take $a = 1$. Then $E = G = \cosh^2 v$. When $v = 0$, $E = G = 1$ and as v gets farther away from 0, E and G get larger. This means that a small square containing the u-parameter curve $v = 0$ will be mapped to a small square of the same size on the catenoid. As v gets farther away from 0, the size of the side lengths of the image square will increase by a factor of $\cosh^2 v$. This can be seen in Figure 2.21 where we have used the **Transparency of surface** and the **Transparency of the frame** sliders to make the u- and v-parameter curves more distinct (the u-parameter curve with $v = 0$ gets mapped to a

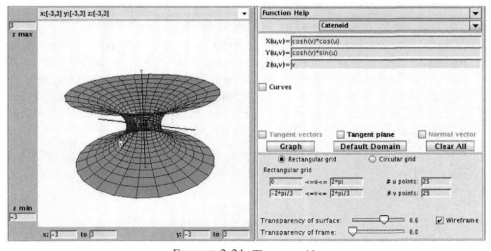

FIGURE 2.21. The catenoid.

parallel on the neck of the catenoid and the u-parameter curve with $v = \frac{2\pi}{3}$ gets mapped to the edge of the catenoid, as displayed in the figure).

To compute the coefficients of the second fundamental form, we need to compute $\mathbf{n} = \frac{\mathbf{x}_u \times \mathbf{x}_v}{|\mathbf{x}_u \times \mathbf{x}_v|}$. Because

$$\mathbf{x}_u \times \mathbf{x}_v = (a^2 \cosh v \cos u, a^2 \cosh v \sin u, -a^2 \cosh v \sinh v),$$

we have

$$|\mathbf{x}_u \times \mathbf{x}_v| = a^2 \cosh^2 v.$$

Hence

$$\mathbf{n} = \left(\frac{\cos u}{\cosh v}, \frac{\sin u}{\cosh v}, -\frac{\sinh v}{\cosh v} \right).$$

We can compute that

$$\mathbf{x}_{uu} = (-a \cosh v \cos u, -a \cosh v \sin u, 0)$$
$$\mathbf{x}_{uv} = (-a \sinh v \sin u, a \sinh v \cos u, 0)$$
$$\mathbf{x}_{vv} = (a \cosh v \cos u, a \cosh v \sin u, 0).$$

Therefore, the coefficients of the second fundamental form are:

$$e = \mathbf{n} \cdot \mathbf{x}_{uu} = -a$$
$$f = \mathbf{n} \cdot \mathbf{x}_{uv} = 0$$
$$g = \mathbf{n} \cdot \mathbf{x}_{vv} = a.$$

What do e, f, and g tell us? Let $(u_0, v_0) \in D$ be a point in the domain, and let $p \in M$ be the image of (u_0, v_0) on the surface. Then at p the vectors $\mathbf{x}_u$ and $\mathbf{x}_v$ create the tangent plane $T_p M$ and the unit normal $\mathbf{n}$. For the catenoid, the u-parameter curve is bending away from $\mathbf{n}$ while the v-parameter curve is bending toward $\mathbf{n}$. Both curves are bending the same amount away from the tangent plane, because $e = -g$.

Exercise 2.32. A torus has the parametrization

$$\mathbf{x}(u, v) = \big((a + b \cos v) \cos u, (a + b \cos v) \sin u, b \sin v \big).$$

(a) Compute the coefficients of the first and the second fundamental forms.

(b) Open *DiffGeomTool* and enter the parametrization with $a = 2$ and $b = 1$. Use the **Rectangular grid** with $0 \leq u \leq 2\pi$ and $0 \leq v \leq 2\pi$. And set the number of u points to 20 and the number of v points to 20. Describe how the results from part (a) match with the image of the torus in *DiffGeomTool*. Use the **Transparency of surface** and the **Transparency of the frame** sliders and click off and on the **Wireframe** button to make the wireframe clearer.

Try it out!

Exercise 2.33. Compute the coefficients of the first and second fundamental forms for Scherk's doubly periodic surface parametrized by

$$\mathbf{x}(u, v) = \left(u, v, \ln \left(\frac{\cos u}{\cos v} \right) \right).$$

Try it out!

We want to express the mean curvature H in terms of the coefficients of the first and second fundamental forms. We will show that

$$H = \frac{Eg + Ge - 2Ff}{2(EG - F^2)}.$$

We will establish the formula by a straightforward calculation from Oprea [22, pp. 40–42]. This approach does not give much insight into the formula. There is a more elegant way to derive this formula, but it requires some concepts that are beyond the scope of what we will need (for a discussion involving this approach, see [3] or [18]).

Let $\mathbf{w}_1, \mathbf{w}_2$ be two perpendicular unit vectors in $T_p M$ and k_1, k_2 be their normal curvatures using the curves $\alpha_1(s) = (u_1(s), v_1(s))$ and $\alpha_2(s) = (u_2(s), v_2(s))$. Let $p_1 = du_1 + i du_2$ and $p_2 = dv_1 + i dv_2$. Then, using the second fundamental form to compute k_1 and k_2, we have

$$2H = k_1 + k_2 = e(du_1^2 + du_2^2) + 2f(du_1 dv_1 + du_2 dv_2) + g(dv_1^2 + dv_2^2)$$
$$= e(p_1 \overline{p_1}) + f(p_1 \overline{p_2} + \overline{p_1} p_2) + g(p_2 \overline{p_2}).$$

We want to further simplify this so that p_1 and p_2 do not appear in the expression. From (2.1)

$$1 = E \, du^2 + 2F \, du dv + G \, dv^2.$$

Also $E du_1 du_2 + F(du_1 dv_2 + du_2 dv_1) + G dv_1 dv_2 = 0$, because $\mathbf{w}_1, \mathbf{w}_2$ are perpendicular and so $\mathbf{w}_1 = \mathbf{w}_2 \alpha_1'(s) \cdot \alpha_2'(s) = 0$. Hence,

$$Ep_1^2 + 2Fp_1 p_2 + Gp_2^2 = E\left[du_1^2 - du_2^2 + i2 du_1 du_2\right]$$
$$+ 2F\left[du_1 dv_1 - du_2 dv_2 + i(du_1 dv_2 + du_2 dv_1)\right]$$
$$+ G\left[dv_1^2 - dv_2^2 + i2 dv_1 dv_2\right]$$
$$= 2i\left[E du_1 du_2 + F(du_1 dv_2 + du_2 dv_1) + G dv_1 dv_2\right]$$
$$+ \left[E du_1^2 + 2F du_1 dv_1 + G dv_1^2\right]$$
$$- \left[E du_2^2 + 2F du_2 dv_2 + G dv_2^2\right]$$
$$= 0 + 1 - 1$$
$$= 0.$$

Thus,

$$p_1 = \frac{-2Fp_2 \pm \sqrt{4F^2 p_2^2 - 4EGp_2^2}}{2E} = \left(-\frac{F}{E} \pm i \frac{\sqrt{EG - F^2}}{E}\right) p_2$$

$$\overline{p_1} = \left(-\frac{F}{E} \mp i \frac{\sqrt{EG - F^2}}{E}\right) \overline{p_2},$$

so

$$p_1 \overline{p_1} = \left(\frac{F^2}{E^2} + \frac{EG - F^2}{E^2}\right) p_2 \overline{p_2} = \frac{G}{E} p_2 \overline{p_2}, \tag{2.2}$$

and

$$p_1 \overline{p_2} + \overline{p_1} p_2 = -\frac{2F}{E} p_2 \overline{p_2}. \tag{2.3}$$

We have

$$2H = k_1 + k_2 = e(p_1 \, \overline{p_1}) + f(p_1 \, \overline{p_2} + \overline{p_1} p_2) + g(p_2 \, \overline{p_2})$$

$$= \left[e\frac{G}{E} + f\left(\frac{-2F}{E} \right) + g \right] p_2 \, \overline{p_2}.$$

To get rid of $p_2 \, \overline{p_2}$, use (2.1) to derive

$$E p_1 \, \overline{p_1} + F(p_1 \, \overline{p_2} + \overline{p_1} p_2) + G p_2 \, \overline{p_2}$$

$$= E(du_1^2 + du_2^2) + 2F(du_1 dv_1 + du_2 dv_2) + G(dv_1^2 + dv_2^2)$$

$$= 1 + 1 = 2.$$

Using (2.2) and (2.3), we have

$$2 = E\left(\frac{G}{E} p_2 \, \overline{p_2} \right) + F\left(\frac{-2F}{E} p_2 \, \overline{p_2} \right) + G p_2 \, \overline{p_2}$$

$$\Rightarrow 2 = \left[2G - \frac{2F^2}{E} \right] p_2 \, \overline{p_2}$$

$$\Rightarrow p_2 \, \overline{p_2} = \frac{E}{EG - F^2}.$$

Therefore,

$$H = \frac{1}{2}\left[e\frac{G}{E} + f\left(\frac{-2F}{E} \right) + g \right] p_2 \, \overline{p_2} = \frac{Eg + eG - 2Ff}{2(EG - F^2)}.$$

2.3   Minimal surfaces

Now that we have some essential ideas from differential geometry, we can begin to explore minimal surfaces. We mentioned that minimal surfaces can be thought of as saddle surfaces. That is, at each point the bending upward in one direction is matched with the bending downward in the orthogonal direction. This can be described mathematically with the definition:

Definition 2.34. A *minimal surface* is a surface M with mean curvature $H = 0$ at all points $p \in M$.

We can use the formula

$$H = \frac{Eg + Ge - 2Ff}{2(EG - F^2)}. \tag{2.4}$$

to show that a surface with a specific parametrization is minimal.

Example 2.35. We will use (2.4) to show that the catenoid is a minimal surface. A catenoid can be parametrized by

$$\mathbf{x}(u, v) = \left(a \cosh v \cos u, a \cosh v \sin u, av \right).$$

From Example 2.31

$$E = \mathbf{x}_u \cdot \mathbf{x}_u = a^2 \cosh^2 v,$$

$$F = \mathbf{x}_u \cdot \mathbf{x}_v = 0,$$

$$G = \mathbf{x}_v \cdot \mathbf{x}_v = a^2 \cosh^2 v,$$

and

$$e = \mathbf{n} \cdot \mathbf{x}_{uu} = -a,$$
$$f = \mathbf{n} \cdot \mathbf{x}_{uv} = 0,$$
$$g = \mathbf{n} \cdot \mathbf{x}_{vv} = a.$$

Hence

$$H = \frac{eG - 2fF + Eg}{2(EG - F^2)} = 0.$$

So the catenoid is a minimal surface.

Exercise 2.36. Using the parametrization for the helicoid

$$\mathbf{x}(u, v) = (a \sinh v \cos u, a \sinh v \sin u, au),$$

prove that it is a minimal surface. ***Try it out!***

Exercise 2.37. Using the parametrization for the torus

$$\mathbf{x}(u, v) = \big((a + b \cos v) \cos u, (a + b \cos v) \sin u, b \sin v\big),$$

prove that it is not a minimal surface. ***Try it out!***

Exercise 2.38. Suppose a surface M is the graph of a function $f(x, y)$ of two variables (see the paragraph before Exercise 2.8). Then M can be parametrized by

$$\mathbf{x}(x, y) = \big(x, y, f(x, y)\big),$$

where its domain is the projection of M onto the xy-plane.

(a) Compute the coefficients of the first and second fundamental forms for M.

(b) A *minimal graph* is a minimal surface that is a graph of a function. Prove M is a minimal graph if and only if

$$f_{xx}\left(1 + f_y^2\right) - 2f_x f_y f_{xy} + f_{yy}\left(1 + f_x^2\right) = 0. \tag{2.5}$$

Try it out!

Before Exercise 2.8, we stated that Scherk's doubly periodic surface is a minimal graph. We will now use (2.5) to prove that. Applying (2.5) is usually not easy, because solving explicitly for f can be complicated. However, we can do this is when f can be separated into two functions, each of which depends on only one variable. Suppose $f(x, y) = g(x) + h(y)$. Then the minimal surface equation becomes

$$g''(x)[1 + (h'(y))^2] + h''(y)[1 + (g'(x))^2] = 0,$$

a separable differential equation that can be solved by separating the terms with x from those with y by putting them on opposite sides, giving

$$-\frac{1 + (g'(x))^2}{g''(x)} = \frac{1 + (h'(y))^2}{h''(y)}. \tag{2.6}$$

If we fix y, the right side remains constant even if we change x in the left side. The same is true if we fix x and vary y. The only way this can occur is if both sides are constant. So we have:

$$-\frac{1 + (g'(x))^2}{g''(x)} = k \qquad \Longrightarrow \qquad 1 + (g'(x))^2 = -kg''(x).$$

To solve, let $\phi(x) = g'(x)$. Then $\frac{d\phi}{dx} = g''(x)$ and so

$$\int dx = -k \int \frac{d\phi}{1 + \phi^2},$$

$$x = -k \arctan \phi + C,$$

$$\phi = -\tan\left(\frac{x + C}{k}\right).$$

For convenience, let $C = 0$ and $k = 1$. Since $\phi = g'$, we can integrate again to get

$$g(x) = \ln[\cos x].$$

Completing the same calculations for the y-side of (2.6) yields

$$h(y) = -\ln[\cos y].$$

Hence

$$f(x, y) = g(x) + h(y) = \ln\left[\frac{\cos x}{\cos y}\right],$$

an equation for Scherk's doubly periodic surface. Using *DiffGeomTool* we display the graph of Scherk's doubly periodic surface (see Figure 2.22). Because $-\frac{\pi}{2} < x, y < \frac{\pi}{2}$, the surface is defined over a square with side lengths π, centered at the origin. By the Schwarz Reflection Principle, we can fit pieces of Scherk's doubly periodic surface together horizontally and vertically to get a checkerboard domain (see Figure 2.23). Because one piece

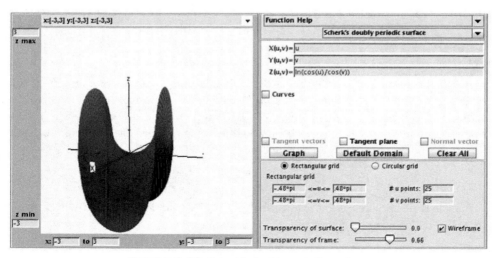

FIGURE 2.22. Scherk's doubly periodic surface.

FIGURE 2.23. A tiling of Scherk's doubly periodic surface.

of the surface can be repeated or tiled in two directions, we get a doubly periodic surface. Periodic minimal surfaces is an active area of research.

Let's summarize the minimal surfaces we have encountered and some other well-known examples.

1. The plane:
    $$\mathbf{x}(u, v) = (u, v, 0).$$

2. Enneper's surface:
    $$\mathbf{x}(u, v) = \left(u - \tfrac{1}{3}u^3 + uv^2, v - \tfrac{1}{3}v^3 + u^2v, u^2 - v^2\right).$$

3. The catenoid:
    $$\mathbf{x}(u, v) = (a \cosh v \cos u, a \cosh v \sin u, av).$$

4. The helicoid:
    $$\mathbf{x}(u, v) = (a \sinh v \cos u, a \sinh v \sin u, au).$$

5. Scherk's doubly periodic surface:
    $$\mathbf{x}(u, v) = \left(u, v, \ln\left(\frac{\cos u}{\cos v}\right)\right).$$

6. Scherk's singly periodic surface:
    $$\mathbf{x}(u, v) = (\operatorname{arcsinh}(u), \operatorname{arcsinh}(v), \arcsin(uv)).$$

7. Henneberg surface:
    $$\mathbf{x}(u, v) = (-1 + \cosh(2u)\cos(2v), -\sinh(u)\sin(v) - \tfrac{1}{3}\sinh(3u)\sin(3v),$$
    $$- \sinh(u)\cos(v) + \tfrac{1}{3}\sinh(3u)\cos(3v)).$$

8. Catalan surface:
    $$\mathbf{x}(u, v) = (1 - \cos(u)\cosh(v), 4\sin(\tfrac{u}{2})\sinh(\tfrac{v}{2}), u - \sin(u)\cosh(v)).$$

In addition to Enneper's surface, the Henneberg surface and the Catalan surface are not embedded.

There is an extensive list of minimal surfaces, but we have no way of listing all of them. So, we often focus on trying to classify minimal surfaces. This means, we try to find results that include all possibilities for minimal surfaces with specific properties. The simplest example of this is Theorem 2.39.

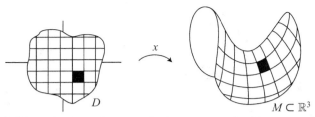

FIGURE 2.24. An isothermal parametrization maps small squares to small squares.

Theorem 2.39. *A nonplanar minimal surface in $\mathbb{R}^3$ that is also a surface of revolution is contained in a catenoid.*

As we have seen, using (2.5) to construct a minimal graph involves solving second order differential equations. We can simplify matters if we use isothermal parametrizations.

Definition 2.40. A parametrization $\mathbf{x}$ is *isothermal* if $E = \mathbf{x}_u \cdot \mathbf{x}_u = \mathbf{x}_v \cdot \mathbf{x}_v = G$ and $F = \mathbf{x}_u \cdot \mathbf{x}_v = 0$.

Because E, F, and G describe how lengths on a surface are distorted as compared to their usual measurements in $\mathbb{R}^3$, if $F = \mathbf{x}_u \cdot \mathbf{x}_v = 0$ then $\mathbf{x}_u$ and $\mathbf{x}_v$ are orthogonal and if $E = G$ then the amount of distortion is the same in the orthogonal directions. Thus, we can think of an isothermal parametrization as mapping a small square in the domain to a small square on the surface. Sometimes an isothermal parametrization is called a *conformal parametrization*, because the angle between a pair of curves in the domain is equal to the angle between the corresponding pair of curves on the surface.

Example 2.41. The parametrization

$$\mathbf{x}(u, v) = (a \cosh v \cos u, a \cosh v \sin u, av)$$

for the catenoid is isothermal, because in Example 2.35 we derived that $E = a^2 \cosh^2 v = G$ and $F = 0$. We can get a geometric sense that it is isothermal by using *DiffGeomTool* to graph it. Open *DiffGeomTool* and enter the parametrization with $a = 1$ and click the **Graph** button. Use the **Transparency of surface** and the **Transparency of the frame** sliders and click off and on the **Wireframe** button to make the curves more distinct. The grid of squares in the domain are mapped to a grid of squares as predicted (see Figure 2.25).

Example 2.42. The parametrization

$$\mathbf{x}(u, v) = \big((a + b \cos v) \cos u, (a + b \cos v) \sin u, b \sin v\big)$$

for the torus is not isothermal (Note: because of Exploration 2.10, we assume that $a > b$). This is because in Exercise 2.32, you derived that

$$E = (a + b \cos v)^2$$
$$F = 0$$
$$G = b^2.$$

Because $F = 0$, $\mathbf{x}_u$ and $\mathbf{x}_v$ are orthogonal on the torus. But $E \neq G$ except if $v = \pi + 2\pi k$, ($k \in \mathbb{Z}$) and $a = 2b$. Thus the image of squares in the domain will be nonsquare

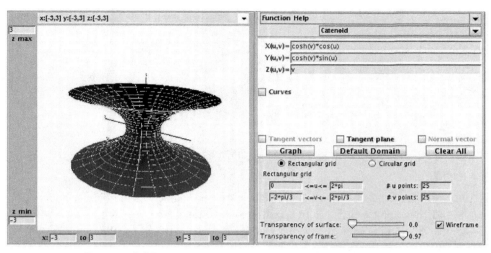

FIGURE 2.25. This parametrization of the catenoid is isothermal.

rectangles whenever $v \neq \pi + 2\pi k$. Open *DiffGeomTool* and enter the parametrization for the torus with $a = 2$ and $b = 1$. Set the **Rectangular grid** values to $0 \leq u \leq 2\pi$ and $0 \leq v \leq 2\pi$. Again, use the **Transparency of surface** and the **Transparency of the frame** sliders and click off and on the **Wireframe** button to make the curves more distinct. The grid of squares in the domain are mapped to a grid of mostly nonsquare rectangles. The ratio, $\frac{length}{height}$, of the sides of the rectangles is largest for the part of the torus farthest away from the origin. This occurs when $v = 0$ (or $v = 2\pi$), resulting in $E = 4$ while $G = 1$. The rectangles are squares for the part of the torus closest to the origin. This occurs when $v = \pi$, resulting in $E = 1$ while $G = 1$. This helps us see why the parametrization is not isothermal (see Figure 2.26).

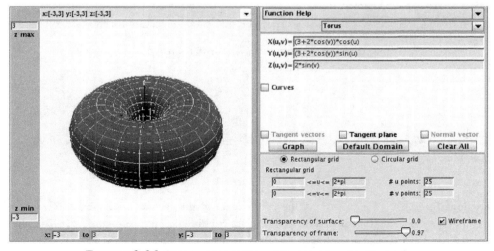

FIGURE 2.26. This parametrization of the torus is not isothermal.

Exercise 2.43. Using Definition 2.40 determine which of the parametrizations of minimal surfaces is isothermal:

(a) Enneper's surface parametrized by
$$\mathbf{x}(u, v) = \left(u - \tfrac{1}{3}u^3 + uv^2, v - \tfrac{1}{3}v^3 + u^2v, u^2 - v^2\right)$$

(b) Scherk's doubly periodic surface parametrized by
$$\mathbf{x}(u, v) = \left(u, v, \ln\left(\frac{\cos u}{\cos v}\right)\right)$$

(c) The helicoid parametized by
$$\mathbf{x}(u, v) = (a \sinh v \cos u, a \sinh v \sin u, au).$$

Try it out!

Exploration 2.44. Use *DiffGeomTool* to check the reasonableness of your answers in Exercise 2.43 by graphing each parametrization as was done in Examples 2.41 and 2.42. Set the number of u points to 20 and the number of v points to 20, and use the following values for the **U/V domain** boxes:

(a) Enneper's surface
$$-\tfrac{\pi}{3} \le u, v \le \tfrac{\pi}{3}$$

(b) Scherk's doubly periodic surface
$$-\tfrac{\pi}{2} + 0.1 \le u, v \le \tfrac{\pi}{2} - 0.1$$

(c) Helicoid
$$-\pi \le u, v \le \pi.$$

From Exercise 2.43 and Exploration 2.44 you have seen there are parametrizations of minimal surfaces that are not isothermal. However, requiring minimal surfaces to have an isothermal parametrization is not a restriction because of the following theorem.

Theorem 2.45. *Every minimal surface in $\mathbb{R}^3$ has an isothermal parametrization.*

Remark 2.46. See [23] for a proof. In fact, every differentiable surface has an isothermal parametrization, an interesting result whose proof is beyond the scope of this text. A proof is given in [2, pp. 15-35].

In Example 2.35, we derived that the isothermal parametrization for the catenoid
$$\mathbf{x}(u, v) = \left(a \cosh v \cos u, a \cosh v \sin u, av\right)$$

has
$$e = -g.$$

In general, we have

Theorem 2.47. *Let M be a surface with isothermal parametrization. Then M is minimal if and only if $e = -g$.*

Exercise 2.48. Prove Theorem 2.47. *Try it out!*

Exploration 2.49. If $e = -g$ for the coefficients of the second fundamental form, the u-parameter curve and the v-parameter curve are bending the same amount away from the normal $\mathbf{n}$ but in different directions. Use *DiffGeomTool* with Theorem 2.47 to geometrically verify which of the following surfaces are minimal:

(a) Enneper's surface
$$\mathbf{x}(u, v) = \left(u - \tfrac{1}{3}u^3 + uv^2, v - \tfrac{1}{3}v^3 + u^2v, u^2 - v^2\right)$$

(b) Cylinder
$$\mathbf{x}(u, v) = \left(\cos u, \sin u, v\right)$$

(c) Helicoid
$$\mathbf{x}(u, v) = (a \sinh v \cos u, a \sinh v \sin u, av).$$

Try it out!

Here is an interesting and important result that uses complex analysis. We know that if $f(z) = x(u, v) + iy(u, v)$ is an analytic function, then the Cauchy-Riemann equations hold for f. That is,
$$x_u = y_v, \qquad x_v = -y_u,$$
and y is called the *harmonic conjugate* of x. If f is analytic, then

$$f'(z) = x_u + iy_u. \tag{2.7}$$

This allows us to relate a minimal surface to another minimal surface, its conjugate minimal surface.

Definition 2.50. Let $\mathbf{x}$ and $\mathbf{y}$ be isothermal parametrizations of minimal surfaces such that their component functions are pairwise harmonic conjugates. That is,

$$\mathbf{x}_u = \mathbf{y}_v \qquad \text{and} \qquad \mathbf{x}_v = -\mathbf{y}_u. \tag{2.8}$$

Then $\mathbf{x}$ and $\mathbf{y}$ are called *conjugate minimal surfaces.*

Example 2.51. Let's find the conjugate surface of the catenoid parametrized by

$$\mathbf{x}(u, v) = (a \cosh v \cos u, a \cosh v \sin u, av).$$

Let $\mathbf{y}(u, v)$ be its parametrization. By the first part of (2.8), we know

$$\mathbf{y}_v = \mathbf{x}_u = (-a \cosh v \sin u, a \cosh v \cos u, 0).$$

Integrating this with respect to v yields

$$\mathbf{y} = (-a \sinh v \sin u + F_1(u), a \sinh v \cos u + F_2(u), F_3(u)),$$

where $F_k(u)$ is a function independent of v. Similarly, by the second part of (2.8), we derive

$$\mathbf{y} = (-a \sinh v \sin u + G_1(v), a \sinh v \cos u + G_2(v), -au + G_3(v)).$$

Equating, we get

$$\mathbf{y} = (-a \sinh v \sin u + K_1, a \sinh v \cos u + K_2, -au + K_3).$$

Using the substitution $u = \widetilde{u} - \frac{\pi}{2}$, $v = \widetilde{v}$, and letting $K_1 = 0$, $K_2 = 0$, and $K_3 = -a\frac{\pi}{2}$, does not affect the geometry of the minimal surface, and yields the parametrization of a helicoid

$$\mathbf{y}(\widetilde{u}, \widetilde{v}) = (a \sinh \widetilde{v} \cos \widetilde{u}, a \sinh \widetilde{v} \sin \widetilde{u}, -a\widetilde{u})$$

given in Exercise 2.24 (The negative sign in the third component function has the effect of reflecting the surface through the xy-plane). Hence, the conjugate surface of the catenoid is the helicoid.

The idea of conjugate minimal surfaces is really interesting. Two conjugate minimal surfaces can be joined through a one-parameter family of minimal surfaces by

$$\mathbf{z} = (\cos t)\mathbf{x} + (\sin t)\mathbf{y},$$

where $t \in \mathbb{R}$. When $t = 0$ we have the minimal surface parametrized by $\mathbf{x}$, and when $t = \frac{\pi}{2}$ we have the minimal surface parametrized by $\mathbf{y}$. So for $0 \le t \le \frac{\pi}{2}$, we have a continuous parameter of minimal surfaces known as *associated surfaces*. We can continuously transform one minimal surface into another minimal surface so that all the in-between surfaces are also minimal. In Example 2.51, we saw that the helicoid and the catenoid are conjugate surfaces. Images of them and associated surfaces are shown in Figure 2.27.

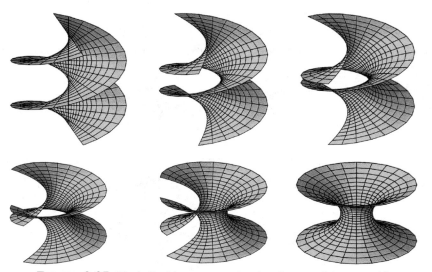

FIGURE 2.27. The helicoid, some associated surfaces, and the catenoid.

This is neat, and it is just the beginning. The rest of this section will explore properties of conjugate surfaces.

Exercise 2.52. Find the conjugate minimal surface for Enneper's surface

$$\mathbf{x}(u, v) = \left(u - \frac{1}{3}u^3 + uv^2, v - \frac{1}{3}v^3 + u^2v, u^2 - v^2\right).$$

If we try to determine the conjugate minimal surface for Scherk's doubly periodic surface with the parametrization

$$\mathbf{x}(u, v) = \left(u, v, \ln\left(\frac{\cos u}{\cos v}\right)\right),$$

the method will not work because the parametrization is not isothermal. Later we will see that Scherk's doubly periodic surface has a conjugate surface, Scherk's singly periodic surface (see Figure 2.28).

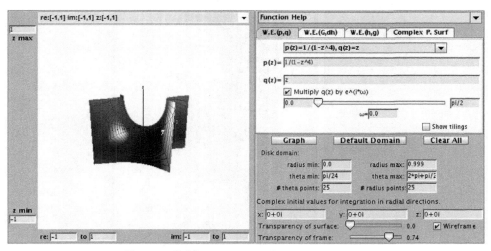

FIGURE 2.28. Scherk's singly periodic surface.

Exploration 2.53. You can see the associated surfaces that occur between Scherk's doubly periodic surface and Scherk's singly periodic surface by using another applet, called *MinSurfTool* that can be used to visualize and explore minimal surfaces in $\mathbb{R}^3$. Open the *MinSurfTool* applet. On the right-hand side near the top, there is a set of tabs for different features. For this exploration, we use the **W.E. (p,q)** feature, so make sure that the **W.E. (p,q)** tab is on top (if it is on top, it should be a different color than the other tabs). In the **Pre-set functions** window, choose $p(z) = 1/(1 - z^4)$, $q(z) = z$. Click on the **Graph** button, and one piece of Scherk's singly periodic surface will appear. Then move the slider arrow above the **Graph** button to the right to see how Scherk's singly periodic surface is transformed by way of the associated surfaces into Scherk's doubly periodic surface. *Try it out!*

Pieces of Scherk's doubly periodic surface can be put together in the xy-plane in a checkerboard fashion. They repeat (or are periodic) in two directions, x and y. Surfaces that repeat in one direction are called singly periodic. Individual pieces of Scherk's singly periodic surface can fit together creating a tower in the z direction. You can visualize adding two pieces together by taking one piece of Scherk's singly periodic surface and adding it to another piece that has been reflected across the xy-plane and shifted up in the z direction. Continuing, you can create a tower of several pieces (see Figure 2.29). The helicoid is a singly periodic surface too.

In Example 2.31 we saw that the coefficients of the first fundamental form for the parametrization of a catenoid are $E = a^2 \cosh^2 v$, $F = 0$, and $G = a^2 \cosh^2 v$. Exercise 2.140 from the Additional Exercises at the end of the chapter shows that for the parametrization of Enneper's surface the coefficients are $E = (1 + u^2 + v^2)^2$, $F = 0$, and $G = (1 + u^2 + v^2)^2$. The coefficients do not match up. However, for two conjugate minimal surfaces and their associated minimal surfaces, the coefficients of the first fundamental form are always the same. The following exercise will help you prove this surprising result.

Exercise 2.54.

(a) Prove that for two conjugate minimal surfaces, **x** and **y**, surfaces of the one-parameter

FIGURE 2.29. Scherk's singly periodic surface.

family

$$\mathbf{z} = (\cos t)\mathbf{x} + (\sin t)\mathbf{y}$$

have the same fundamental form: $E = \mathbf{x}_u \cdot \mathbf{x}_u = \mathbf{y}_u \cdot \mathbf{y}_u$, $F = 0$, $G = \mathbf{x}_v \cdot \mathbf{x}_v = \mathbf{y}_v \cdot \mathbf{y}_v$.

(b) Prove that the surfaces $\mathbf{z}$ are minimal for all $t \in \mathbb{R}$.

Try it out!

The normal vector $\mathbf{n}$ at a point on a surface points orthogonally away from the surface. Since minimal surfaces have different shapes, there is no reason to suppose that the normal vectors on one surface will be related to the normal vectors on another surface. However, for conjugate minimal surfaces and their associated minimal surfaces there is a strong connection. For any point in the domain, the corresponding surface normal points in the same direction on all these minimal surfaces. The next theorem establishes this.

Theorem 2.55. *Let $\mathbf{x}, \mathbf{y} : D \to \mathbb{R}^3$ be isothermal parametrizations of conjugate minimal surfaces. Then for $(u_0, v_0) \in D$, the surface unit normal is the same for the associated surfaces.*

Proof. Let $(u_0, v_0) \in D$. Let $\mathbf{n}^{\mathbf{x}}$ and $\mathbf{n}^{\mathbf{y}}$ represent the surface normal for $\mathbf{x}$ and for $\mathbf{y}$, respectively. Then by the definition of conjugate surfaces, $\mathbf{x}$ and $\mathbf{y}$ have the same unit normal, because

$$\mathbf{n}^{\mathbf{x}} = \frac{\mathbf{x}_u \times \mathbf{x}_v}{\left| \mathbf{x}_u \times \mathbf{x}_v \right|} = \frac{\mathbf{y}_v \times -\mathbf{y}_u}{\left| \mathbf{y}_v \times -\mathbf{y}_u \right|} = \frac{\mathbf{y}_u \times \mathbf{y}_v}{\left| \mathbf{y}_u \times \mathbf{y}_v \right|} = \mathbf{n}^{\mathbf{y}}.$$

To show that this is true for the associated surfaces, let $\mathbf{z} = (\cos t)\mathbf{x} + (\sin t)\mathbf{y}$ be the parametrization of the associated surfaces. Then

$$\begin{aligned}
\mathbf{z}_u \times \mathbf{z}_v &= \left(\cos t \, \mathbf{x}_u + \sin t \, \mathbf{y}_u \right) \times \left(\cos t \, \mathbf{x}_v + \sin t \, \mathbf{y}_v \right) \\
&= \cos^2 t \left(\mathbf{x}_u \times \mathbf{x}_v \right) + \cos t \sin t \left(\mathbf{x}_u \times \mathbf{y}_v \right) + \cos t \sin t \left(\mathbf{y}_u \times \mathbf{x}_v \right) + \sin^2 t \left(\mathbf{y}_u \times \mathbf{y}_v \right) \\
&= \cos^2 t \left(\mathbf{x}_u \times \mathbf{x}_v \right) + \cos t \sin t \left(\mathbf{x}_u \times \mathbf{x}_u \right) + \cos t \sin t \left(-\mathbf{x}_v \times \mathbf{x}_v \right) + \sin^2 t \left(-\mathbf{x}_v \times \mathbf{x}_u \right) \\
&= \cos^2 t \left(\mathbf{x}_u \times \mathbf{x}_v \right) + \cos t \sin t \left(\mathbf{0} \right) + \cos t \sin t \left(\mathbf{0} \right) + \sin^2 t \left(\mathbf{x}_u \times \mathbf{x}_v \right) \\
&= \mathbf{x}_u \times \mathbf{x}_v.
\end{aligned}$$

$\square$

The following example and exploration helps us visualize this idea.

Example 2.56. Using *DiffGeomTool* we can graph the catenoid and its conjugate surface, the helicoid, whose parametrizations are given in Example 2.51. If we plot the normal **n** at $(\frac{\pi}{3}, -\frac{\pi}{4})$ on the conjugate surfaces, we see that both normals point in the same direction as guaranteed by Theorem 2.55 (see Figure 2.30 and Figure 2.31).

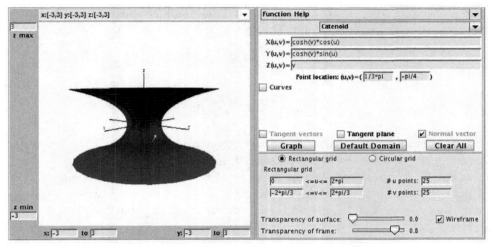

FIGURE 2.30. The catenoid with **n** at $(\frac{\pi}{3}, -\frac{\pi}{4})$.

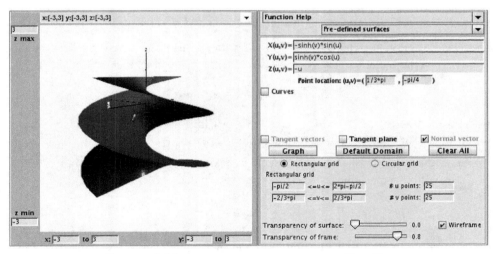

FIGURE 2.31. The conjugate helcoid with **n** at $(\frac{\pi}{3}, -\frac{\pi}{4})$.

Exploration 2.57. Open two windows of *DiffGeomTool*. In one plot the catenoid parametrized by

$$\mathbf{x}(u, v) = (\cosh v \cos u, \cosh v \sin u, v),$$

where $0 \leq u \leq 2\pi$ and $-\frac{2\pi}{3} \leq v \leq \frac{2\pi}{3}$. In the other plot its conjugate surface, the helicoid, with parametrization

$$\mathbf{y}(u, v) = (-\sinh v \sin u, \sinh v \cos u, -u),$$

where $-\frac{\pi}{2} \leq u \leq 2\pi - \frac{\pi}{2}$ and $-\frac{2\pi}{3} \leq v \leq \frac{2\pi}{3}$ (The u values for the helicoid are different than the values for the catenoid because we used the substitution $u = \widetilde{u} - \frac{\pi}{2}$ in Example 2.51). Make sure that the x, y, z-axes are positioned in the same directions. Then plot the following unit normals, $\mathbf{n}$, on each surface at the following points and observe that $\mathbf{n}$ points in the same direction as prescribed by Theorem 2.55:

(a) at $(\frac{\pi}{3}, 0)$ (b) at $(\frac{\pi}{4}, -\frac{\pi}{2})$ (c) at $(\frac{\pi}{6}, \frac{\pi}{2})$.

Try it out!

2.4 Weierstrass representation

At the end of the last section, we saw that we could apply complex analysis to minimal surface theory to define conjugate surfaces. In this section we will again use complex analysis to learn more about minimal surfaces. First, we will use inherent properties of an isothermal parametrization to give a necessary and sufficient condition for a surface to be minimal. As we will see, the condition is useful for finding and classifying minimal surfaces.

Theorem 2.58. *If the parametrization* $\mathbf{x}$ *is isothermal, then*

$$\mathbf{x}_{uu} + \mathbf{x}_{vv} = 2EH\mathbf{n},$$

where E is a coefficient of the first fundamental form and H is the mean curvature.

Exercise 2.59 (Proof of Theorem 2.58). The set $\{\mathbf{x}_u, \mathbf{x}_v, \mathbf{n}\}$ forms a basis for $\mathbb{R}^3$. Assume $F = 0$. Then the vector $\mathbf{x}_{uu}$ can be expressed in terms of the basis vectors. That is,

$$\mathbf{x}_{uu} = \Gamma^u_{uu}\mathbf{x}_u + \Gamma^v_{uu}\mathbf{x}_v + e\mathbf{n},$$

where the coefficients, Γ^u_{uu} and Γ^v_{uu}, are known as Christoffel symbols and e comes from the coefficient of the second fundamental form: $e = \mathbf{n} \cdot \mathbf{x}_{uu}$.

(a) Show that $\Gamma^u_{uu} = \frac{E_u}{2E}$ and $\Gamma^v_{uu} = -\frac{E_v}{2G}$ by taking the inner product of $\mathbf{x}_{uu}$ with the basis vectors. In a similar manner, it can be shown that

$$\mathbf{x}_{vv} = -\frac{G_u}{2E}\mathbf{x}_u + \frac{G_v}{2G}\mathbf{x}_v + g\mathbf{n}.$$

(b) Use the mean curvature equation (2.4) and the results from (a) to show that if the parametrization $\mathbf{x}$ is isothermal, then

$$\mathbf{x}_{uu} + \mathbf{x}_{vv} = 2EH\mathbf{n}.$$

Try it out!

Now, where do we go from here? For a minimal surface, $H \equiv 0$, so Theorem 2.58 tells us that $\mathbf{x}_{uu} + \mathbf{x}_{vv} \equiv 0$. What does that equation represent? It is Laplace's equation and relates to harmonic functions. Recall that $\varphi(u, v)$ is a real-valued *harmonic function* if $\varphi_{uu} + \varphi_{vv} = 0$ (for example, $\varphi(u, v) = u^2 - v^2$ is harmonic). This leads to the condition for a surface to be minimal that we referred to in the beginning of this section. Corollary 2.60 will allow us to explicitly describe minimal surfaces using the Weierstrass representation, which we will derive in this section.

Corollary 2.60. *A surface M with an isothermal parametrization $\mathbf{x}(u, v) = \big(x_1(u, v), x_2(u, v), x_3(u, v)\big)$ is minimal if and only if x_1, x_2, and x_3 are harmonic.*

We need an isothermal parametrization for our surface, but this is not a difficulty because of Theorem 2.45. Then this result tells us we will have a minimal surface if and only if its coordinate functions are harmonic. This provides us another way to prove a surface is minimal.

Proof. ($\Rightarrow$) If M is minimal, then $H = 0$ and so by Theorem 2.58 $\mathbf{x}_{uu} + \mathbf{x}_{vv} = 0$, and hence the coordinate functions are harmonic. ($\Leftarrow$) Suppose x_1, x_2, and x_3 are harmonic. Then $\mathbf{x}_{uu} + \mathbf{x}_{vv} = 0$. So by Theorem 2.58 we have that $2(\mathbf{x}_u \cdot \mathbf{x}_u)H\mathbf{n} = 0$. But $\mathbf{n} \neq 0$ and $E = \mathbf{x}_u \cdot \mathbf{x}_u \neq 0$. Hence $H = 0$ and M is minimal. $\qquad\square$

Exercise 2.61. Given the parametrization for Enneper's surface

$$\mathbf{x}(u, v) = \left(u - \frac{1}{3}u^3 + uv^2, v - \frac{1}{3}v^3 + u^2 v, u^2 - v^2\right),$$

use Corollary 2.60 to prove that it is a minimal surface. ***Try it out!***

The importance of Corollary 2.60 is not in proving specific surfaces are minimal but lies in establishing a formula that will guarantee that a surface created by it is minimal. The formula is known as the Weierstrass representation for minimal surfaces. It provides us with a simple way to construct many examples of minimal surfaces using functions from complex analysis. After stating the Weierstrass representation in Theorem 2.66 we will create more minimal surfaces. However, not all minimal surfaces are of equal interest. So in section 2.5 we will investigate properties that make minimal surfaces interesting.

We now derive this important formula, the Weierstrass representation. Suppose M is a minimal surface with an isothermal parametrization $\mathbf{x}(u, v)$. Let $z = u + iv$ be a point in the complex plane, so $\bar{z} = u - iv$. Solving for u, v in terms of $z, \bar{z}$ we get

$$u = \frac{z + \bar{z}}{2} \qquad \text{and} \qquad v = \frac{z - \bar{z}}{2i}.$$

The parametrization of the minimal surface M can be written in terms of the complex variables z and $\bar{z}$ as

$$\mathbf{x}(z, \bar{z}) = \big(x_1(z, \bar{z}), x_2(z, \bar{z}), x_3(z, \bar{z})\big).$$

Exercise 2.62. Let $f(u, v) = x(u, v) + iy(u, v)$ be a complex function. Using $u = \frac{z+\bar{z}}{2}$ and $v = \frac{z-\bar{z}}{2i}$, we can express f in terms of z and $\bar{z}$. In this exercise you will prove that f is analytic if and only if f can be written in terms of $z = u + iv$ alone, without using $\bar{z} = u - iv$.

(a) Using the chain rule, derive

$$\frac{\partial f}{\partial z} = \frac{1}{2}\left(\frac{\partial x}{\partial u} + \frac{\partial y}{\partial v}\right) + \frac{i}{2}\left(\frac{\partial y}{\partial u} - \frac{\partial x}{\partial v}\right),$$

$$\frac{\partial f}{\partial \overline{z}} = \frac{1}{2}\left(\frac{\partial x}{\partial u} - \frac{\partial y}{\partial v}\right) + \frac{i}{2}\left(\frac{\partial y}{\partial u} + \frac{\partial x}{\partial v}\right).$$

(b) Show that f is analytic $\iff \frac{\partial f}{\partial \overline{z}} = 0$.

Try it out!

Example 2.63. The function $f_1(z) = z^2$ is analytic because $\frac{\partial f_1}{\partial \overline{z}}(z^2) = 0$. However, $f_2(z) = |z|^2 = z\overline{z}$ is not analytic because $\frac{\partial f_1}{\partial \overline{z}} = z \neq 0$.

Exercise 2.64. Prove that

$$4\left(\frac{\partial}{\partial z}\left(\frac{\partial f}{\partial \overline{z}}\right)\right) = f_{uu} + f_{vv}. \tag{2.9}$$

Try it out!

The next theorem uses Corollary 2.60 to establish the Weierstrass representation for minimal surfaces.

Theorem 2.65. *Let M be a surface with parametrization* $\mathbf{x} = (x_1, x_2, x_3)$ *and let* $\boldsymbol{\phi} = (\varphi_1, \varphi_2, \varphi_3)$, *where* $\varphi_k = \frac{\partial x_k}{\partial z}$. *Let* $\boldsymbol{\phi}^2$ *denote* $(\varphi_1)^2 + (\varphi_2)^2 + (\varphi_3)^2$. *Then* $\mathbf{x}$ *is isothermal* $\iff \boldsymbol{\phi}^2 \equiv 0$. *If* $\mathbf{x}$ *is isothermal, then M is minimal* $\iff$ *each φ_k is analytic.*

Before we prove Theorem 2.65, let's apply it to an example to help us understand what the theorem says. Suppose we have the parametrization $\mathbf{x} = (x_1, x_2, x_3) = (z - \frac{1}{3}z^3, -i(z + \frac{1}{3}z^3), z^2)$. Then $\varphi_1 = \frac{\partial x_1}{\partial z} = 1 - z^2$, $\varphi_2 = \frac{\partial x_2}{\partial z} = -i(1 + z^2)$, and $\varphi_3 = \frac{\partial x_3}{\partial z} = 2z$. Because $\boldsymbol{\phi}^2 = [1 - z^2]^2 + [-i(1 + z^2)]^2 + [2z]^2 = 0$, the parametrization is isothermal by the theorem. Each φ_k is a polynomial and hence analytic. So $\mathbf{x}$ is a parametrization of a minimal surface (it is Enneper's surface). Make sure you understand how this example relates to Theorem 2.65 before you read the proof of the theorem.

Proof. Applying the complex differential operator $\frac{\partial f}{\partial z}$ from Exercise 2.62 and squaring the terms, we have

$$(\varphi_k)^2 = \left(\frac{\partial x_k}{\partial z}\right)^2 = \left[\frac{1}{2}\left(\frac{\partial x_k}{\partial u} - i\frac{\partial x_k}{\partial v}\right)\right]^2 = \frac{1}{4}\left[\left(\frac{\partial x_k}{\partial u}\right)^2 - \left(\frac{\partial x_k}{\partial v}\right)^2 - 2i\frac{\partial x_k}{\partial u}\frac{\partial x_k}{\partial v}\right].$$

Also,

$$\mathbf{x_u} \cdot \mathbf{x_u} = \left(\frac{\partial x_1}{\partial u}\right)^2 + \left(\frac{\partial x_2}{\partial u}\right)^2 + \left(\frac{\partial x_3}{\partial u}\right)^2 = \sum_{k=1}^{3}\left(\frac{\partial x_k}{\partial u}\right)^2$$

and similarly $\mathbf{x}_v \cdot \mathbf{x}_v = \sum_{k=1}^{3} (\frac{\partial x_k}{\partial v})^2$. Hence,

$$\phi^2 = (\varphi_1)^2 + (\varphi_2)^2 + (\varphi_3)^2$$

$$= \frac{1}{4} \left[\sum_{k=1}^{3} \left(\frac{\partial x_k}{\partial u} \right)^2 - \sum_{k=1}^{3} \left(\frac{\partial x_k}{\partial v} \right)^2 - 2i \sum_{k=1}^{3} \frac{\partial x_k}{\partial u} \frac{\partial x_k}{\partial v} \right]$$

$$= \frac{1}{4} \left(\mathbf{x}_u \cdot \mathbf{x}_u - \mathbf{x}_v \cdot \mathbf{x}_v - 2i (\mathbf{x}_u \cdot \mathbf{x}_v) \right)$$

$$= \frac{1}{4} (E - G - 2iF).$$

Thus, $\mathbf{x}$ is isothermal $\Longleftrightarrow E = G$, and $F = 0 \Longleftrightarrow \phi^2 \equiv 0$.

Suppose that $\mathbf{x}$ is isothermal. By Corollary 2.60, it suffices to show that for each k, x_k is harmonic $\Longleftrightarrow \varphi_k$ is analytic. Using (2.9) and Exercise 2.62 this follows because

$$\frac{\partial^2 x_k}{\partial u \partial u} + \frac{\partial^2 x_k}{\partial v \partial v} = 4 \left(\frac{\partial}{\partial \bar{z}} \left(\frac{\partial x_k}{\partial z} \right) \right) = 4 \left(\frac{\partial}{\partial \bar{z}} \left(\varphi_k \right) \right) = 0.$$

$\square$

Let's apply this theorem. Suppose we have analytic functions φ_k and we want to find the functions x_k. If $\mathbf{x}$ is isothermal, then

$$|\phi|^2 = \left| \frac{\partial x_1}{\partial z} \right|^2 + \left| \frac{\partial x_2}{\partial z} \right|^2 + \left| \frac{\partial x_3}{\partial z} \right|^2 = \frac{1}{4} \left(\sum_{k=1}^{3} \left(\frac{\partial x_k}{\partial u} \right)^2 + \sum_{k=1}^{3} \left(\frac{\partial x_k}{\partial v} \right)^2 \right)$$

$$= \frac{1}{4} (\mathbf{x}_u \cdot \mathbf{x}_u + \mathbf{x}_v \cdot \mathbf{x}_v) = \frac{1}{4} (E + G) = \frac{E}{2}.$$

So if $|\phi|^2 \equiv 0$, then the coefficients of the first fundamental form are zero and M is a point.

We need to solve $\varphi_k = \frac{\partial x_k}{\partial z}$ for x_k since the parametrization of the surface is given as $\mathbf{x} = (x_1, x_2, x_3)$. The difficulty is that x_k is a function of two variables, z and $\bar{z}$, and we want to have a representation where we have to integrate with respect to only one variable. To overcome this, we will use some ideas about differentials (see [26] for an introduction to differentials). Since x_k is a function of the two variables u and v, we can write

$$dx_k = \frac{\partial x_k}{\partial u} du + \frac{\partial x_k}{\partial v} dv. \tag{2.10}$$

Also, $dz = du + i dv$. Using Exercise 2.62 we have

$$\varphi_k dz = \frac{\partial x_k}{\partial z} dz = \frac{1}{2} \left(\frac{\partial x_k}{\partial u} - i \frac{\partial x_k}{\partial v} \right) (du + i dv)$$

$$= \frac{1}{2} \left[\frac{\partial x_k}{\partial u} du + \frac{\partial x_k}{\partial v} dv + i \left(\frac{\partial x_k}{\partial u} dv - \frac{\partial x_k}{\partial v} du \right) \right],$$

$$\overline{\varphi_k dz} = \overline{\varphi_k} \overline{dz} = \overline{\frac{\partial x_k}{\partial z}} \overline{dz} = \frac{1}{2} \left(\frac{\partial x_k}{\partial u} + i \frac{\partial x_k}{\partial v} \right) (du - i dv)$$

$$= \frac{1}{2} \left[\frac{\partial x_k}{\partial u} du + \frac{\partial x_k}{\partial v} dv - i \left(\frac{\partial x_k}{\partial u} dv - \frac{\partial x_k}{\partial v} du \right) \right].$$

Adding we get

$$\frac{\partial x_k}{\partial u}du + \frac{\partial x_k}{\partial v}dv = \varphi_k dz + \overline{\varphi_k dz} = 2\operatorname{Re}\{\varphi_k dz\}. \tag{2.11}$$

Combining (2.10) and (2.11), we have

$$dx_k = 2\operatorname{Re}\{\varphi_k dz\}.$$

Therefore, $x_k = 2\operatorname{Re}\int \varphi_k dz + c_k$. Since adding c_k translates the image by a constant amount and multiplying a coordinate function by 2 scales the the surface, the constants do not affect the geometric shape of the surface. Hence, we do not need them and we will let our coordinate function be

$$x_k = \operatorname{Re}\int \varphi_k dz.$$

Summary: If we have analytic functions $'_k$ $(k = 1, 2, 3)$ such that $\phi^2 \equiv 0$ and $|\phi|^2 \not\equiv 0$ and is finite, then the parametrization

$$\mathbf{x} = \left(\operatorname{Re}\int \varphi_1(z)dz, \operatorname{Re}\int \varphi_2(z)dz, \operatorname{Re}\int \varphi_3(z)dz\right) \tag{2.12}$$

defines a minimal surface.

For example, consider the functions $p(z)$ and $q(z)$ such that

$$\varphi_1 = p(1 + q^2)$$
$$\varphi_2 = -ip(1 - q^2)$$
$$\varphi_3 = -2ipq.$$

Then

$$\begin{aligned}
\phi^2 &= [p(1 + q^2)]^2 + [-ip(1 - q^2)]^2 + [-2ipq]^2 \\
&= [p^2 + 2p^2q^2 + p^2q^4] - [p^2 - 2p^2q^2 + p^2q^4] - [4p^2q^2] \\
&= 0,
\end{aligned}$$

and

$$\begin{aligned}
|\phi|^2 &= |p(1 + q^2)|^2 + |-ip(1 - q^2)|^2 + |-2ipq|^2 \\
&= |p|^2[(1 + q^2)(1 + \overline{q}^2) + (1 - q^2)(1 - \overline{q}^2) + 4q\,\overline{q}] \\
&= |p|^2[2(1 + 2q\,\overline{q} + q^2\,\overline{q}^2)] \\
&= 4|p|^2(1 + |q|^2)^2 \neq 0 \qquad (\text{if } p = 0, \text{ then } \varphi_k = 0 \text{ for all } k).
\end{aligned}$$

For φ_k to be analytic, p, pq^2, and pq have to be analytic. If p is analytic with a zero of order $2m$ at z_0, then q can have a pole of order no larger than m at z_0. A function that is analytic in a domain D except possibly at poles is a *meromorphic function* in D. This leads to the following result.

Theorem 2.66 (Weierstrass Representation (p, q)). *Every regular minimal surface has a local isothermal parametric representation of the form*

$$\mathbf{x} = (x_1(z), x_2(z), x_3(z))$$
$$= \left(\operatorname{Re} \left\{ \int_a^z p(1 + q^2) dz \right\}, \operatorname{Re} \left\{ \int_a^z -ip(1 - q^2) dz \right\}, \operatorname{Re} \left\{ \int_a^z -2ipq dz \right\} \right),$$

where p is an analytic function and q is a meromorphic function in a domain $\Omega \subset \mathbb{C}$, having the property that where q has a pole of order m, p has a zero of order at least $2m$, and $a \in \Omega$ is a constant.

Example 2.67. For $p(z) = 1, q(z) = iz$, we get

$$\mathbf{x} = \left(\operatorname{Re} \left\{ \int_0^z (1 - z^2) \, dz \right\}, \operatorname{Re} \left\{ \int_0^z -i(1 + z^2) \, dz \right\}, \operatorname{Re} \left\{ \int_0^z 2z \, dz \right\} \right)$$
$$= \left(\operatorname{Re} \left\{ z - \frac{1}{3} z^3 \right\}, \operatorname{Re} \left\{ -i \left(z + \frac{1}{3} z^3 \right) \right\}, \operatorname{Re} \left\{ z^2 \right\} \right).$$

Letting $z = u + iv$, this yields

$$\mathbf{x}(u, v) = \left(u - \frac{1}{3} u^3 + uv^2, v - \frac{1}{3} v^3 + u^2 v, u^2 - v^2 \right),$$

which gives Enneper's surface.

You can use the applet *MinSurfTool* to graph an image of this surface using p and q. After opening *MinSurfTool*, make sure that the **W.E. (p, q)** tab is on top. In the appropriate boxes, put $p(z) = 1$ and $q(z) = iz$. Then click on the **Graph** button. You can increase the size of the image of the surface by clicking on the left button on the mouse, and you can decrease it by clicking on the right mouse button. You can rotate the surface by placing the cursor arrow on its image and clicking, holding the left button on the mouse as you move the cursor.

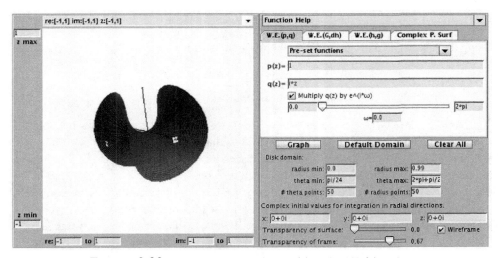

FIGURE 2.32. Enneper's surface using $p(z) = 1$ and $q(z) = iz$.

Example 2.68. Let $p(z) = 1$ and $q(z) = 1/z$ on the domain $\mathbb{C} - \{0\}$. We see that q is meromorphic with a pole of order 1 at $z_0 = 0$ and p does not have a zero of order 2 at $z_0 = 0$. This does not violate the conditions of Theorem 2.66, because the domain is $\mathbb{C} - \{0\}$. We will show that p and q generate a helicoid. Using the Weierstrass representation (p, q) and letting $z = u + iv$, we get $\mathbf{x}(u, v) = (x_1, x_2, x_3)$, where

$$x_1 = \operatorname{Re} \int_1^z \left(1 + \frac{1}{z^2}\right) dz = \operatorname{Re}\left(z - \frac{1}{z}\right) = u - \frac{u}{u^2 + v^2}$$

$$x_2 = \operatorname{Re} \int_1^z -i\left(1 - \frac{1}{z^2}\right) dz = \operatorname{Im}\left(z + \frac{1}{z}\right) = v - \frac{v}{u^2 + v^2}$$

$$x_3 = \operatorname{Re} \int_1^z -2i\frac{1}{z}\, dz = 2\operatorname{Im}(\log z) = 2 \arg z = 2 \arctan\left(\frac{v}{u}\right).$$

The parametrization is different than the parametrization we have been using for the helicoid,

$$\widetilde{x}(\widetilde{u}, \widetilde{v}) = (\widetilde{x_1}, \widetilde{x_2}, \widetilde{x_3}) = (a \sinh \widetilde{v} \cos \widetilde{u}, a \sinh \widetilde{v} \sin \widetilde{u}, a\widetilde{u}).$$

To show that $\mathbf{x}$ also gives an image of the helicoid, we find a substitution that will change $\mathbf{x}$ into $\widetilde{x}$. We have

$$x_1^2 + x_2^2 = (u^2 + v^2) - 2 + \frac{1}{u^2 + v^2}$$

$$\widetilde{x_1}^2 + \widetilde{x_2}^2 = a^2 \sinh^2 \widetilde{v} = a^2 \left(\frac{e^{\widetilde{v}} - e^{-\widetilde{v}}}{2}\right)^2.$$

Equating the right-hand sides and letting $a = 2$, we get that

$$u^2 + v^2 = e^{2\widetilde{v}}.$$

With $x_3 = \widetilde{x_3}$, we see that

$$\frac{v}{u} = \tan \widetilde{u}.$$

Using the last two equations we can solve for u and v to get

$$u = e^{\widetilde{v}} \cos \widetilde{u} \quad \text{and} \quad v = e^{\widetilde{v}} \sin \widetilde{u}.$$

If we substitute u and v into $\mathbf{x}(u, v)$ we get the parametrization $\widetilde{x}(\widetilde{u}, \widetilde{v})$ for the helicoid.

Using the **W.E. (p,q)** tab in *MinSurfTool*, we can get a graph of the helicoid by choosing from the **Pre-set functions** the values $p(z) = 1$ and $q(z) = 1/z$. Since the domain is $\mathbb{C} \setminus \{0\}$, the **Disk domain: radius min:** box is set to 0.2. Because of the singularity at $z = 0$, there needs to be functions entered into the x, y and z boxes in the **Complex initial values for integration in radial direction** section. These functions come from explicitly solving the integrals for $x_1(z)$, $x_2(z)$, and $x_3(z)$ in the Weierstrass representation when $p(z) = 1$ and $q(z) = 1/z$ and substituting in $re^{i\theta}$ for z. If you use a choice from the **Pre-set functions**, the functions will be entered automatically. However, if you enter your own p and q values, you will need to compute the functions and enter them into these boxes to get the correct minimal surface image.

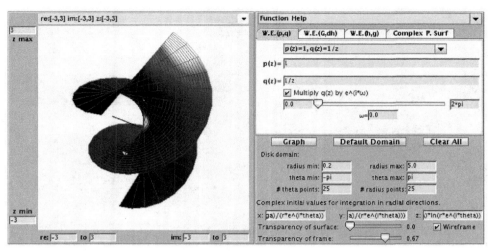

FIGURE 2.33. The helicoid using $p(z) = 1$ and $q(z) = \frac{1}{z}$.

Exercise 2.69. Show that the minimal surface generated by using $p(z) = 1$ and $q(z) = 0$ on the domain $\mathbb{C}$ in the Weierstrass representation is the plane. *Try it out!*

Exercise 2.70. Show that the minimal surface generated by using $p(z) = 1$ and $q(z) = i/z$ on the domain $\mathbb{C} - \{0\}$ in the Weierstrass representation is the catenoid. Use the appropriate **Pre-set function** in the **W.E. (p,q)** tab of *MinSurfTool* to graph an image of the surface. *Try it out!*

Exploration 2.71. Enneper's surface can be constructed with $p(z) = 1$ and $q(z) = z$. It has four leaves (two pointing up and two pointing down). The number of leaves can be increased.

(a) Using $p(z) = 1$ and $q(z) = z^2$ on the domain $\mathbb{C}$ in the Weierstrass representation gives Enneper's surface with six leaves (see Figure 2.34). Compute its parametrization $\mathbf{x}(u, v)$.

(b) Conjecture the values of p and q for Enneper's surface with n leaves.

(c) Use *MinSurfTool* with the **W.E. (p,q)** tab to check your conjecture.

Try it out!

Exploration 2.72. Use *MinSurfTool* with the **W.E. (p,q)** tab to graph an image of the surface generated by the **Pre-set functions** $p(z) = \frac{1}{1-z^4}$ and $q(z) = z$.

(a) What minimal surface is it?

(b) Click on the box "Multiply $q(z)$ by e∧(i*ω)" and move the slider to generate a family of minimal surfaces. They are associated surfaces (see the paragraph after Definition 2.50). When $\theta = \frac{\pi}{2}$ you get the conjugate surface. What is it?

(c) Experiment with *MinSurfTool* to view the associated family and find the conjugate surface of the various minimal surfaces discussed above.

Try it out!

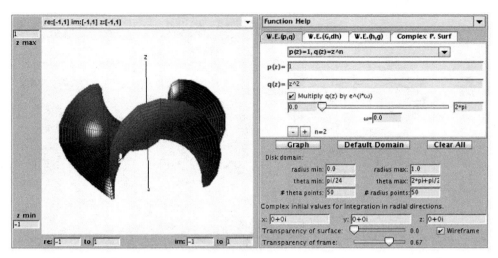

FIGURE 2.34. Enneper's surface with six leaves using $p(z) = 1$ and $q(z) = z^2$.

Exploration 2.73. Scherk's doubly periodic surface is generated with $p(z) = \frac{1}{1-z^4}$ and $q(z) = iz$. Use the appropriate **Pre-set function** in the **W.E. (p,q)** tab of *MinSurfTool* to graph an image of the surface generated by $p(z) = \frac{1}{1-z^{2n}}$ and $q(z) = iz^{n-1}$ for various values of $n = 2, 3, 4, \ldots$. The $-$ and $+$ boxes above the **Graph** button can be used to increase or decrease n.

(a) What happens to the surface as n increases?

(b) The surface has leaves that alternate between going up and going down. How is n related to the number of leaves?

(c) What is the image of the projection of the surface onto the x_1x_2-plane for each n?

(d) Using the previous parts conjecture how many leaves the surface has if $p(z) = \frac{1}{1-z^5}$. Why could such a surface not exist?

Try it out!

Exploration 2.74. Use the appropriate **Pre-set function** in the **W.E. (p,q)** tab of *MinSurfTool* to graph an image of the surface generated by $p(z) = \frac{1}{(1-z^4)^2}$ and $q(z) = iz^3$. This surface is known as the 4-noid (see Figure 2.35).

(a) Try to create a 3-noid by changing the values of p and q and graphing the result in *MinSurfTool*.

(b) Conjecture the values of p and q that will generate an n-noid.

Try it out!

While the Weierstrass representation will generate a minimal surface, there is no guarantee that it will be embedded. In the introduction of this chapter that we said that a surface is embedded if it has no self-intersections. The plane, the catenoid, the helicoid, and Scherk's doubly periodic surface are embedded minimal surfaces, but Enneper's surface is not embedded. In Exploration 2.11 you saw that Enneper's surface intersects itself when the domain contains a disk centered at the origin of radius $R \geq \sqrt{3}$.

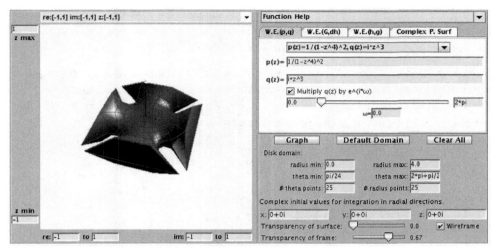

FIGURE 2.35. Image of the 4-noid minimal surface.

Exploration 2.75. Using *MinSurfTool* with the **W.E. (p,q)** tab, get three sets of functions p and q that create minimal surfaces that are not embedded. If the p and q functions result in a singularity in the Weierstrass representation, then you will need to enter functions into the x, y, and z boxes in the **Complex initial values for integration in radial direction** section. They come from explicitly solving the integrals for $x_1(z)$, $x_2(z)$, and $x_3(z)$ in the Weierstrass representation for $p(z)$ and $q(z)$ and then substituting in $re^{i\theta}$ for z. *Try it out!*

An important part of minimal surface theory is the study of complete (boundaryless) embedded minimal surfaces. The following theorem tells us that a minimal surface without boundary cannot be closed and bounded.

Theorem 2.76. *If M is a complete minimal surface in $\mathbb{R}^3$, then M is not compact.*

Proof. By Theorem 2.45, we can assume that M has an isothermal parametrization. If M were compact, then each coordinate function would attain a maximum. Since the real part of an analytic function is harmonic, the coordinate functions in this parametrization are harmonic by Theorem 2.65. Because harmonic functions attain their maximums on the boundary of the set, M must have a boundary, which contradicts M being complete. $\square$

2.5 The Gauss map, G, and height differential, dh

We can use other representations for $\boldsymbol{\phi} = (\varphi_1, \varphi_2, \varphi_3)$ to form different Weierstrass representations as long as $\boldsymbol{\phi}^2 = 0$ and $|\boldsymbol{\phi}|^2 \neq 0$ (see the Summary on page 120). An important representation employs the function known as the Gauss map, G, and the height differential, dh. This representation is useful because G and dh describe the geometry of the minimal surface. To develop the representation, we need some background about the Gauss map.

The curvature of a unit speed curve, α, at a point s is $|\alpha''(s)|$. That is, the curvature of a curve is described by the rate of change of its tangent vector. Similarly, the curvature of a surface is related to the rate of change of its tangent plane. Since a tangent plane is determined by its unit normal vector, **n**, we can investigate the curvature of a surface by studying the variation of **n**. This is the idea behind the Gauss map.

Definition 2.77. Let $M : \Omega \to \mathbb{R}^3$ be a surface with a chosen orientation (that is, a differentiable field of unit normal vectors **n**). The *Gauss map*, $\mathbf{n}_p$, translates the unit normal on M at a point p to the unit vector at the origin pointing in the same direction as the unit normal and thus corresponds to a point on the unit sphere S^2.

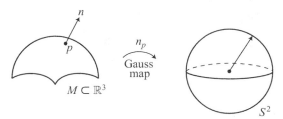

FIGURE 2.36. The Gauss map.

Example 2.78. Let's determine the image of the Gauss map for the catenoid. A meridian is a curve formed by a vertical slice on the surface. Take a meridian on the entire catenoid (the image in Figure 2.37 is only part of a catenoid that extends on forever). Its Gauss map, $\mathbf{n_p}$, will be a meridian on S^2 from the north pole $(0, 0, 1)$ to the south pole $(0, 0, -1)$ that excludes the end points. They are excluded because no matter how far the catenoid extends, the unit normal **n** never points exactly straight up or exactly straight down. Since the catenoid is a surface of revolution, if we revolve the meridian, we get that the image of the Gauss map for the catenoid is $S^2 \setminus \{(0, 0, 1), (0, 0, -1)\}$.

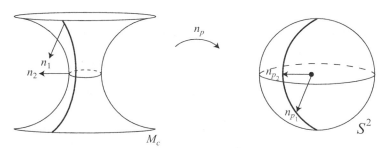

FIGURE 2.37. Image of a meridian on a catenoid under the Gauss map.

Exercise 2.79. Describe the image of the Gauss map for

(a) a right circular cylinder

(b) a torus

(c) Enneper's surface defined only on $\mathbb{D}$

(d) a helicoid

(e) Scherk's doubly periodic surface.

Try it out!

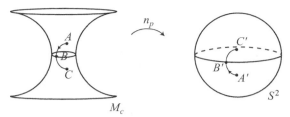

FIGURE 2.38. $\mathbf{n_p}$ is orientation reversing.

Theorem 2.80 ([22]). *Let M be a minimal surface with an isothermal parametrization. Then the Gauss map of M preserves angles.*

Though the Gauss map preserves angles, it reverses orientation. Such maps are called *anticonformal*. To help visualize the reversal, take three points A, B, and C on a curved path near the neck of the catenoid (see Figure 2.38). Since A is above the neck, the outward pointing unit normal at A will be pointing downward and hence the Gauss map will put it below the equator on S^2 at A'. The point B is on the neck and so the outward pointing unit normal at B will be horizontal. The Gauss map will put it on the equator of S^2 at B'. Similarly, the normal at C will be mapped to C'. Thus, the curved path from A to B to C in the positive direction on the catenoid gets sent by the Gauss map to a curve from A' to B' to C' in the negative direction on S^2. That is, we have an orientation-reversing map.

Since the Gauss map associates a point on M with a point on S^2, we can also associate the original point with a point in the complex plane $\mathbb{C}$ by using stereographic projection. Stereographic projection takes a point on S^2 to a point in the extended complex plane, $\mathbb{C} \cup \infty$. As in Figure 2.39, we place the complex plane through the equator of the sphere and take a line connecting the north pole, $(0, 0, 1) \in S^2$, with the point $(x_1, x_2, x_3) \in S^2$. This line intersects the extended complex plane at a point, $z = x + iy$. In this setting the unit sphere is known as the *Riemann sphere*.

Exercise 2.81. Describe the projections of the following sets on the Riemann sphere onto the extended complex plane

(a) meridians

(b) parallels

(c) circles

(d) circles that contain (i.e., touch) the point $(0, 0, 1)$

(e) antipodal points (i.e., diametrically opposite points).

Try it out!

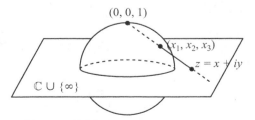

FIGURE 2.39. Stereographic projection.

Let $\overline{\sigma}$ be the projection of $(x_1, x_2, x_3) \in S^2$ to the point $x - iy \in \mathbb{C}$ by stereographic projection of (x_1, x_2, x_3) to $z = x + iy$ followed by reflection across the real axis to $\overline{z} = x - iy$. Note that $\overline{\sigma}$ is an anticonformal map.

Let $G : D \subset \mathbb{C} \to \mathbb{C}$ be the map defined by $G = \overline{\sigma} \circ \mathbf{n} \circ \mathbf{x}$. It preserves angles because $\overline{\sigma}$, $\mathbf{n}$, and $\mathbf{x}$ preserve angles and so the composition of these maps preserves angles. G is orientation preserving because $\overline{\sigma}$ and $\mathbf{n}$ are orientation reversing and so their composition is orientation preserving. Thus, G is a meromorphic function. It is also called the Gauss map. While there are two different maps called the Gauss map, we will mean G when we refer to "the Gauss map" in this chapter. See Figure 2.40

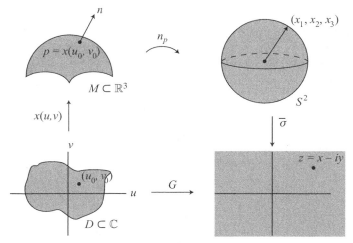

FIGURE 2.40. The map G.

Example 2.82. Using the geometry of Enneper's surface, M_E, we can determine specific values of a Gauss map, G, on M_E even though we do not know what function G is. What is

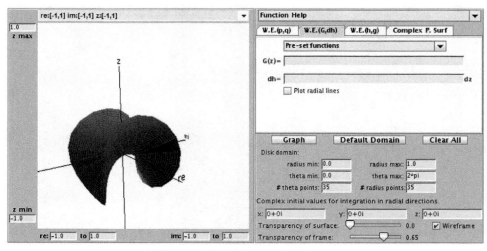

FIGURE 2.41. Enneper's surface.

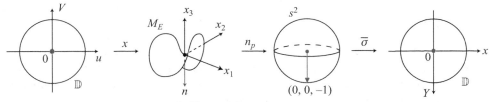

FIGURE 2.42. $G(0) = 0$ for Enneper's surface.

$G(0)$? Enneper's surface is formed by bending a disk into a saddle surface. The point $0 \in \mathbb{C}$ should get mapped to the point in the center of Enneper's surface. There are two normal vectors on the surface. We will chose the downward pointing normal $\mathbf{n}$. Hence, the unit normal at the center of M_E points straight down, and thus mapping it to S^2 under $\mathbf{n}_p$ gives the vector pointing at $(0, 0, -1)$ (see Figure 2.42). Taking the stereographic projection, σ, results in the point $z = 0 \in \mathbb{C}$ and reflecting it across the real axis does not change 0, so $\bar{\sigma}(0, 0, -1) = 0$. Hence, $G(0) = 0$.

What is $G(r)$ when $r \in [0, 1]$? The points r get mapped under $\mathbf{x}$ to a curve moving upward along one of the upward pointing leaves of Enneper's surface. The corresponding downward pointing unit normal, $\mathbf{n}_r$, stays in the x_1x_3-half plane (where $x_1 \geq 0$) also moving upward (i.e., x_3 is increasing). As r approaches 1, $\mathbf{n}_r$ approaches being parallel to the x_1 axis (see Figure 2.43). Thus, mapping the unit normals to S^2, the curve $\{r \in \mathbb{D} : 0 \leq r < 1\}$ traces a meridian on S^2 from $(0, 0, -1)$ to $(1, 0, 0)$. Its stereographic projection onto the complex plane gives $\{r \in \mathbb{D} : 0 \leq r < 1\}$ and reflecting this across the real axis does not change the values. Hence, $G(r) = r$, where $0 \leq r \leq 1$.

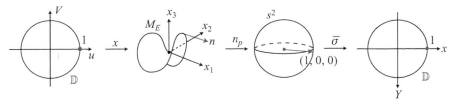

FIGURE 2.43. $G(1) = 1$ for Enneper's surface.

What is $G(e^{i\theta})$ for $0 \leq \theta \leq \frac{\pi}{2}$? If we restrict the domain of Enneper's surface to $\mathbb{D}$, these points get mapped to the edge of our image of Enneper's surface in a positive direction. At $\theta = 0$, the unit normal is pointing outward (i.e., away from the opposite leaf) and under the Gauss map, $\mathbf{n}_p$, this corresponds to $(1, 0, 0) \in S^2$. As θ moves from 0 to $\frac{\pi}{2}$, the unit normal moves from pointing outward to pointing inward (i.e., toward the opposite leaf). At $\theta = \frac{\pi}{2}$, the unit normal is mapped under n_p to $(0, -1, 0) \in S^2$ which projects under σ to $-i \in \mathbb{C}$ (see Figure 2.44). Reflecting across the real axis gives

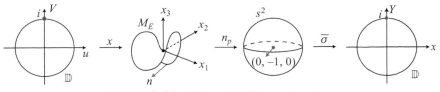

FIGURE 2.44. $G(i) = i$ for Enneper's surface.

$\overline{\sigma}(0, -1, 0) = \overline{-i} = i$. A similar argument shows that the same thing happens for all $\theta \in [0, \frac{\pi}{2}]$. That is, $G(e^{i\theta}) = e^{i\theta}, (0 \le \theta \le \frac{\pi}{2})$.

Example 2.83. Let's determine some values for the Gauss map G for the singly periodic Scherk surface with six leaves, M_S. The domain used is $\mathbb{D}$. The leaves are centered at rays from the origin through each of the sixth roots of unity (i.e., $e^{i\pi k/3}, (k = 0, ..., 5)$) with the leaf centered at the positive real axis pointing upward and the subsequent leaves alternating between pointing downward and upward (see Figure 2.45).

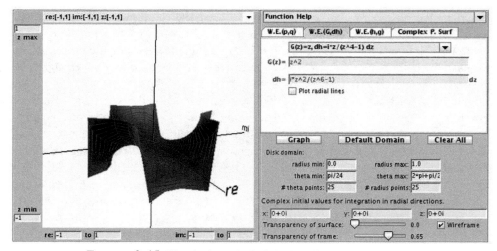

FIGURE 2.45. The singly periodic Scherk surface with six leaves.

As in the previous example, we will use the downward pointing unit normal, and so $G(0) = 0$. Because the leaves are centered at the sixth roots of unity, let's look at $G(e^{i\pi k/3})$, where $k = 0, ..., 5$. The point 1 gets mapped under $\mathbf{x}$ to the edge of M_S above the positive real axis. By looking at the graph of M_S, we see that this is in the middle of an upward pointing leaf, and the corresponding unit normal $\mathbf{n}$ lies above the positive real axis and points away from the origin in a plane parallel to the horizontal $x_1 x_2$-plane. Mapping this normal under $\mathbf{n}_p$ and then $\overline{\sigma}$, results in the point 1. Hence, $G(1) = 1$. Now, consider $G(e^{i\pi/3})$. The point $e^{i\pi/3}$ gets mapped to the edge of M_S above the line $re^{i\pi/3}, r > 0$. Since the leaves alternate between pointing upward and downward, this is in the middle of a downward pointing leaf and points toward the origin. Mapping it under $\mathbf{n}_p$ and then $\overline{\sigma}$, results in the point $\overline{e^{i4\pi/3}} = e^{i2\pi/3}$. Hence, $G(e^{i\pi/3}) = e^{i2\pi/3}$.

In a similar way, we get

$$G(1) = 1, \qquad G\left(e^{i\frac{\pi}{3}}\right) = e^{i\frac{2\pi}{3}}, \qquad G\left(e^{i\frac{2\pi}{3}}\right) = e^{i\frac{4\pi}{3}},$$

$$G(-1) = 1, \qquad G\left(e^{i\frac{4\pi}{3}}\right) = e^{i\frac{2\pi}{3}} = e^{i\frac{8\pi}{3}}, \qquad G\left(e^{i\frac{5\pi}{3}}\right) = e^{i\frac{4\pi}{3}} = e^{i\frac{10\pi}{3}}.$$

Exercise 2.84. A picture of the half catenoid on its side defined on $\mathbb{D}$ is shown in Figure 2.46 with the positive real axis on the right.

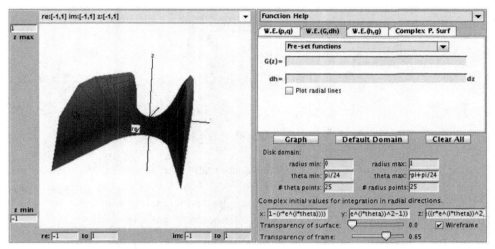

FIGURE 2.46. A view of the half catenoid on its side.

For the catenoid on its side determine

(a) $G(0)$ (b) $G(1)$ (c) $G(-1)$ (d) $G(i)$ (e) $G(-i)$.

Try it out!

Exercise 2.85. For the 4-noid (see Figure 2.35), determine:

(a) $G(0)$ (b) $G(1)$ (c) $G(-1)$ (d) $G(i)$ (e) $G(-i)$.

Try it out!

We will use the Gauss map G to form another Weierstrass representation of a parametrized minimal surface. In doing so, we will need the height differential, dh, which is so called because it is locally (though not globally) the differential of the height coordinate. We will not get into the definition of differential forms. They are in [26]. It is worth mentioning that at points where the Gauss map is vertical (i.e., $G = 0$ or $G = \infty$), the height function ought to have local minimums and maximums. Hence, dh ought to have a zero at these points (for example, Figure 2.42).

Theorem 2.86 (Weierstrass Representation (G,dh)). *Every regular minimal surface has a local isothermal parametric representation of the form*

$$\mathbf{x} = \operatorname{Re} \int_a^z \left(\frac{1}{2}\left(\frac{1}{G} - G \right), \frac{i}{2}\left(\frac{1}{G} + G \right), 1 \right) dh, \qquad (2.13)$$

where G is the Gauss map, dh is the height differential, and $a \in \Omega$ is a constant.

Proof. From the Summary on page 120, we need

$$\phi^2 = 0 \quad \text{and} \quad |\phi|^2 \neq 0 \text{ and be finite.}$$

Comparing (2.13) and (2.12), we have that

$$\varphi_1 \, dz = \frac{1}{2}\left(\frac{1}{G} - G \right) dh, \qquad \varphi_2 \, dz = \frac{i}{2}\left(\frac{1}{G} + G \right) dh, \qquad \varphi_3 \, dz = dh.$$

In Exercise 2.87 you will show that

$$\boldsymbol{\phi}^2 = 0 \qquad \text{and} \qquad |\boldsymbol{\phi}|^2 \neq 0.$$

$\square$

Exercise 2.87. Prove that $\boldsymbol{\phi}^2 = 0$ and $|\boldsymbol{\mathfrak{E}}|^2 \neq 0$ in the proof of Theorem 2.86. ***Try it out!***

We have

$$G = \frac{\varphi_1 + i\varphi_2}{-\varphi_3} \qquad \text{and} \qquad dh = \varphi_3 \, dz.$$

One advantage of using the Weierstrass representation with the Gauss map and height differential is that the complex analytic properties of G and dh are related to the geometry of a minimal surface. We will discuss this later, but first we will look at some examples. Here is a list of the Weierstrass data for some common minimal surfaces.

(a) Enneper's surface $\qquad\qquad\qquad G(z) = z \qquad dh = z \, dz \qquad$ on $\mathbb{C}$.

(b) The catenoid $\qquad\qquad\qquad\qquad G(z) = z \qquad dh = \frac{1}{z} \, dz \qquad$ on $\mathbb{C} \setminus \{0\}$.

(c) The helicoid $\qquad\qquad\qquad\qquad G(z) = z \qquad dh = \frac{i}{z} \, dz \qquad$ on $\mathbb{C} \setminus \{0\}$.

(d) Scherk's doubly periodic surface $\qquad G(z) = z \qquad dh = \frac{z}{z^4-1} \, dz \qquad$ on $\mathbb{D}$.

(e) Scherk's singly periodic surface $\qquad G(z) = z \qquad dh = \frac{iz}{z^4-1} \, dz \qquad$ on $\mathbb{D}$.

(f) Polynomial Enneper $\qquad\qquad\quad G(z) = p(z) \quad dh = p(z) \, dz \qquad$ on $\mathbb{C}$.

(g) Wavy plane $\qquad\qquad\qquad\qquad G(z) = z \qquad dh = dz \qquad$ on $\mathbb{C} \setminus \{0\}$.

Example 2.88. For $G(z) = z^k$ and $dh = z^k \, dz$, where $k = 1, 2, \ldots$, we get

$$\mathbf{x} = \mathrm{Re} \int_0^z \left(\frac{1}{2}\left(\frac{1}{z^k} - z^k \right), \frac{i}{2}\left(\frac{1}{z^k} + z^k \right), 1 \right) z^k \, dh$$

$$= \left(\mathrm{Re} \, \frac{1}{2}\left\{ z - \frac{1}{2k+1} z^{2k+1} \right\}, \mathrm{Re} \, \frac{1}{2}\left\{ -i\left(z + \frac{1}{2k+1} z^{2k+1} \right) \right\}, \mathrm{Re} \left\{ \frac{z^{k+1}}{k+1} \right\} \right).$$

This is Enneper's surface with $2k + 2$ leaves described in Exploration 2.71.

Exercise 2.89. The Weierstrass data for the catenoid and the helicoid, which are conjugate surfaces (see Definition 2.50), have the same Gauss map, G, while the height diffferentials dh differ by a multiple of i. Prove that this is true for any conjugate surfaces. ***Try it out!***

Exercise 2.90. Let $G(z) = z^4$ and $dh = z^2 \, dz$.

(a) Using (2.13), compute the parametrization.

(b) This minimal surface has a planar end (i.e., looks like a plane) and an Enneper end. To graph the surface, use the entry under the **Pre-set functions** in the **W.E. (G, dh)** tab of *MinSurfTool*.

Try it out!

The catenoid and the surface in Exercise 2.90 are examples of minimal surfaces with ends. Loosely, an *end* of a minimal surface is a piece that goes on forever, or, more precisely, leaves all compact subsets of the minimal surface. From Theorem 2.76, all complete minimal surfaces in $\mathbb{R}^3$ are not compact, and hence they must possess at least one end.

Exercise 2.91. Determine the number of ends of

(a) the catenoid   (b) the plane   (c) the helicoid   (d) Enneper's surface.
Try it out!

Ends occur in a deleted neighborhood (i.e., a disk with the center removed) centered at a singularity. Three common types of ends for minimal surfaces are Enneper ends, catenoidal ends, and flat or planar ends. In discussing ends, we represent ds, the metric (i.e., a way to measure distance) on a minimal surface, in terms of G and dh. Using $ds^2 = |\phi|^2$ and (2.13), we derive

$$ds = \frac{1}{\sqrt{2}}\left(|G| + \frac{1}{|G|}\right)|dh|. \tag{2.14}$$

An Enneper end has $ds \sim |z^k| \cdot |dz|$, so the metric grows in the same manner as a polynomial as we approach the end. A catenoidal end and a planar end have $ds \sim |dz|$, so the metric becomes Euclidean. A catenoidal end differs from a planar end because the residue of dh is logarithmic.

Exercise 2.92. Prove (2.14). *Try it out!*

Example 2.93. We know that the catenoid has two catenoidal ends. Let's show how we could prove. Using the Weierstrass data, $G(z) = z$ and $dh = \frac{1}{z} dz$, for the catenoid, we have the parametrization

$$\mathbf{x}(z) = \left(\frac{1}{2}\operatorname{Re}\left(-\frac{1}{z} - z\right), \frac{i}{2}\operatorname{Re}\left(-\frac{1}{z} + z\right), \operatorname{Re}\left(\log z\right)\right).$$

There is a singularity or pole of order 1 at 0 and a pole of order 1 at ∞. To see that there is a singularity at ∞, we replace z with $\frac{1}{w}$ and look at the limit as w goes to 0. Thus, the catenoid will have two ends (one at 0 and one at ∞). To determine their types, we look at ds at these points. Because

$$ds = \frac{1}{\sqrt{2}}\left(|z| + \frac{1}{|z|}\right)\frac{1}{|z|}|dz|$$

as $z \to \infty$, $ds \sim |dz|$ and because x_3 is logarithmic, we have a catenoidal end.

At $z = 0$, substituting 0 does not work, so we let $w = \frac{1}{z}$ and consider $w \to \infty$. We have

$$dh = \frac{1}{z} dz = w\left(\frac{1}{w}\right)' = w\left(-\frac{1}{w^2} dw\right) = -\frac{1}{w} dw.$$

Therefore

$$ds = \frac{1}{\sqrt{2}}\left(|w| + \frac{1}{|w|}\right)\frac{1}{|w|}|dw|.$$

As $w \to \infty$, $ds \sim |dw|$ and because x_3 is logarithmic, we have a catenoidal end.

Example 2.94. The surface in Exercise 2.90 has $G(z) = z^4$ and $dh = z^2 dz$, and the corresponding parametrization is

$$\mathbf{x}(z) = \left(\frac{1}{2}\operatorname{Re}\left(\frac{1}{7}z^7 - \frac{1}{z}\right), \frac{i}{2}\operatorname{Re}\left(\frac{1}{7}z^7 + \frac{1}{z}\right), \operatorname{Re}\left(\frac{1}{3}z^3\right)\right).$$

There are singularities at 0 and at ∞ and

$$ds = \frac{1}{\sqrt{2}}\left(|z|^4 + \frac{1}{|z|^4}\right)|z|^2|dz|.$$

As $z \to \infty$, $ds \sim |z|^6\,|dz|$ and so we have an Enneper end.

At $z = 0$, we let $w = \frac{1}{z}$ and consider $w \to \infty$. Then $G(w) = w^4$ and $dh = -\frac{1}{w^4}\,dw$. Hence,

$$ds = \left(|w|^4 + \frac{1}{|w|^4}\right)\frac{1}{|w|^4}|dw|.$$

As $w \to \infty$, $ds \sim |dw|$, but because there is no logarithmic term, we have a planar end.

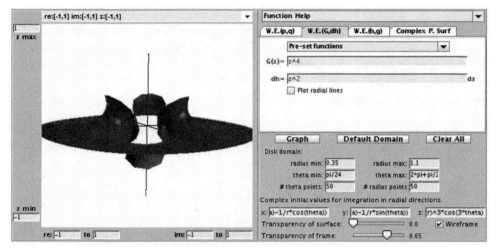

FIGURE 2.47. A minimal surface with a planar end and an Enneper end.

Exercise 2.95. Let $G(z) = \frac{z^2+3}{z^2-1}$ and $dh = \frac{z^2+3}{z^2-1}\,dz$. Show that the minimal surface has one planar end and two catenoidal ends.

Try it out!

The Gauss map and height differential also tell us about two important types of curves on a minimal surface, the asymptotic lines and the curvature lines. To understand what these lines are, we will review normal curvature and principal directions discussed in Section 2. Let p be a point on a curve on a minimal surface M. The tangent vector $\mathbf{w}$ and normal vector $\mathbf{n}$ at p form a plane that intersects the surface in a curve, say α (see Figure 2.20). The normal curvature in the direction $\mathbf{w}$ is $\alpha'' \cdot \mathbf{n}$ and measures how much the surface bends toward $\mathbf{n}$ as you move in the direction of $\mathbf{w}$ at point p. An *asymptotic line* is a curve that is tangent to a direction in which the normal curvature is zero.

As we rotate the plane through the normal $\mathbf{n}$, we get a set of curves on the surface each of which has a value for its curvature. The directions in which the normal curvature attains its absolute maximum and absolute minimum values are known as the principal directions. *Curvature lines* are curves that are always tangent to a principal direction.

The relationship between these lines and the Weierstrass data is

$$\text{A curve } z(t) \text{ is an asymptotic line} \iff \frac{dG}{G}(z) \cdot dh(z) \in i\mathbb{R}$$

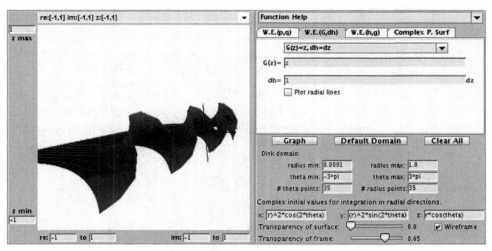

FIGURE 2.48. The wavy plane.

$$A \text{ curve } z(t) \text{ is a curvature line} \iff \frac{dG}{G}(z) \cdot dh(z) \in \mathbb{R}.$$

Often these equations are difficult to solve, but there are times when we can find the asymptotic lines and curvature lines at specific points.

Example 2.96. Let $G(z) = z$ and $dh(z) = dz$. This is a parametrization of the wavy plane. Computing the Weierstrass representation, we get the parametrization

$$(x_1(z), x_2(z), x_3(z)) = \left(\mathrm{Re}\left\{ \frac{1}{2}\log(z) - \frac{1}{4}z^2 \right\}, \mathrm{Re}\left\{ \frac{i}{2}\log(z) + \frac{i}{4}z^2 \right\}, \mathrm{Re}\left\{ z \right\} \right).$$

We can plot an image of it using the corresponding function under **Pre-set functions** in the **W.E. (G,dh)** tab in *MinSurfTool*.

For the wavy plane

$$\frac{dG}{G}(z) \cdot dh(z) = \frac{dz}{z} \cdot dz.$$

If we let $z = e^{i\theta}$ (since we let the maximum radius be 1), then $dz = i e^{i\theta} d\theta$ and

$$\frac{dG}{G}(z) \cdot dh(z) \bigg|_{z=e^{i\theta}} = -e^{i\theta}(d\theta)^2.$$

So, we get that for $k \in \mathbb{Z}$ asymptotic lines occur at $z = re^\theta$ as r varies with $\theta = \frac{\pi}{2} + k\pi$ and curvature lines occur at $z = re^\theta$ as r varies with $\theta = k\pi$.

If we use *MinSurfTool* to plot the wavy plane with the minimum and maximum θ values of $-\frac{\pi}{2}$ and $\frac{\pi}{2}$, we see that the asymptotic lines lie in the x_1x_2-plane. If we plot the minimum and maximum θ values of 0 and π, we see that the curvature lines are reflection lines through which the wavy plane can be reflected as if through a mirror to get a smooth continuation of the minimal surface.

Exercise 2.97. Using $G(z) = z$ and $dh = zdz$ for Enneper's surface, show that asymptotic lines occur at $z = re^\theta$ as r varies with $\theta = \frac{\pi}{4} + \frac{k\pi}{2}$, where $k \in \mathbb{Z}$, and curvature lines occur at $z = re^\theta$ as r varies with $\theta = \frac{k\pi}{2}$, where $k \in \mathbb{Z}$. Use *MinSurfTool* to plot the lines on Enneper's surface. ***Try it out!***

Exercise 2.98. Prove that for conjugate surfaces, the asymptotic lines (curvature lines) of one surface are the curvature lines (asymptotic lines) of the other surface. ***Try it out!***

As mentioned, an advantage of using the Gauss map and height differential is that the complex analytic properties of G and dh are related to the geometry of a minimal surface. So, let's look at the complex analytic properties of G and dh. From the Summary on page 120, we need $|\phi|^2$ to be finite and nonzero. Because of (2.14), we have the following proposition.

Proposition 2.99. *At a nonsingular point, G has a zero or pole of order n if and only if dh has a zero of order n.*

The integrals in the Weierstrass representation in (2.13) might depend upon the path of integration if the domain of G and dh is not simply connected. If the representation does not depend on the path of integration, then all closed paths γ in the domain satisfy the three conditions

$$\text{Re}\left(\frac{1}{2}\int_\gamma \left(\frac{1}{G} - G\right) dh\right) = 0$$

$$\text{Re}\left(\frac{i}{2}\int_\gamma \left(\frac{1}{G} + G\right) dh\right) = 0$$

$$\text{Re}\int_\gamma dh = 0.$$

If a curve has the property that one of these quantities is a nonzero constant, then the surface will have periodic behavior. Example 2.101 illustrates this.

The three equations can be reduced to the two period conditions

$$(i) \int_\gamma G\, dh = \overline{\int_\gamma \frac{1}{G}\, dh} \quad \text{(horizontal period condition)},$$

$$(ii)\ \text{Re}\int_\gamma dh = 0 \quad \text{(vertical period condition)}$$

(2.15)

for closed paths γ in the domain.

Exercise 2.100. Show that the conditions

$$\text{Re}\left(\frac{1}{2}\int_\gamma \left(\frac{1}{G} - G\right) dh\right) = 0 \text{ and } \text{Re}\left(\frac{i}{2}\int_\gamma \left(\frac{1}{G} + G\right) dh\right) = 0$$

are equivalent to

$$\int_\gamma G\, dh = \overline{\int_\gamma \frac{1}{G}\, dh}.$$

Try it out!

The period conditions are useful in determining horizontal periods (e.g., Scherk's doubly periodic surface), vertical periods (e.g., Scherk's singly periodic surface), and in determining constant values in G and dh.

Example 2.101. Scherk's doubly periodic surface has the Weierstrass data $G(z) = z$ and $dh(z) = \frac{z}{z^4-1}dz$. The horizontal period condition is $\int_\gamma G \, dh = \overline{\int_\gamma \frac{1}{G} \, dh}$ for all closed paths γ in the domain. Both integrands are meromorphic with poles of order 1 at $\pm 1, \pm i$. So, the only paths that concern us are ones that enclose one, two, or three of these poles. To calculate the integrals we use the residue theorem that states if γ is a simple closed positively-oriented contour and f is analytic inside and on γ except at the points $z_1, \ldots, z_n$ inside γ, then

$$\int_\gamma f(z) \, dz = 2\pi i \sum_{j=1}^n \operatorname{Res}(f, z_j).$$

For poles of order 1,

$$\operatorname{Res}(f, z_j) = \lim_{z \to z_j} (z - z_j) f(z).$$

Thus, for $\int_\gamma G \, dh$, we have

$$\operatorname{Res}(G \, dh, z_j) = \lim_{z \to z_j} \frac{z^3 - z_j z^2}{z^4 - 1} = \lim_{z \to z_j} \frac{3z^2 - 2z_j z}{4z^3} = \lim_{z \to z_j} \frac{3z^3 - 2z_j z^2}{4z^4} = \frac{z_j^3}{4}.$$

In particular,

$$\operatorname{Res}(G \, dh, 1) = \frac{1}{4} \qquad\qquad \operatorname{Res}(G \, dh, i) = \frac{-i}{4}$$

$$\operatorname{Res}(G \, dh, -1) = \frac{-1}{4} \qquad\qquad \operatorname{Res}(G \, dh, -i) = \frac{i}{4}.$$

Similarly, we can compute that

$$\operatorname{Res}\left(\frac{1}{G} \, dh, 1\right) = \frac{1}{4} \qquad\qquad \operatorname{Res}\left(\frac{1}{G} \, dh, i\right) = \frac{i}{4}$$

$$\operatorname{Res}\left(\frac{1}{G} \, dh, -1\right) = \frac{-1}{4} \qquad\qquad \operatorname{Res}\left(\frac{1}{G} \, dh, -i\right) = \frac{-i}{4}.$$

Now, if the path γ_1 contains only the pole at $z_1 = 1$, then the horizontal period conditions result in

$$\int_{\gamma_1} G \, dh = 2\pi i \operatorname{Res}(G \, dh, 1) = \frac{i\pi}{2}, \qquad \overline{\int_{\gamma_1} \frac{1}{G} \, dh} = \overline{2\pi i \operatorname{Res}\left(\frac{1}{G} \, dh, 1\right)} = \frac{-i\pi}{2}.$$

The integrals should be equal, which occurs if the minimal surface is periodic in the imaginary direction with period π. Likewise, if we take a path γ_2 that contains only the pole $z_2 = i$, then we get

$$\int_{\gamma_2} G \, dh = \frac{\pi}{2}, \qquad \overline{\int_{\gamma_2} \frac{1}{G} \, dh} = \frac{-\pi}{2},$$

and the minimal surface is periodic in the real direction with period π. All other paths γ are covered by these two cases. If we look at the vertical period condition, we get that the condition is automatically true for all paths γ and so the minimal surface is not periodic in the vertical direction. This matches the image of Scherk's doubly periodic surface from *MinSurfTool*.

Exercise 2.102. Show that the period conditions given in (2.15) result in Scherk's singly periodic surface being periodic in the vertical direction. ***Try it out!***

Let's look at an example of how this can help us use the geometry of a minimal surface to determine G and dh.

Example 2.103. From the Weierstrass data on page 132, we know that $G(z) = z$ and $dh = z\, dz$ for Enneper's surface. We want to show how it can be determined by using the geometric shape of the surface. Let's determine a plausible candidate for G by making a guess based on the value of G at a few points. From Example 2.82, we know that $G(0) = 0$, $G(r) = r$ for $0 \le r \le 1$, and $G(e^{i\theta}) = e^{i\theta}$ for $0 \le \theta \le \frac{\pi}{2}$. Therefore, it seems plausible to let $G(z) = z$. With this G, let's determine dh. Because ds must be finite in (2.14) and Enneper's surface has no ends in $\mathbb{C}$, dh cannot have poles in $\mathbb{C}$. However, from the sentence before Exercise 2.91, we know that Enneper's surface must have at least one end. It corresponds to the point at infinity, $z = \infty$, and so dh has a pole at ∞. Thus, $dh = \rho z^n\, dz$ for some $n \in \mathbb{N}$ and $\rho \in \mathbb{C}$. Since $G(z) = z$ has a zero of order 1 at 0, by Proposition 2.99 dh must have a zero of order 1 at 0 and no other zeros. Thus, $dh = \rho z\, dz$. For simplicity, we let $\rho = 1$. The period conditions in (2.15) hold because there are no poles in $\mathbb{C}$, and so every integral along any closed path γ will equal 0. Hence, the Weierstrass data

$$G(z) = z, \qquad dh = z\, dz$$

generate a minimal surface.

Example 2.104. The singly periodic Scherk surface, M_S, with six leaves that go off to infinity (see Figure 2.45) will have six poles. Because of symmetry, we choose them to be at the sixth roots of unity (i.e., $e^{i\pi k/3}, (k = 0, ..., 5)$). This means that dh will have $z^6 - 1$ in its denominator. However, we will need to determine G first to know what should be in the numerator of dh. From the results in Example 2.83, it seems reasonable that $G(z) = z^2$. Since $G(z) = z^2$ has a zero of order 2 at 0, by Proposition 2.99 dh must also have a zero of order 2 at 0 and no other zeros. Thus, we so far have

$$G(z) = z^2, \qquad dh = \rho \frac{z^2}{z^6 - 1}\, dz,$$

where $\rho \in \mathbb{C}$. To determine values of ρ, consider the period conditions in (2.15). There are poles of order 1 at $e^{ik\pi/3}, k = 0, \ldots, 5$. We compute that

$$\operatorname{Res}(G\, dh, z_j) = \frac{\rho z_j^5}{6}, \qquad \operatorname{Res}\left(\frac{1}{G}\, dh, z_j\right) = \frac{\rho z_j}{6}.$$

Hence, if γ contains the pole z_j, then the horizontal period condition requires

$$\int_\gamma G\, dh = 2\pi i \operatorname{Res}(G\, dh, z_j) = \frac{\rho \pi i z_j^5}{3}$$

and

$$\overline{\int_\gamma \frac{1}{G}\, dh} = \overline{2\pi i \operatorname{Res}\left(\frac{1}{G}\, dh, z_j\right)} = \frac{-\overline{\rho} \pi i \overline{z}_j}{3}$$

to be equal (since there is no periodicity of M_S in the horizontal direction).The integrals will be equal for the poles $z_j^5 = \overline{z}_j$ if $\rho = -\overline{\rho}$. That is, ρ is purely imaginary. Without loss of generality, we let $\rho = i$ and check that the vertical period condition holds. Hence, we have that the Weierstrass data for singly periodic Scherk surface with six leaves is

$$G(z) = z^2, \qquad dh = \frac{iz^2}{z^6 - 1} \, dz.$$

Exercise 2.105. Let M be Enneper's surface with eight leaves. Using the approach of Example 2.103 determine G and dh. ***Try it out!***

Exercise 2.106. Let M be the 3-noid with ends symmetrically placed so that if the surface is rotated by $\frac{2\pi}{3}$ it is unchanged. Determine G and dh. ***Try it out!***

Small Project 2.107. Let M be a minimal surface that has six symmetrically-placed catenoidal ends with four ends along the side (like a 4-noid), one end on the top, and one end on the bottom. So, M will look the same if it is rotated horizontally by $\frac{\pi}{2}$ or if it is rotated vertically by $\frac{\pi}{2}$. Determine G and dh. **Optional**

Small Project 2.108. For Scherk's singly periodic surface the four ends are symmetrically placed so that if the surface is rotated by $\frac{\pi}{2}$ you will get the same image. This is because the denominator of dh is $z^4 - 1$, which has zeros equally spaced on the unit circle. It is possible to create a variation of Scherk's singly periodic surface that has four ends with rotational symmetry of π. That is, if the ends are labelled $E_1, \ldots, E_4$, then E_2 will be closer to E_1 than to E_3 (and likewise, E_4 will be closer to E_3 than to E_1) and if the surface is rotated by π you will get the same image. Determine G and dh. **Optional**

Large Project 2.109. Describe and classify the minimal surfaces where one of the coordinates of the parametrization is fixed while the other two vary. For example, if x_3 is fixed to a specific function, what are the possible coordinate functions for x_1 and x_2? Generalize this approach. **Optional**

Large Project 2.110. Describe and classify the minimal surfaces with $G(z) = z^m$ and $dh = z^n \, dz$, for all $n, m \in \mathbb{N}$ (see Example 2.88 and Example 2.94). There are several cases to consider. Determine how to separate m, n into the cases remembering to discuss types of surface, types of ends, lines of symmetry, etc. **Optional**

2.6 Minimal surfaces and harmonic univalent mappings

In the Summary on page 120, we learned that each coordinate function of the parametrization $\mathbf{x}$ of a minimal surface had the form $x_k = \text{Re} \int \varphi_k \, dz$ with φ_k analytic. Since the real part of an analytic function is a harmonic function, we see that each x_k is harmonic. From the Summary, we have that $(\varphi_1)^2 + (\varphi_2)^2 + (\varphi_3)^2 = 0$. This means that if we know φ_1 and φ_2, we can determine φ_3. So, a way to get a Weierstrass representation for minimal surfaces is to use two harmonic functions x_1 and x_2. Thus, we can investigate minimal surfaces by studying harmonic mappings in the complex plane. They are known as *planar harmonic mappings* and have been studied independently of minimal surfaces.

In this section we will develop another Weierstrass representation. We will use planar harmonic mappings instead of p and q as in Section 2.4 or G and dh as in Section

2.5. What benefits do we obtain from this new approach? First, we can use results about planar harmonic mappings to prove results about minimal surfaces. Since these two areas of mathematics developed independently, this approach can be fruitful. Second, in the study of planar harmonic mappings, much of the work deals with one-to-one functions. Geometrically, this means that the image $f(G)$ will not overlap or intersect itself. When one-to-one functions are lifted from the complex plane into $\mathbb{R}^3$, the resulting surfaces are minimal graphs and hence have no self-intersections. This can be useful in establishing the embeddedness of the minimal surfaces. For example, the one-to-one planar harmonic function given by

$$f(z) = h(z) + \overline{g(z)} = \text{Re}\left[\frac{i}{2}\log\left(\frac{i+z}{i-z}\right)\right] + i\,\text{Im}\left[\frac{1}{2}\log\left(\frac{1+z}{1-z}\right)\right]$$

maps the unit disk onto a square that is the projection of Scherk's doubly periodic surface onto the plane. So, we can lift f from $\mathbb{C}$ into $\mathbb{R}^3$ to get Scherk's doubly periodic surface. Planar harmonic mappings that are one-to-one are also known as harmonic univalent mappings, which can be studied on their own as is done in Chapter 4.

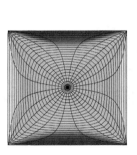

FIGURE 2.49. The image of $f(\mathbb{D})$ and Scherk's doubly periodic surface.

Exercise 2.111.

(a) Although minimal graphs are embedded, the converse is not true. Give an example of an embedded minimal surface that is not a minimal graph.

(b) Suppose you have a nonunivalent harmonic mapping. Why could it not be the projection of a minimal graph?

Try it out!

A planar harmonic mapping is a function $f = u(x, y) + iv(x, y)$ where u and v are real harmonic functions. Planar harmonic mappings are more general than analytic functions, because we do not require u and v to be harmonic conjugates. The next theorem allows us to relate planar harmonic mappings to analytic functions. We will assume that the domain of f is the unit disk, $\mathbb{D}$.

Theorem 2.112. *Let a function $f = u + iv$, where u and v are real harmonic functions. If D is a simply-connected domain and $f : D \to \mathbb{C}$, then there exist analytic functions h and g such that $f = h + \overline{g}$.*

Exercise 2.113.

(a) Show that $f(x, y) = u(x, y) + iv(x, y) = \left(\frac{1}{3}x^3 - xy^2 + x\right) + i\left(\frac{1}{3}y^3 - x^2y + y\right)$ is complex-valued harmonic by showing that u and v are real harmonic functions.

(b) Using $x = \frac{1}{2}(z + \overline{z})$ and $y = \frac{1}{2i}(z - \overline{z})$, write $f(x, y) = \left(\frac{1}{3}x^3 - xy^2 + x\right) + i\left(\frac{1}{3}y^3 - x^2y + y\right)$ in terms of z and $\overline{z}$.

(c) Determine the analytic functions h and g such that $f = h + \overline{g}$.

Try it out!

Example 2.114. In the previous exercise, we saw that the planar harmonic map $f : \mathbb{D} \to \mathbb{C}$ defined by

$$f(x, y) = u(x, y) + iv(x, y) = \left(\frac{1}{3}x^3 - xy^2 + x\right) + i\left(\frac{1}{3}y^3 - x^2y + y\right)$$

can be written as

$$f(z) = h(z) + \overline{g}(z) = z + \frac{1}{3}\overline{z}^3.$$

The image of $\mathbb{D}$ under f is a hypocycloid with four cusps. This can be seen by considering $f(e^{i\theta}) = u(\theta) + iv(\theta)$ and comparing the component functions, $u(\theta)$ and $v(\theta)$, to the parametrized equation for a hypocycloid with four cusps. To help us visualize the image, we can use the applet *ComplexTool*. To graph the image of $\mathbb{D}$ under the harmonic function $f(z) = z + \frac{1}{3}\overline{z}^3$, enter f in *ComplexTool* in the form $z + 1/3$ **conj** $(z \wedge 3)$ (see Figure 2.50). We will later show that f is related to a minimal graph.

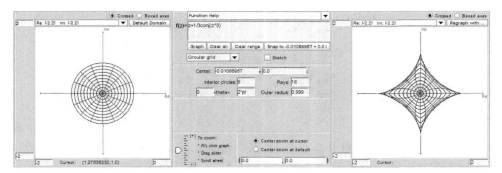

FIGURE 2.50. Image of $\mathbb{D}$ under the harmonic function $f(z) = z + \frac{1}{3}\overline{z}^3$.

The harmonic function $f(z) = h(z) + \overline{g}(z)$ can also be written as

$$f(z) = \text{Re}\{h(z) + g(z)\} + i\,\text{Im}\{h(z) - g(z)\} \tag{2.16}$$

because

$$\text{Re}\{h + g\} = \frac{1}{2}[(h + g) + \overline{(h + g)}] \quad \text{and} \quad \text{Im}\{h - g\} = \frac{1}{2i}[(h - g) - \overline{(h - g)}].$$

Hence, in the previous example $f(z) = z + \frac{1}{3}\bar{z}^3$ can be written as $f(z) = \text{Re}\left\{z + \frac{1}{3}z^3\right\} + i\,\text{Im}\left\{z - \frac{1}{3}z^3\right\}$.

We are interested in harmonic functions that are one-to-one or univalent, because this is a necessary condition to lift the harmonic mapping to a minimal graph. Theorem 2.119 establishes univalency, but it requires background material.

Definition 2.115. The *dilatation* of $f = h + \bar{g}$ is $\omega(z) = g'(z)/h'(z)$.

Theorem 2.116 (Lewy [16]). *$f = h + \bar{g}$ is locally univalent and orientation-preserving if and only if $|g'(z)/h'(z)| < 1$, for all $z \in \mathbb{D}$.*

See Chapter 4 for more background on the dilatation and on orientation-preserving functions.

Exercise 2.117. Show that if $z \in \mathbb{D}$, then $|\omega(z)| < 1$ for

(a) $\omega_1(z) = e^{i\theta}z$, where $\theta \in \mathbb{R}$

(b) $\omega_2(z) = z^n$, where $n = 1, 2, 3, \ldots$

(c) $\omega_3(z) = \dfrac{z + a}{1 + \bar{a}z}$, where $|a| < 1$

(d) $\omega_4(z)$, the composition of any of ω_1, ω_2, and ω_3.

Try it out!

Creating nontrivial examples of harmonic univalent mappings that lift to minimal graphs is not easy. One way to do this is to use the shearing technique of Clunie and Sheil-Small. Before we proceed, we need to discuss a certain type of domain.

Definition 2.118. A domain Ω is *convex in the direction of the real axis* (or *convex in the horizontal direction*, CHD) if every line parallel to the real axis has a connected intersection with Ω.

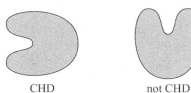

CHD   not CHD

Theorem 2.119 (Clunie and Sheil-Small). *A harmonic function $f = h + \bar{g}$ locally univalent in $\mathbb{D}$ is a univalent mapping of $\mathbb{D}$ onto a CHD domain $\iff$ $h - g$ is an analytic univalent mapping of $\mathbb{D}$ onto a CHD domain.*

Remark 2.120. This is known as the shear method, or shearing a function. In our situation, suppose $F = h - g$ is an analytic univalent function convex in the real direction. Then the corresponding harmonic shear is

$$f = h + \bar{g} = h - g + g + \bar{g} = h - g + 2\,\text{Re}\{g\}.$$

The harmonic shear differs from the analytic function by the addition of a real function. Geometrically, you can think of this as taking F, the original analytic univalent function convex in the real direction, and cutting it into thin horizontal slices that are then translated or scaled in a continuous way to form the corresponding harmonic function, f. This is why the method is called shearing. Since F is univalent and convex in the real direction and we are only adding a continuous real function to it, the univalency is preserved.

Example 2.121. Let

$$h(z) - g(z) = \frac{1}{2} \log \left(\frac{1+z}{1-z} \right), \tag{2.17}$$

which is an analytic function that maps $\mathbb{D}$ onto a horizontal strip convex in the direction of the real axis (see Figure 2.51).

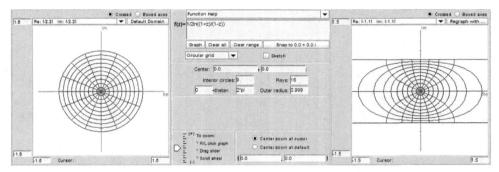

FIGURE 2.51. Image of $\mathbb{D}$ under the analytic function $\frac{1}{2} \log \left(\frac{1+z}{1-z} \right)$.

Let

$$\omega(z) = g'(z)/h'(z) = -z^2.$$

Applying the shearing method from Theorem 2.119 with the substitution $g'(z) = -z^2 h'(z)$, we have

$$h'(z) - g'(z) = \frac{1}{1-z^2} \Rightarrow h'(z) + z^2 h'(z) = \frac{1}{1-z^2}$$

$$\Rightarrow h'(z) = \frac{1}{1-z^4} = \frac{1}{4} \left[\frac{1}{1+z} + \frac{1}{1-z} + \frac{i}{i+z} + \frac{i}{i-z} \right].$$

Integrating $h'(z)$ and normalizing so that $h(0) = 0$ yields

$$h(z) = \frac{1}{4} \log \left(\frac{1+z}{1-z} \right) + \frac{i}{4} \log \left(\frac{i+z}{i-z} \right). \tag{2.18}$$

We can use the same method to solve for normalized $g(z)$, where $g(0) = 0$. We can also find $g(z)$ by using (4.11) and (4.12). Either way, we get

$$g(z) = -\frac{1}{4} \log \left(\frac{1+z}{1-z} \right) + \frac{i}{4} \log \left(\frac{i+z}{i-z} \right).$$

So

$$f(z) = h(z) + \overline{g(z)} = \text{Re} \left[\frac{i}{2} \log \left(\frac{i+z}{i-z} \right) \right] + i \, \text{Im} \left[\frac{1}{2} \log \left(\frac{1+z}{1-z} \right) \right].$$

What is $f(\mathbb{D})$? We have

$$f(z) = \left[-\frac{1}{2} \arg\left(\frac{i+z}{i-z}\right) \right] + i\left[\frac{1}{2} \arg\left(\frac{1+z}{1-z}\right) \right] = u + iv.$$

Let $z = e^{i\theta} \in \partial\mathbb{D}$. Then

$$\frac{i+z}{i-z} = \frac{i+e^{i\theta}}{i-e^{i\theta}} \frac{-i-e^{-i\theta}}{-i-e^{-i\theta}} = \frac{1 - i(e^{i\theta}+e^{-i\theta}) - 1}{1 + i(e^{i\theta}-e^{-i\theta}) + 1} = -i\frac{\cos\theta}{1-\sin\theta}.$$

Thus,

$$u = -\frac{1}{2}\arg\left(\frac{i+z}{i-z}\right)\Bigg|_{z=e^{i\theta}} = \begin{cases} \frac{\pi}{4} & \text{if } \cos\theta > 0, \\ -\frac{\pi}{4} & \text{if } \cos\theta < 0. \end{cases}$$

Likewise, we can show that

$$v = \begin{cases} \frac{\pi}{4} & \text{if } \sin\theta > 0, \\ -\frac{\pi}{4} & \text{if } \sin\theta < 0. \end{cases}$$

Therefore, we have that $z = e^{i\theta} \in \partial\mathbb{D}$ is mapped to

$$u + iv = \begin{cases} z_1 = \frac{\pi}{2\sqrt{2}} e^{i\frac{\pi}{4}} = \frac{\pi}{4} + i\frac{\pi}{4} & \text{if } \theta \in (0, \frac{\pi}{2}) \\ z_2 = \frac{\pi}{2\sqrt{2}} e^{i\frac{3\pi}{4}} = -\frac{\pi}{4} + i\frac{\pi}{4} & \text{if } \theta \in (\frac{\pi}{2}, \pi) \\ z_3 = \frac{\pi}{2\sqrt{2}} e^{i\frac{5\pi}{4}} = -\frac{\pi}{4} - i\frac{\pi}{4} & \text{if } \theta \in (\pi, \frac{3\pi}{2}) \\ z_4 = \frac{\pi}{2\sqrt{2}} e^{i\frac{7\pi}{4}} = \frac{\pi}{4} - i\frac{\pi}{4} & \text{if } \theta \in (\frac{3\pi}{2}, 2\pi). \end{cases}$$

Thus, the harmonic function maps $\mathbb{D}$ onto the interior of the region bounded by a square with vertices at z_1, z_2, z_3, and z_4.

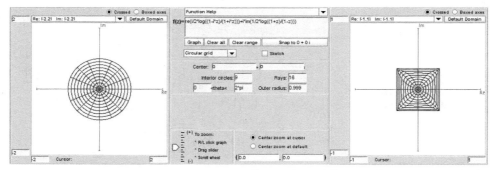

FIGURE 2.52. Image of $\mathbb{D}$ under $f(z) = \text{Re}\left[\frac{i}{2}\log\left(\frac{i+z}{i-z}\right)\right] + i\,\text{Im}\left[\frac{1}{2}\log\left(\frac{1+z}{1-z}\right)\right]$.

Exercise 2.122. Verify that shearing $h(z) - g(z) = z - \frac{1}{3}z^3$ with $\omega(z) = z^2$ yields $f(z) = z + \frac{1}{3}\bar{z}^3$. **_Try it out!_**

To find the minimal graph that is associated with specific types of harmonic univalent mappings we need to develop the appropriate Weierstrass representation as outlined in

(2.12). It must satisfy $\phi^2 = 0$ and $|\phi|^2 \neq 0$, and we want it to use planar harmonic mappings. A natural choice is to take

$$x_1 = \text{Re}(h + g) = \text{Re} \int (h' + g') \, dz = \text{Re} \int \varphi_1 \, dz$$

$$x_2 = \text{Im}(h - g) = \text{Re} \int -i(h' - g') \, dz = \text{Re} \int \varphi_2 \, dz$$

$$x_3 = \text{Re} \int \varphi_3 \, dz$$

and then solve for φ_3.

Exercise 2.123. Derive $\varphi_3 = 2ih'\sqrt{g'/h'} = 2i\sqrt{g'h'}$. *Try it out!*

We need φ_3 to be analytic and so we require the dilatation $\omega = g'/h'$ to be a perfect square.

Theorem 2.124 (Weierstrass Representation (h, g)). *If $f = h + \overline{g}$ is a sense-preserving harmonic univalent mapping of $\mathbb{D}$ onto a domain $\Omega \subset \mathbb{C}$ with dilatation $\omega = q^2$ for a function q analytic in $\mathbb{D}$, then the isothermal parametrization*

$$\mathbf{x}(u, v) = (x_1, x_2, x_3)$$
$$= \left(\text{Re}\{h(z) + g(z)\}, \text{Im}\{h(z) - g(z)\}, 2\,\text{Im}\left\{ \int_0^z \sqrt{g'(\zeta)h'(\zeta)}d\zeta \right\} \right)$$

defines a minimal graph whose projection onto the complex plane is f. Conversely, if a minimal graph $\mathbf{x}(u, v) = \{(u, v, F(u, v)) : u + iv \in \Omega\}$ is parametrized by sense-preserving isothermal parameters $z = x + iy \in \mathbb{D}$, then the projection onto its base plane defines a harmonic univalent mapping $f(z) = u + iv = \text{Re}\{h(z) + g(z)\} + i\,\text{Im}\{h(z) - g(z)\}$ of $\mathbb{D}$ onto Ω whose dilatation is the square of an analytic function.

Summary: Let $f = h + \overline{g}$ defined on $\mathbb{D}$ be a harmonic univalent mapping such that the dilatation $\omega = g'/h'$ is the square of an analytic function and $|\omega(z)| < 1$ for all $z \in \mathbb{D}$. Then f lifts to a minimal graph using the Weierstrass formula given in Theorem 2.124.

Example 2.125. In Example 2.114 we had the harmonic univalent mapping

$$f(z) = z + \frac{1}{3}\overline{z}^3 = \text{Re}\left(z + \frac{1}{3}z^3 \right) + i\,\text{Im}\left(z - \frac{1}{3}z^3 \right),$$

so $h(z) = z$ and $g(z) = \frac{1}{3}z^3$. Also, $\omega(z) = z^2$, the square of an analytic function. Hence this harmonic mapping lifts to a minimal graph. We compute that

$$x_3 = 2\,\text{Im} \int_0^z \sqrt{h'(\zeta)g'(\zeta)} \, d\zeta = \text{Im}\left(z^2 \right).$$

This yields a parametrization of a surface that is the conjugate of Enneper's surface given in Example 2.67

$$\mathbf{x} = \left(\text{Re}\left\{ z + \frac{1}{3}z^3 \right\}, \text{Im}\left\{ z - \frac{1}{3}z^3 \right\}, \text{Im}\left\{ z^2 \right\} \right)$$

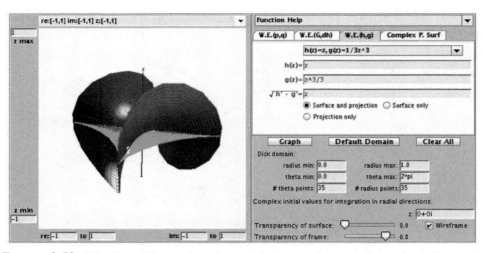

FIGURE 2.53. Side view of Enneper's surface and the image of the unit disk under the harmonic map.

and hence yields Enneper's surface. The projection of Enneper's surface onto the x_1x_2-plane is the image of $\mathbb{D}$ under the harmonic mapping f. While Enneper's surface is not a graph over $\mathbb{C}$, it is a graph over $\mathbb{D}$ as this result proves. You can see this by using *MinSurfTool* with the **W.E. (h,g)** tab. Enter the functions $h(z) = z$, $g(z) = \frac{1}{3}z^3$, and $\sqrt{h' \cdot g'} = z$. Make sure the **Surface and projection** button is clicked allowing you to see the minimal surface and its projection that is related to the harmonic mapping. For the **Disk domain** use the unit disk (i.e., radius min = 0; radius max = 1; theta min = 0; theta max = 2π). The minimal surface is colored purple while $f(\mathbb{D})$ is colored green (see Figure 2.53). As you move the image so that it is viewed from the top, the projection of the minimal surface matches the image of $f(\mathbb{D})$ (see Figure 2.54).

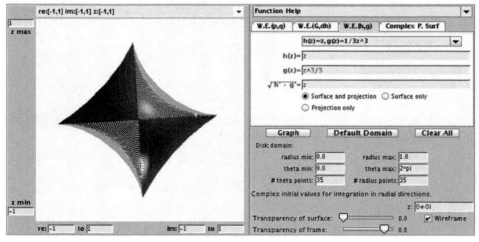

FIGURE 2.54. The projection of Enneper's surface is the image of the unit disk under the harmonic map.

Exploration 2.126. In the Weierstrass representation (h, g), we require that $\omega = g'/h'$ be the square of an analytic function. This is necessary because $\varphi_3 = ih'\sqrt{g'/h'}$, and if g'/h' were not the square of an analytic function, then there would be two branches of the square root. Geometrically, we can see that this is necessary. Use *MinSurfTool* with the **W.E. (h,g)** tab to graph the following images and describe why the geometry of the functions $f = h + \bar{g}$ in the left column lift to a minimal graph while those in the right column do not. To enter $\sqrt{z}$ into the applet type `sqrt(z)`.

(a) $z + \frac{1}{3}\bar{z}^3$
($\omega = z^2$ and $\sqrt{h' \cdot g'} = z$)

(b) $z + \frac{1}{2}\bar{z}^2$
($\omega = z$ and $\sqrt{h' \cdot g'} = \sqrt{z}$)

(c) $z - \frac{1}{5}\bar{z}^5$
($\omega = -z^4$ and $\sqrt{h' \cdot g'} = iz^2$)

(d) $z - \frac{1}{4}\bar{z}^4$.
($\omega = -z^5$ and $\sqrt{h' \cdot g'} = i\sqrt{z^5}$)

Example 2.127. Consider the harmonic univalent mapping from Example 2.121 given by

$$f(z) = h(z) + \overline{g(z)} = \text{Re}\left[\frac{i}{2}\log\left(\frac{i+z}{i-z}\right)\right] + i\,\text{Im}\left[\frac{1}{2}\log\left(\frac{1+z}{1-z}\right)\right].$$

Because $\omega(z) = -z^2$ is the square of an analytic function, we can lift it to a minimal graph. We compute that

$$x_3 = 2\,\text{Im}\int_0^z \frac{iz}{1-z^4}\,d\zeta = \frac{1}{2}\,\text{Im}\left\{i\log\left(\frac{1+z^2}{1-z^2}\right)\right\}.$$

This yields a parametrization of Scherk's doubly periodic minimal surface:

$$\mathbf{x} = \left(\text{Re}\left[\frac{i}{2}\log\left(\frac{i+z}{i-z}\right)\right], \text{Im}\left[\frac{1}{2}\log\left(\frac{1+z}{1-z}\right)\right], \frac{1}{2}\,\text{Im}\left[i\log\left(\frac{1+z^2}{1-z^2}\right)\right]\right).$$

We can use **Pre-set functions** from the **W.E. (h,g)** tab on *MinSurfTool* to plot the minimal graph and the image of the unit disk under the planar harmonic mapping, where

$$h(z) = \frac{1}{4}\log\left(\frac{1+z}{1-z}\right) + \frac{i}{4}\log\left(\frac{i+z}{i-z}\right),$$
$$g(z) = -\frac{1}{4}\log\left(\frac{1+z}{1-z}\right) + \frac{i}{4}\log\left(\frac{i+z}{i-z}\right).$$

Because of the singularities at $\pm 1, \pm i$, the radius max is set to 0.999. The projection of Scherk's doubly periodic surface onto the x_1x_2-plane is a square that is the image of $\mathbb{D}$ under the harmonic mapping f.

Exploration 2.128. Use the **Pre-set functions** from the **W.E. (h,g)** tab on *MinSurfTool* to plot the minimal graphs associated with the functions h and g for the planar harmonic mappings. Determine which minimal surfaces these are.

(a) $h(z) = z$, $g(z) = \frac{1}{2n+1}z^{2n+1}$ $(n = 1, 2, 3, \ldots)$

(b) $h(z) = z$, $g(z) = \frac{1}{z}$

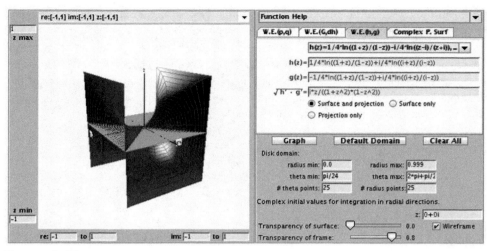

FIGURE 2.55. Side view of Scherk's doubly periodic surface and the image of the unit disk under the harmonic map.

(c) $h(z) = z, g(z) = -\frac{1}{z}$

(d) $h(z) = \frac{1}{5}z^5, g(z) = -\frac{1}{z}$

(e) $h(z) = \frac{1}{4}\log\left(\frac{i+z}{i-z}\right) - \frac{i}{4}\log\left(\frac{1+z}{1-z}\right), g(z) = \frac{1}{4}\log\left(\frac{i+z}{i-z}\right) + \frac{i}{4}\log\left(\frac{1+z}{1-z}\right).$

Try it out!

If we start with a harmonic univalent mapping that has a perfect square dilatation we can use Theorem 2.124 to find the parametrization of the corresponding minimal graph. In Example 2.125, we saw that the minimal graph corresponding to the harmonic univalent map $f(z) = z + \frac{1}{3}\bar{z}^3$ is Enneper's surface, and in Example 2.127 the minimal graph corresponding to the harmonic univalent map $f(z) = \text{Re}[\frac{i}{2}\log(\frac{i+z}{i-z})] + i \, \text{Im}[\frac{1}{2}\log(\frac{1+z}{1-z})]$ is Scherk's doubly periodic surface. There have been several research papers that have used Theorem 2.124 to create minimal graphs from harmonic univalent mappings (e.g., [8], [9], [10], [11], [17]). However, many of them have not identified the specific minimal graph created, because it is often not as easy to as it was in Examples 2.125 and 2.127.

Question: Given a harmonic univalent mapping we can use Theorem 2.124 to find the parametrization of a minimal graph. Can we determine which minimal graph this is?

The following examples show how we can determine the minimal graph when the parametrization is not a standard parametrization for a known minimal surface.

Example 2.129. The analytic function

$$F(z) = \frac{z}{1-z}$$

is an important and interesting function in complex analysis. It maps $\mathbb{D}$ onto the right

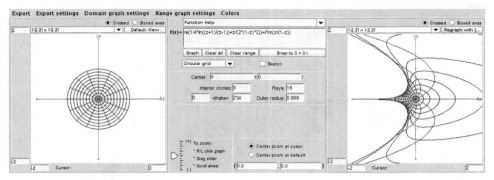

FIGURE 2.56. Image of $\mathbb{D}$ under $f = \mathrm{Re}\left(\frac{1}{4}\log\left(\frac{z+1}{z-1}\right) + \frac{z}{2(1-z)^2}\right) + \mathrm{Im}\left(\frac{z}{1-z}\right)$.

half-plane $\{z \in \mathbb{C} \,|\, \mathrm{Re}\{z\} > 0\}$. If we shear

$$h(z) - g(z) = \frac{z}{1-z} \quad \text{with} \quad \omega(z) = g'(z)/h'(z) = z^2$$

then we get the harmonic univalent mapping $f = h + \overline{g}$, where

$$h = \frac{1}{8}\log\left(\frac{z+1}{z-1}\right) + \frac{3z - 2z^2}{4(1-z)^2} \quad \text{and} \quad g = \frac{1}{8}\log\left(\frac{z+1}{z-1}\right) - \frac{z - 2z^2}{4(1-z)^2}.$$

The image of $\mathbb{D}$ under $f = h + \overline{g}$ is shown in Figure 2.56.

Because $\omega(z) = z^2$, we can use Theorem 2.124 to find the parametrization of the corresponding minimal graph:

$$\mathbf{x} = \left(\mathrm{Re}\left\{\frac{1}{4}\log\left(\frac{z+1}{z-1}\right) + \frac{z}{2(1-z)^2}\right\}, \mathrm{Im}\left\{\frac{z}{1-z}\right\}, \mathrm{Im}\left\{\frac{1}{4}\log\left(\frac{z+1}{z-1}\right) - \frac{z}{2(1-z)^2}\right\}\right).$$

What minimal graph is this? It is difficult to tell. The coordinate functions do not look like a standard parametrization for a known minimal surface. Let us use a substitution to rewrite the parametrization into a form that allows us to identify the minimal graph. If we use a Möbius transformation for the substitution, it will not affect the geometry of the minimal graph. Letting $z \mapsto \frac{\hat{z}+1}{\hat{z}-1}$, we get the parametrization

$$\hat{\mathbf{x}} = \left(\frac{1}{4}\mathrm{Re}\left\{\log(\hat{z}) + \frac{1}{2}\hat{z}^2 - \frac{1}{2}\right\}, -\frac{1}{2}\mathrm{Im}\{\hat{z}\}, -\frac{1}{4}\mathrm{Im}\left\{\log(\hat{z}) - \frac{1}{2}\hat{z}^2 + \frac{1}{2}\right\}\right).$$

This is useful, because it simplifies the log terms in x_1 and x_3. By switching the coordinate functions and factoring out $\frac{1}{2}$ we have

$$\widetilde{x} = \left(-\frac{1}{2}\left[\frac{1}{2}\mathrm{Im}\left\{\log(\widetilde{z}) - \frac{1}{2}\widetilde{z}^2\right\}\right], \frac{1}{2}\left[\frac{1}{2}\mathrm{Re}\left\{\log(\widetilde{z}) + \frac{1}{2}\widetilde{z}^2\right\}\right], -\frac{1}{2}[\mathrm{Im}\{\widetilde{z}\}]\right), \tag{2.19}$$

and this looks close to the standard parametrization for the wavy plane. In Figure 2.57 we have used the **Complex P. Surf** tab in *MinSurfTool* to graph the image of the surface with this parametrization.

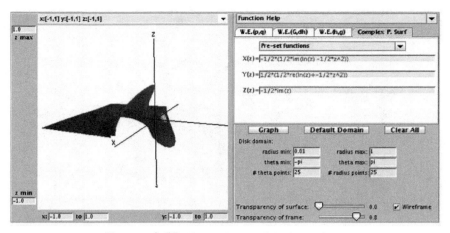

FIGURE 2.57. The conjugate of the wavy plane.

The coordinate functions correspond to the conjugate surface of the wavy plane scaled by $\frac{1}{2}$. This is clear given the actual coordinates of the wavy plane

$$\mathbf{W} = \left(\frac{1}{2}\,\mathrm{Re}\left\{ \log(z) - \frac{1}{2}z^2 \right\} , -\frac{1}{2}\,\mathrm{Im}\left\{ \log(z) + \frac{1}{2}z^2 \right\} , \mathrm{Re}\{z\} \right).$$

Since the wavy plane is its own conjugate surface, this means that we can describe our original minimal graph as the wavy plane.

How does this match with the image in Figure 2.56 of $\mathbb{D}$ under the original harmonic univalent map? In deriving the parametrization of the wavy plane, we applied the transformation $z \mapsto \frac{\hat{z}+1}{\hat{z}-1}$, switched the coordinate functions, and took the conjugate function. These changes are equivalent to altering the original domain $\mathbb{D}$, so we are looking at a piece of the wavy plane coming from a different domain G. The projection of the wavy plane coming from G is shown in Figure 2.58. There is similarity between the image in Figure 2.56 and the image in Figure 2.58.

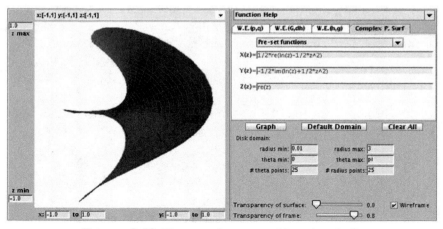

FIGURE 2.58. The wavy plane mapped from domain G.

Exercise 2.130. An important function in complex analysis is the Koebe function $\frac{z}{(1-z)^2}$. By shearing

$$h(z) - g(z) = \frac{z}{(1-z)^2} \quad \text{with} \quad \omega(z) = z^2,$$

we derive the harmonic univalent mapping $f = h + \overline{g}$, where

$$h = \frac{z - z^2 + \frac{1}{3}z^3}{(1-z)^3} \quad \text{and} \quad g = \frac{\frac{1}{3}z^3}{(1-z)^3}.$$

(a) Use Theorem 2.124 to find the parametrization of the minimal graph that f lifts to.

(b) Use the approach of Example 2.129 to show analytically that the minimal graph is Enneper's surface.

(c) Use *ComplexTool* to graph the image of $\mathbb{D}$ under f and *MinSurfTool* to sketch the corresponding minimal graph.

Try it out!

Example 2.131. By shearing $h(z) - g(z) = \frac{1}{2}\log\left(\frac{1+z}{1-z}\right)$ with $\omega(z) = g'(z)/h'(z) = m^2 z^2$, where $|m| \leq 1$, it was shown in [9] that the harmonic function $f = h + \overline{g}$ is univalent, where

$$h(z) = \frac{1}{2(1-m^2)}\log\left(\frac{1+z}{1-z}\right) + \frac{m}{2(m^2-1)}\log\left(\frac{1+mz}{1-mz}\right)$$

$$g(z) = \frac{m^2}{2(1-m^2)}\log\left(\frac{1+z}{1-z}\right) + \frac{m}{2(m^2-1)}\log\left(\frac{1+mz}{1-mz}\right).$$

When $m = e^{i\frac{\pi}{2}}$, the function f is the same as in Example 2.121 and the image of $\mathbb{D}$ under $f = h + \overline{g}$ is a square. For every m with $|m| = 1$, the image of $\mathbb{D}$ under $f = h + \overline{g}$ is a parallelogram.

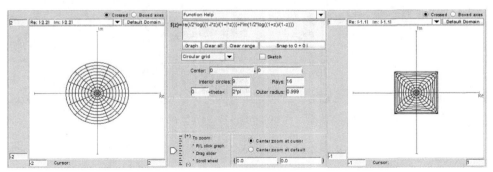

FIGURE 2.59. Image of $\mathbb{D}$ under $f = h + \overline{g}$ when $m = e^{i\frac{\pi}{2}}$.

Since $\omega(z) = g'(z)/h'(z) = m^2 z^2$, we can lift f to a minimal graph. We compute that

$$x_3 = \text{Im}\left\{\frac{m}{1-m^2}\log\left(\frac{1-m^2 z^2}{1-z^2}\right)\right\}.$$

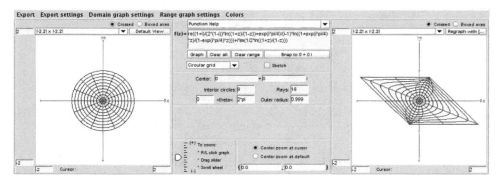

FIGURE 2.60. Image of $\mathbb{D}$ under $f = h + \overline{g}$ when $m = e^{i\frac{\pi}{4}}$.

Hence, the corresponding parametrization of the minimal graph is

$$\mathbf{x} = \left(\mathrm{Re}\left\{ \frac{1+m^2}{2(1-m^2)} \log\left(\frac{1+z}{1-z}\right) + \frac{m}{(m^2-1)} \log\left(\frac{1+mz}{1-mz}\right) \right\}, \right.$$

$$\left. \mathrm{Im}\left\{ \frac{1}{2} \log\left(\frac{1+z}{1-z}\right) \right\}, \mathrm{Im}\left\{ \frac{m}{1-m^2} \log\left(\frac{1-m^2z^2}{1-z^2}\right) \right\} \right).$$

When $m = e^{i\frac{\pi}{2}}$, the minimal graph is Scherk's doubly periodic surface (see Figure 2.61).

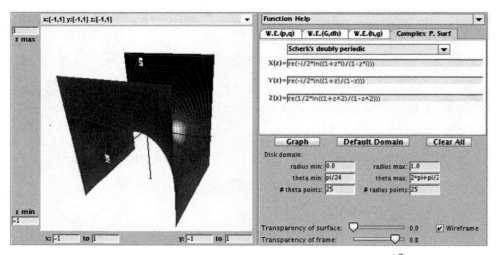

FIGURE 2.61. Scherk's doubly periodic square surface ($m = e^{i\frac{\pi}{2}}$).

For $m = e^{i\theta}, (0 < \theta < \frac{\pi}{2})$, the minimal graphs are slanted Scherk's surfaces (see Figure 2.62).

What is the minimal graph when $m = 1$? In the limit (i.e., $\theta = 0$) we have the equation

$$\mathbf{x} = \left(\mathrm{Re}\left\{ \frac{z}{1-z^2} \right\}, \mathrm{Re}\left\{ -\frac{i}{2} \log\left(\frac{1+z}{1-z}\right) \right\}, \mathrm{Re}\left\{ \frac{-iz^2}{1-z^2} \right\} \right).$$

This parametrization does not look like a standard parametrization of a known minimal surface. We can use a substitution to rewrite it in a form that allows us to identify the

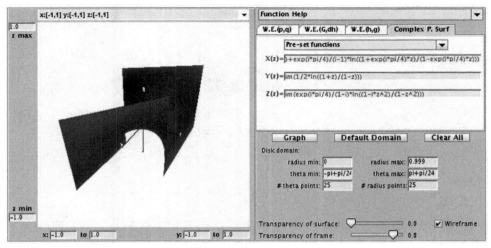

FIGURE 2.62. A Scherk's doubly periodic parallelogram surface ($m = e^{i\frac{\pi}{4}}$).

minimal graph. Using the substitution $z \longmapsto \frac{e^z - 1}{e^z + 1}$ and the fact that $\mathrm{Re}\left\{ \dfrac{-iz^2}{1 - z^2} \right\} =$
$\mathrm{Re}\left\{ \dfrac{1}{2i} \dfrac{1 + z^2}{1 - z^2} \right\}$, the equation is equivalent to

$$X = \left(\frac{1}{2} \sinh u \cos v, \frac{1}{2} v, \frac{1}{2} \sinh u \sin v \right),$$

which is an equation of a helicoid. Thus, we have a family of minimal surfaces that transform Scherk's doubly periodic surface continuously into the helicoid. This is neat!

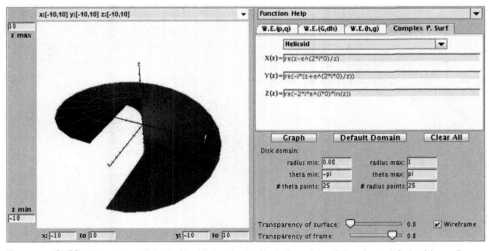

FIGURE 2.63. Side view of the helicoid that is the limit function of the slanted Scherk's surfaces.

Exercise 2.132. Show that $\mathrm{Re}\left\{ \dfrac{-iz^2}{1 - z^2} \right\} = \mathrm{Re}\left\{ \dfrac{1}{2i} \dfrac{1 + z^2}{1 - z^2} \right\}$. *Try it out!*

Exercise 2.133. Consider the harmonic univalent map $f(z) = h(z) + \overline{g}(z)$, where

$$h = \frac{1}{4} \log \left(\frac{1+z}{1-z} \right) + \frac{\frac{1}{2}z}{1-z^2} \quad \text{and} \quad g = \frac{1}{4} \log \left(\frac{1+z}{1-z} \right) - \frac{\frac{1}{2}z}{1-z^2}.$$

(a) Use Theorem 2.124 to find the parametrization of the minimal graph that f lifts to.

(b) Use *ComplexTool* to graph the image of $\mathbb{D}$ under f and *MinSurfTool* with tab **W.E. (h,g)** to sketch the corresponding minimal graph.

(c) Use the approach of Example 2.131 to show analytically that the minimal graph is the catenoid.

Try it out!

Large Project 2.134. The analytic function $F(z) = z$ maps the unit disk $\mathbb{D}$ onto itself. Shear $h(z) - g(z) = z$ with various perfect-square dilatations, ω, that satisfy the condition $|\omega| < 1$ for all $z \in \mathbb{D}$ (e.g., $\omega = z^{2n} (n \in \mathbb{N}), \omega = e^{i\theta} z^2 (\theta \in \mathbb{R}), \omega = \left(\frac{z-a}{1-\overline{a}z} \right)^2 (|a| < 1))$. Determine the corresponding minimal graphs. **Optional**

Large Project 2.135. The analytic function $F(z) = \frac{z}{1-z}$ maps the unit disk $\mathbb{D}$ onto a right half-plane and is an important function. Shear $h(z) + g(z) = \frac{z}{1-z}$ with various perfect-square dilatations, ω, that satisfy the condition $|\omega| < 1$ for all $z \in \mathbb{D}$ (e.g., $\omega = z^{2n} (n \in \mathbb{N}), \omega = e^{i\theta} z^2 (\theta \in \mathbb{R}), \omega = \left(\frac{z-a}{1-\overline{a}z} \right)^2 (|a| < 1))$. Determine the corresponding minimal graphs. **Optional**

2.7   Conclusion

We have presented an introduction to minimal surfaces and described a few topics that you can explore using the exercises, the exploratory problems, and the projects along with the applets. For a deeper and thorough explanation of differential geometry consult [7], [18], or [21] for beginners, and [3] for intermediates. Also, you should consider Spivak's five volume work [24]. Oprea's book [22] is a nice source for an introduction to minimal surfaces. For more background on minimal surfaces we recommend [25], [14], [15], [6], [23], and [20].

2.8   Additional exercises

Differential geometry

Exploration 2.136. An oblique cylinder can be parametrized by

$$\mathbf{x}(u, v) = (\cos u, \sin u + v \cos \theta, v \sin \theta),$$

where $\theta \in \left(0, \frac{\pi}{2} \right)$ is a fixed value. Use *DiffGeomTool* to explore what happens to it as θ varies between 0 and $\frac{\pi}{2}$.

Exercise 2.137. Use *DiffGeomTool* to graph the surface parametrized by

$$\mathbf{x}(u, v) = \left(\cos u\left(1 + v \sin\left(\frac{1}{2}u\right)\right), \sin u\left(1 + v \sin\left(\frac{1}{2}u\right)\right), v \cos\left(\frac{1}{2}u\right)\right),$$

where $-\pi < u < \pi$, $-\frac{1}{2} < v < \frac{1}{2}$. This surface is a Möbius strip and is nonorientable; that is, the normal vector can change from pointing outward to pointing inward as it travels along a closed path on the surface. You can see this in *DiffGeomTool* by clicking on the **Normal vector** box and setting the **Point locator: (u,v) =** to $(\pi - 0.1, 0)$. Next, change the u coordinate to $u = \pi - 0.1 - 1$, $u = \pi - 0.1 - 2$, $u = \pi - 0.1 - 3$, $u = \pi - 0.1 - 4$, $u = \pi - 0.1 - 5$, and $u = \pi - 0.1 - 6$. Observe that $\mathbf{n}$ will make nearly a complete path along a closed curve but it will change its direction.

Exercise 2.138. Describe the u-parameter and v-parameter curves on Enneper's surface.

Exercise 2.139. In Exploration 2.11(c), you determined the largest value of r for which Enneper's surface has no self-intersections assuming that the intersection occurs on the x_3-axis. In this exercise, prove the result without making that assumption.

Exercise 2.140. Compute the coefficients of the first and the second fundamental forms for Enneper's surface whose parametrization is

$$\mathbf{x}(u, v) = \left(u - \frac{1}{3}u^3 + uv^2, v - \frac{1}{3}v^3 + u^2v, u^2 - v^2\right).$$

Exercise 2.141. A *CMC (Constant Mean Curvature) surface* is a surface that has the same mean curvature everywhere. Minimal surfaces are a subset of CMC surfaces. Using *DiffGeomTool* sketch the following surfaces and determine which are CMC surfaces:

(a) $\mathbf{x}(u, v) = (u - v, u + v, 2(u^2 + v^2))$, where $-1 < u < 1, -1 < v < 1$

(b) $\mathbf{x}(u, v) = (\cos u, \sin u, v)$, where $-\pi < u < \pi, -2 < v < 2$

(c) $\mathbf{x}(u, v) = \left((2 + \cos v)\cos u, (2 + \cos v)\sin u, \sin v\right)$, where $0 < u, v < 2\pi$

(d) $\mathbf{x}(u, v) = (\sqrt{1 - u^2}\cos v, \sqrt{1 - u^2}\sin v, u)$, where $-1 < u < 1, -\pi < v < \pi$.

Minimal surfaces

Exercise 2.142. Use (2.4) to show that Enneper's surface parametrized by

$$\mathbf{x}(u, v) = \left(u - \frac{1}{3}u^3 + uv^2, v - \frac{1}{3}v^3 + u^2v, u^2 - v^2\right)$$

is a minimal surface.

Exercise 2.143. Prove Theorem 2.39 when the surface of revolution has the parametrization

$$\mathbf{x}(u, v) = \left(f(v)\cos u, f(v)\sin u, v\right).$$

Exercise 2.144. An oblique cylinder is a cylinder whose side forms an angle θ with the x_1x_2-plane, where $0 < \theta \le \frac{\pi}{2}$. For a fixed θ it can be parametrized by

$$\mathbf{x}(u, v) = (\cos u, \sin u + v \cos \theta, v \sin \theta).$$

Determine the values of θ for which $\mathbf{x}$ is isothermal.

Exercise 2.145. Show that the parametrization

$$\mathbf{x}(u, v) = \left(\arctan\left(\frac{2u}{1 - (u^2 + v^2)} \right), \arctan\left(\frac{-2v}{1 - (u^2 + v^2)} \right), \right.$$
$$\left. \frac{1}{2} \ln\left(\frac{(u^2 - v^2 + 1)^2 + 4u^2 v^2}{(u^2 - v^2 - 1)^2 + 4u^2 v^2} \right) \right)$$

is an isothermal parametrization of Scherk's doubly periodic surface (that is, show that it is isothermal and that there is transformation that maps it to the parametrization given in Exercise 2.43(b) for Scherk's doubly periodic surface).

Exercise 2.146. Let X be a minimal surface that is not a plane and let Y be its conjugate minimal surface. Is it possible that the plane is one of the associated surfaces for X and Y? If so, describe the geometry of X. If not, explain why.

Weierstrass representation

Exercise 2.147. In Example 2.67, show the details in going from Enneper's surface parametrization

$$\mathbf{x} = \left(\operatorname{Re}\left\{ z - \frac{1}{3}z^3 \right\}, \operatorname{Re}\left\{ -i\left(z + \frac{1}{3}z^3 \right) \right\}, \operatorname{Re}\left\{ z^2 \right\} \right)$$

to the parametrization

$$\mathbf{x}(u, v) = \left(u - \frac{1}{3}u^3 + uv^2, v - \frac{1}{3}v^3 + vu^2, u^2 - v^2 \right)$$

that is also for Enneper's surface.

Exercise 2.148. Compute the parametrization for the minimal surfaces generated by using $p(z) = \frac{1}{2z}$ and $q(z) = iz$ on the domain $\mathbb{C} - \{0\}$ in the Weierstrass representation. Use *MinSurfTool* with the **W.E. (p,q)** tab to graph an image of the surface, which is known as the *wavy plane*. [Use radius min=0.001, radius max=1.3, θ min=$-\pi$, θ max=π with initial values $x = \operatorname{Re}(1/2 \log(z) - 1/4z^2)$, $y = \operatorname{Im}(1/2 \log(z) + 1/4z^2)$, and $z = \operatorname{Re}(z)$.]

Exercise 2.149. Compute the parametrization for the minimal surfaces generated by using $p(z) = z^2$ and $q(z) = \frac{i}{z^2}$ on the domain $\mathbb{C} - \{0\}$ in the Weierstrass representation. Use *MinSurfTool* with the **W.E. (p,q)** tab to graph an image of this surface, which is known as *Richmond's surface*. [Use radius min=0.1, radius max=1, θ min=$\pi/24$, θ max=$2\pi + \pi/24$ with initial values $x = \operatorname{Re}(1/3z^3 + 1/z)$, $y = \operatorname{Im}(1/3z^3 - 1/z)$, and $z = \operatorname{Re}(2z)$.]

The Gauss map, G, and height differential, dh

Exercise 2.150. Show that if $z = x + iy$ is the projection of (x_1, x_2, x_3) on the Riemann sphere onto to complex plane, then

$$x = \frac{x_1}{1 - x_3}, \qquad y = \frac{x_2}{1 - x_3}.$$

Exercise 2.151. For Scherk's doubly periodic surface find

   (a) $G(0)$ (b) $G(1)$ (c) $G(-1)$ (d) $G(i)$ (e) $G(-i)$.

Exercise 2.152. The Weierstrass data for a 4-noid are

$$G(z) = z^3 \quad \text{and} \quad dh = \frac{z^3}{(z^4 - 1)^2}\, dz.$$

Show that the ends of the 4-noid are catenoidal.

Exercise 2.153. Determine asymptotic and curvature lines for Scherk's doubly periodic surface with $G(z) = z$ and $dh(z) = \frac{iz}{z^4-1} dz$.

Exercise 2.154. Determine the period conditions for the wavy plane with $G(z) = z$ and $dh(z) = dz$.

Exercise 2.155. Let M be the Scherk doubly periodic surface with six ends. Using the approach of Example 2.104 determine G and dh.

Minimal surfaces and harmonic univalent mappings

Exercise 2.156. Prove that if $f = u + iv$ is harmonic in a simply-connected domain G, then $f = h + \overline{g}$, where h and g are analytic.

Exercise 2.157. Prove that the representations $f(z) = h(z) + \overline{g}(z)$ and $f(z) = \mathrm{Re}\,\{h(z) + g(z)\} + i\,\mathrm{Im}\,\{h(z) - g(z)\}$ are equivalent.

Exercise 2.158. Shear $h(z) - g(z) = \frac{z}{1-z}$ with $\omega(z) = z^2$ to get the harmonic univalent function $f = h + \overline{g}$ given in Example 2.129, where

$$h = \frac{1}{8}\log\left(\frac{z+1}{z-1}\right) + \frac{3z - 2z^2}{4(1-z)^2} \quad \text{and} \quad g = \frac{1}{8}\log\left(\frac{z+1}{z-1}\right) - \frac{z - 2z^2}{4(1-z)^2}.$$

Exercise 2.159. Shear $h(z) - g(z) = \frac{z}{(1-z)^2}$ with $\omega(z) = z^2$ to get the harmonic univalent function $f = h + \overline{g}$ given in Exercise 2.130, where

$$h = \frac{1}{8}\log\left(\frac{z+1}{z-1}\right) + \frac{3z - 2z^2}{4(1-z)^2} \quad \text{and} \quad g = \frac{1}{8}\log\left(\frac{z+1}{z-1}\right) - \frac{z - 2z^2}{4(1-z)^2}.$$

Exercise 2.160. Show that the parametrization

$$\mathbf{x} = \left(\mathrm{Re}\left[\frac{i}{2}\log\left(\frac{i+z}{i-z}\right)\right], \mathrm{Im}\left[\frac{1}{2}\log\left(\frac{1+z}{1-z}\right)\right], \frac{1}{2}\mathrm{Im}\left[i\log\left(\frac{1+z^2}{1-z^2}\right)\right]\right)$$

is equivalent to the parametrization in Exercise 2.145 that gives Scherk's doubly periodic minimal surface.

Exercise 2.161. For the harmonic univalent map $f(z) = h(z) + \overline{g}(z)$, where

$$h = \frac{3}{16}\log\left(\frac{1+z}{1-z}\right) - \frac{3i}{16}\log\left(\frac{1+iz}{1-iz}\right) + \frac{1}{4}\frac{z}{1-z^4}$$

$$g = -\frac{3}{16}\log\left(\frac{1+z}{1-z}\right) - \frac{3i}{16}\log\left(\frac{1+iz}{1-iz}\right) + \frac{1}{4}\frac{z^3}{1-z^4}$$

(a) Use Theorem 2.124 to find the parametrization of the minimal graph that f lifts to.

(b) Use *ComplexTool* to graph the image of $\mathbb{D}$ under f and *MinSurfTool* with tab **W.E.(h,g)** to sketch the corresponding minimal graph.

(c) This minimal surface has four helicoidial ends and is not embedded (i.e., it does have self-intersections). However, Theorem 2.124 states that it should be a minimal graph and hence have no self-intersections. Explain why this is not a contradiction.

Large Project 2.162. The analytic function $F(z) = \frac{z}{(1-z)^2}$ maps the unit disk $\mathbb{D}$ onto $\mathbb{C} \setminus (-\infty, -\frac{1}{4})$ and is an important function. Shear $h(z) - g(z) = \frac{z}{(1-z)^2}$ with various perfect-square dilatations, ω, that satisfy the condition $|\omega| < 1$ for all $z \in \mathbb{D}$ (e.g., $\omega = z^{2n}(n = \mathbb{N})\omega = e^{i\theta}z^2, (\theta \in \mathbb{R}), \omega = \left(\frac{z-a}{1-\bar{a}z}\right)(|a| < 1)$). Determine the corresponding minimal graphs.

2.9 Bibliography

[1] Fred J.Almgren and Jean Taylor, The geometry of soap films and soap bubbles, *Sci. Am.* **235** (1976), 82–93.

[2] Lipman Bers, *Riemann Surfaces*, New York Univ., Institute of Mathematical Sciences, New York, 1957–1958.

[3] William M. Boothby, *An Introduction to Differentiable Manifolds and Riemannian Geometry*, Second edition. Pure and Applied Mathematics, vol 120. Academic Press, Inc., Orlando, FL, 1986.

[4] James Clunie and Terry Sheil-Small, Harmonic univalent functions, *Ann. Acad. Sci. Fenn. Ser. A.I Math.* **9** (1984), 3–25.

[5] Richard Courant and Herbert Robbins, *What Is Mathematics?*, Oxford University Press, 1941.

[6] Ulrich Dierkes, Stefan Hildebrandt, Albrecht Kster, and Ortwin Wohlrab, *Minimal Surfaces I*, Grundlehren der Mathematischen Wissenschaften, 295, Springer-Verlag, Berlin, 1992.

[7] Manfredo do Carmo, *Differential Geometry of Curves and Surfaces*, translated from the Portuguese, Prentice-Hall, Inc., Englewood Cliffs, NJ, 1976.

[8] Michael Dorff, Minimal graphs in $\mathbb{R}^3$ over convex domains, *Proc. Amer. Math. Soc.* **132** (2004), no. 2, 491–498.

[9] Michael Dorff and Jan Szynal, Harmonic shears of elliptic integrals, *Rocky Mountain J. Math.* **35** (2005), no. 2, 485–499.

[10] Kathy Driver and Peter Duren, Harmonic shears of regular polygons by hypergeometric functions, *J. Math. Anal. App.* **239** (1999), 72–84.

[11] Peter Duren and William Thygerson, Harmonic mappings related to Scherk's saddle-tower minimal surfaces, *Rocky Mountain J. Math.* **30** (2000), no. 2, 555–564.

[12] Walter Hengartner and Glenn Schober, On schlicht mappings to domains convex in one direction, *Comment. Math. Helv.* **45** (1970), 303–314.

[13] Stefan Hildebrandt and Anthony Tromba, *Mathematics and Optimal Form*, Scientific American Library, 1985.

[14] Karcher, Hermann, Construction of minimal surfaces, *Survey in Geometry*, Univ. of Tokyo, 1989.

[15] Hermann Karcher, Introduction to the complex analysis of minimal surfaces, Lecture notes given at the NCTS, Taiwan, 2003.

[16] Hans Lewy, On the non-vanishing of the Jacobian in certain one-to-one mappings, *Bull. Amer. Math. Soc.* **42** (1936), no. 10, 689–692.

[17] Jane McDougall and Lisabeth Schaubroeck, Minimal surfaces over stars, *J. Math. Anal. Appl.* **340** (2008), no. 1, 721–738.

[18] Richard Millman and George Parker, *Elements of Differential Geometry*, Prentice-Hall Inc., Englewood Cliffs, NJ, 1977.

[19] Frank Morgan, Minimal surfaces, crystals, shortest networks, and undergraduate research, *Math. Intel.*, **14** (1992), 37–44.

[20] Johannes C. C. Nitsche, *Lectures on Minimal Surfaces*, vol. 1, Cambridge U. Press, 1989.

[21] John Oprea, *Differential Geometry and Its Applications.*, 2nd ed., Classroom Resource Materials Series. Math. Assoc. of America, Washington, DC, 2007.

[22] ——, *The Mathematics of Soap Films: Explorations with Maple*, Student Mathematical Library, vol 10, Amer. Math. Soc., Providence, RI, 2000.

[23] Robert Osserman, *A Survey of Minimal Surfaces*, 2nd ed. Dover Publications, Inc., New York, 1986.

[24] Michael Spivak, *A Comprehensive Introduction to Differential Geometry*, vols 1–5, 3rd ed., Publish or Perish, Inc., Houston, TX, 2005.

[25] Matthias Weber, Classical minimal surfaces in Euclidean space by examples: geometric and computational aspects of the Weierstrass representation in *Global Theory of Minimal Surfaces*, 19–63, Clay Math. Proc., 2, Amer. Math. Soc., Providence, RI, 2005.

[26] Steven Weintraub, *Differential Forms: A Complement to Vector Calculus*, Academic Press, Inc., San Diego, CA, 1997.

Applications to Flow Problems

Michael A. Brilleslyper (text),
James S. Rolf (software)

3.1 Introduction

This chapter developed from a series of lectures prepared for an undergraduate mathematical physics course. The lectures were designed to show students applications of complex function theory that connected to familiar topics from calculus and physics. Two dimensional flows of ideal fluids was a natural topic. Many ideas from vector calculus are used and there are numerous applications of the methods that are developed. Modeling ideal fluid flow is a standard application of conformal mappings and is readily found in most undergraduate complex analysis texts (see [5] or [1]). However, in preparing the notes it became apparent that there was a need for a unified treatment that included a variety of applications and extensions such as including sources and sinks in the flow or accounting for the role played by sources or sinks at infinity. The emphasis in many texts is on the analytic aspects of the subject and not on the geometric or visual aspects of the flows. This chapter combines both analytic and geometric aspects of flow problems. It is self-contained and relies only on basic results from vector calculus and a standard first course in complex variables. We created the easy-to-use applet *FlowTool* to encourage discovery and experimentation.

FlowTool plots the streamlines for the flow of an ideal fluid. It permits the user to select the number and location of sources and sinks on the boundary or in the interior of the region and then allows the user to dynamically vary the strength of those sources and sinks. The applet shows steady state fluid flow in four preset regions: the entire plane, the half plane, the quadrant, and the strip.

The applet is limited to pre-set regions and particular types of sources. To deal with the wide variety of problems that may be encountered in practice, we also use powerful computer algebra systems such as Mathematica.

Vector fields arise naturally in many applications. They are used to model physical phenomena such as the velocity field of a fluid flowing in a region or the electric force field generated by a collection of charges. The geometry of the region, along with sources or sinks in the field, determines the nature of the resulting vector field. Finding descriptions of these fields is the focus of this chapter. It is also interesting to find the integral curves of a vector field, which are curves that are everywhere tangent to the field. Integral curves have different names depending on the context including *flow lines*, *streamlines*, or *lines of force*.

In this chapter we study physical situations in which complex function theory can be used to help solve the problem of finding such vector fields and their integral curves. We focus mainly on using techniques of complex analysis to solve flow problems. We start with an example, omitting many of the details, but providing a glimpse at what lies ahead.

Example 3.1. Imagine an infinitely long, very shallow, channel in which fluid flows. Because the channel is shallow we assume the flow is a two dimensional. Real flows are three dimensional, but we assume the flow is identical on all parallel planes. Orient the channel on the complex plane so that one edge runs along the real axis and the other along the horizontal line Im $z = \pi$, so the channel has width π. Assume that fluid is being pumped into the channel and drained from it at various points along the edges. For the time being we assume that all pumps and drains operate at the same constant rate. Our goal is to describe the velocity of the fluid in the channel. We assume that the flow is in a steady state, meaning that the velocity at a point of the domain does not change with time. Let the fluid be pumped into the channel at equal rates at $z = 0$ and $z = 2$, and suppose fluid is drained from the channel at the same rate at the point $z = -2 + \pi i$. There is more fluid being pumped into the channel than is being drained. See Figure 3.1. The curves shown are the integral curves of the underlying vector field and they represent the path that a drop of dye would follow if placed into the flow. A critical assumption in this model is that we are dealing with an *ideal fluid*. This means that the fluid is incompressible, non-viscous, and there is no loss of energy due to friction between the walls of the channel and the fluid. Such fluids do not physically exist, but they frequently provide good approximations to physical situations and they have mathematical properties that lend themselves to analysis and modeling.

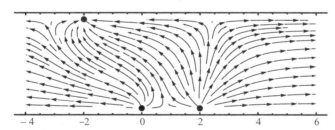

FIGURE 3.1. Flow lines for an ideal fluid in a channel with sources and sinks.

The plot in Figure 3.1 and most of the other plots in this chapter were generated using the computer algebra system Mathematica. Software packages such as Mathematica, Maple, and Matlab are powerful and can be useful in generating complicated graphics. However, they may not be easy to use or to adapt to a particular problem. To facilitate understanding of this chapter we have provided the *FlowTool* applet. It allows you to experiment with complicated flows for a fixed number of domains. We use it throughout the chapter and in numerous exercises. We also provide the code in two examples for generating these types of plots in Mathematica.

Before moving on you should open the *FlowTool* applet and generate your own version of Figure 3.1. Open the applet and select the strip domain from the drop down menu. Uncheck the box marked **equipotential** lines since we wish to view only the flow lines in this example. Use the mouse to place a source at $z = 0$, $z = 2$, and $z = -2 + \pi i$ by clicking once near each of the three locations. A dialog box with a slider opens in the right-hand panel each time a source is placed on the boundary of the channel. The sliders are used to change the strength of a source. If the slider is moved to the left, the strength will eventually become negative, signifying that the source has changed to a sink. Move the slider to create a sink of strength -1 at $z = -2 + \pi i$. The other two locations are sources of strength 1. Figure 3.2 shows a graph similar to what you should see on the applet. You can experiment further by adding or deleting sources and using the sliders to vary the strengths. Observe how the flow lines run parallel to the edges of the channel, indicating that the boundary of the region acts like a frictionless barrier to the flow.

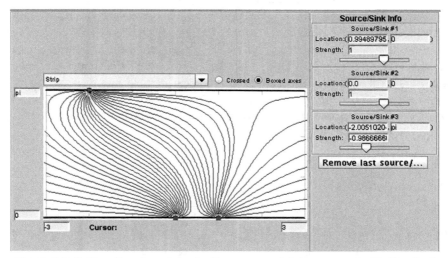

FIGURE 3.2. Flow lines in a channel with sources and sinks generated with *FlowTool*.

There is interesting mathematics behind the plot of the flow lines in Figures 3.1 and 3.2. This chapter develops the key ideas needed to solve flow problems in various regions with combinations of sources and sinks along the boundary. In addition, later sections extend the ideas and methods to more complicated settings.

Section 3.2 starts with a review of basic ideas from vector calculus including the important notions of the divergence and curl of a vector field and the definition of a harmonic function. This material should be familiar to most students.

Section 3.3 provides a review of the main results needed from a standard course in complex variables. It makes the critical connection between planar vector fields and complex functions. We also introduce the Polya field and give a mathematical description of sources and sinks.

Section 3.4 introduces the complex potential function that is the main tool for constructing ideal fluid flows. We make several connections between the complex potential function and the underlying vector field. This leads directly to the construction of uniform flows in various regions in Section 3.5. We continue the development in Section 3.6 by

allowing sources and sinks along the boundary of the region. We then look at the uniform
flow in a channel in Section 3.7. This development sheds light on the important role played
by sources or sinks at infinity. Section 3.8 puts the previous material together by discussing
flows in any region with combinations of sources or sinks along the boundary.

The final five sections focus on applications or extensions of the material. Topics include flows inside disks, interval sources, steady state temperature problems, sources or
sinks in the interior of regions, and electric fields generated by dipoles and multipoles.

3.2   Background and Fundamental Results

We need several fundamental concepts from multivariable calculus. We briefly review the
main ideas. More background regarding vector fields and their associated operations can
be found in any standard calculus text such as [4].

We represent a 2-dimensional or planar vector field in Cartesian coordinates using two
real-valued functions of position: $\vec{F}(x, y) = \; < P(x, y), Q(x, y) >$. Examples include
constant vector fields such as $\vec{F}(x, y) = \; < 3, 4 >$ and the field tangent to concentric
circles about the origin given by $\vec{G}(x, y) = \; < -y, x >$. Other examples include slope
fields for first-order differential equations or the magnetic field in a plane perpendicular
to a wire with a current flowing through it. Vector fields are represented graphically in
the plane by drawing arrows indicating the direction of the field at selected points. The
magnitude of the field at a point is

$$|\vec{F}| = \sqrt{P(x, y)^2 + Q(x, y)^2}$$

and the direction θ satisfies

$$\tan \theta = \frac{Q(x, y)}{P(x, y)}$$

for $P(x, y) \neq 0$. If $Q(x, y) \neq 0$ and $P(x, y) = 0$, then the direction is $\pm \frac{\pi}{2}$ at (x, y), the
choice being determined by the sign of $Q(x, y)$. The vector $\langle 0, 0 \rangle$ has no defined direction.
When graphing vector fields, it is difficult to draw vectors with their true magnitudes, so

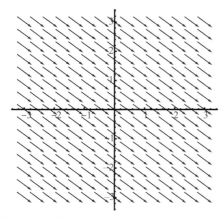

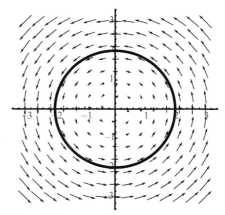

FIGURE 3.3. The constant vector field given by   FIGURE 3.4. Magnetic field generated by a cur-
$\vec{F}(x, y) = \langle 4, -3 \rangle$.   rent in a wire. The circle represents an integral
                                                                                      curve of the field.

graphical representations frequently scale their length down to a more manageable size. Figure 3.3 shows a constant vector field and Figure 3.4 shows the magnetic field generated by a current running perpendicular to the plane through the origin. We also show an integral curve of the field.

The *curl* and *divergence* of $\vec{F}$ are defined in terms of the vector differential operator $\nabla = \langle \partial_x, \partial_y \rangle$:

$$\nabla \times \vec{F} = (Q_x - P_y)\hat{k} \tag{3.1}$$

$$\nabla \cdot \vec{F} = P_x + Q_y \tag{3.2}$$

The curl operator acts on 3-dimensional vector fields and yields another vector field orthogonal to the original field. However, for 2-dimensional planar fields we assume that the $\hat{k}$-component of the vector field $\vec{F}$ is zero, implying that the curl will always point in the $\hat{k}$-direction. Thus, it is sufficient to compute the scalar component of $\nabla \times \vec{F}$, namely $Q_x - P_y$.

A vector field is called *irrotational* at a point if the curl is zero and it is called *incompressible* at a point if the divergence is zero. The standard physical description of a fluid with zero curl is that an infinitesimally small paddle wheel placed horizontally into the flow would not rotate. That is, the flow contains no vortices. The physical description of zero divergence is that the fluid is incompressible. A consequence of this is that the amount of fluid entering a region must be equal to the amount of fluid leaving it.

In this chapter we study vector fields that are both irrotational and incompressible throughout a region. This may seem restrictive, but it includes many important examples in addition to modeling ideal fluid flow. One example is the electric field generated in a plane perpendicular to an infinitely long wire with a uniform current density. In this case, the electric field is identical on all planes perpendicular to the wire. The electric field in this scenario has magnitude inversely proportional to the distance from the wire. This is different from the electric field generated by a point charge in space in which the magnitude of the field is inversely proportional to the square of the distance from the wire, the familiar inverse square law. A spherically symmetric field is not identical on parallel planes and thus does not readily lend itself to analysis using complex variable methods. When we discuss electric fields in this chapter, we will always assume the field is generated by uniform density current in a wire running perpendicular to the plane.

The requirement that both the curl and the divergence be zero imposes conditions on the component functions of the vector field $\vec{F}(x, y) = <P(x, y), Q(x, y)>$. From (3.1) and (3.2) we obtain a pair of partial differential equations relating the components of $\vec{F}$:

$$P_x = -Q_y \tag{3.3}$$

$$P_y = Q_x.$$

These equations are similar to the Cauchy-Riemann equations satisfied by analytic functions. If $f(z) = u(x, y) + iv(x, y)$ is an analytic function, then the real and imaginary parts satisfy the Cauchy-Riemann equations:

$$u_x = v_y \tag{3.4}$$

$$u_y = -v_x.$$

Exercise 3.2. Determine if the following vector fields are incompressible, irrotational, or both.

(a) $\vec{F}(x, y) = \langle ax + by, \, cx + dy \rangle$, where $a, b, c,$ and d are real

(b) $\vec{G}(x, y) = \langle e^x \cos y, \, e^x \sin y \rangle$

Try it out!

Given a vector field $\vec{F}(x, \, y)$ and a path C, the line integral of $\vec{F}$ along C is denoted by $\int_C \vec{F} \cdot d\vec{R}$ and measures the work done by the field on a particle traversing the path C. The line integral is computed by letting $r(t) = (x(t), \, y(t))$, be a parameterization of C, where $a \le t \le b$, and computing $\int_a^b \vec{F}(r(t)) \cdot r'(t) \, dt$.

A vector field $\vec{F}$ is said to be *conservative* in a simply connected domain D if the line integral $\int_C \vec{F} \cdot d\vec{R}$ between any two fixed points in D is independent of the chosen path C, provided the path lies entirely in D. There are numerous equivalent conditions that guarantee that a field is conservative. See [4] for a complete discussion.

Theorem 3.3. *Let $\vec{F}$ be a vector field with component functions that are continuous and have continuous first order partial derivatives throughout a simply connected region D. The following are equivalent:*

(a) $\vec{F}$ is conservative.

(b) There exists a differentiable potential function ϕ such that $\nabla \phi = \vec{F}$, where $\nabla \phi = \langle \phi_x, \phi_y \rangle$.

(c) $\int_C \vec{F} \cdot d\vec{R} = 0$ for every closed loop C in D.

(d) $\nabla \times \vec{F} = 0$.

We make extensive use of the result that a curl-free vector field has a potential function. The condition $\nabla \phi = \vec{F}$ implies the system of partial differential equations

$$\phi_x = P(x, \, y) \qquad\qquad (3.5)$$
$$\phi_y = Q(x, \, y)$$

is satisfied. The equations allow us to construct potential functions through partial integration. The function ϕ is unique up to an additive constant (why?) and is called the *real potential function* of $\vec{F}$.

Another definition that plays an important role in our work is that of a *harmonic function*, a solution of Laplace's equation. Harmonic functions play a critical role in much of applied mathematics. They arise frequently as steady-state solutions to various physical problems. Harmonic functions and their properties are closely tied to the theory of analytic functions in complex analysis. In solving flow problems, harmonic functions play a central role.

Definition 3.4. Let $u : D \subseteq \mathbb{R}^2 \to \mathbb{R}$ have continuous second order partial derivatives. If u satisfies Laplace's equation $\Delta u = \nabla \cdot \nabla u = u_{xx} + u_{yy} = 0$, then u is a *harmonic function* in D.

The following exercises explore vector fields, potential functions, and harmonic functions. Exercise 3.6 shows that the potential function for an irrotational and incompressible field is always harmonic.

Exercise 3.5. Let $\vec{F}(x, y) = \langle x^3 - 3xy^2, y^3 - 3x^2y \rangle$. Compute the curl and divergence of $\vec{F}$. If $\vec{F}$ is conservative, then find a potential function for it. If $\vec{F}$ is incompressible, show that the potential function is harmonic. (Optional) Use a program such as Mathematica or Matlab to graph $\vec{F}$ and several level curves of the potential function. What key geometrical observation connects the direction of $\vec{F}$ with the tangents to the level curves of the potential function? *Try it out!*

Exercise 3.6. Suppose that $\vec{F}$ is irrotational and incompressible. Let ϕ be a real potential function of $\vec{F}$. Show that ϕ is a harmonic function. *Try it out!*

The result of Exercise 3.6 is particularly important. It is the irrotational feature of the vector field that implies the existence of the potential function and the incompressibility then implies that the potential function is harmonic.

Exercise 3.7. Consider an attracting force at the origin of the xy-plane whose magnitude at a point is inversely proportional to the distance from the origin to the point. Determine a formula for the vector field representing the force field and determine if the field is irrotational and incompressible in a region not containing the origin. If possible, find a potential function for the vector field. Change the field by assuming the force of attraction at a point is inversely proportional to the square of the distance to the point. Compute the curl and divergence of this field. How do your answers compare? *Try it out!*

3.3 Complex Functions and Vector Fields

In this section we make several connections between vector fields and complex functions. A vector field that is irrotational and incompressible is closely related to the concept of an analytic function.

A complex function of the complex variable $z = x + i\,y$ can be expressed in terms of its real and imaginary parts as $f(z) = u(x, y) + i\,v(x, y)$, where u and v are real-valued functions of x and y. This is accomplished by setting $z = x + i\,y$ and separating into real and imaginary parts. The function f is said to be *analytic* at $z = z_0$ if the complex derivative $f'(z)$ exists for every z in a neighborhood of z_0. An immediate and far-reaching consequence is that the functions u and v must satisfy the Cauchy-Riemann equations:

$$u_x = v_y \tag{3.6}$$
$$u_y = -v_x.$$

A basic result about analytic functions is

Theorem 3.8. *Let $f(z) = u(x, y) + iv(x, y)$ be analytic in a domain D. Then the functions u and v are harmonic in D.*

This result follows directly from the Cauchy-Riemann equations and the reader is strongly encouraged to prove it.

Exercise 3.9. Prove Theorem 3.8. *Try it out!*

An equally important result is that a harmonic function on a simply connected domain is the real part of an analytic function. For a complete discussion, see [1].

Theorem 3.10. *Let $u(x, y)$ be a harmonic function defined on a simply connected domain D. Then, there exists a harmonic function $v(x, y)$ defined on D called the harmonic conjugate of u, such that $f(z) = u(x, y) + i v(x, y)$ is analytic on D. The harmonic conjugate is uniquely determined up to an additive constant.*

Exercise 3.11. Find a harmonic conjugate for $u(x, y) = 3x^2 - 2y - 3y^2$. ***Try it out!***

Exercise 3.12. Let $u(x, y) = \frac{1}{2}\ln(x^2 + y^2)$. Show that u is harmonic on the punctured plane. Next, omit an infinite ray emanating from the origin and construct a harmonic conjugate for u on the plane minus the ray. ***Try it out!***

Next we establish a correspondence between the set of complex functions and the set of planar vector fields. Since both complex functions and planar vector fields can be represented by a pair of real-valued functions, a complex function may be thought of as a vector field and vice versa. We note the correspondence:

$$V(z) = P(x, y) + i Q(x, y) \quad \Longleftrightarrow \quad \vec{F}(x, y) = \langle P(x, y),\ Q(x, y) \rangle \qquad (3.7)$$

The correspondence may appear to be only a renaming of one object as another. However, looking at complex functions as vector fields sometimes yields insights that are not otherwise apparent. When plotting the vector field of a complex function, the magnitude of the vector field at z is $|V(z)|$ and the direction is $\arg(V(z))$. Figure 3.5 shows the vector field corresponding to the complex function $V(z) = z^2$

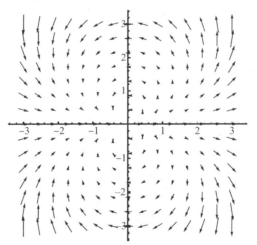

FIGURE 3.5. Vector field representation of $V(z) = z^2$.

For the remainder of this chapter we will use the notation $V(z)$ to refer to both complex functions and their corresponding vector field representations.

The correspondence in (3.7), though simple, is not the most useful vector field representation of a complex function. For reasons that will become clear, we instead associate a complex function with its Polya vector field.

Definition 3.13. Let $V(z) = u(x, y) + i\ v(x, y)$ be a complex function. The vector field given by the conjugate $\overline{V(z)} = u(x, y) - i\ v(x, y)$ is called the *Polya vector field* of $V(z)$.

Note that when plotting $\overline{V(z)}$, we attach the arrow to the point z, not $\bar{z}$.

The Polya vector field of $V(z) = z^2$ is $\overline{V(z)} = \bar{z}^2$ and is shown in Figure 3.6. Contrast this field with the one shown in Figure 3.5. We will see the value of the Polya field representation when we look at the vector field corresponding to an analytic function.

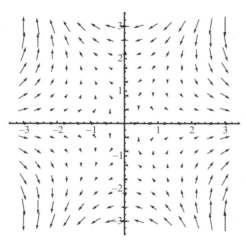

FIGURE 3.6. The irrotational, incompressible Polya field $\overline{V(z)} = \bar{z}^2$.

The next theorem shows that if a complex function is analytic, then its Polya field is irrotational and incompressible. The proof is a straightforward calculation involving the Cauchy-Riemann equations and is left as an exercise.

Theorem 3.14. *Let $V(z)$ be a complex function that is twice differentiable on a domain D. $V(z)$ is analytic on D if and only if its Polya vector field $\overline{V(z)}$ is irrotational and incompressible on D.*

Exercise 3.15. Prove Theorem 3.14. *Try it out!*

Theorem 3.14 establishes a bijection between the set of irrotational, incompressible planar vector fields and the set of analytic functions. It is this connection that allows us to use the techniques of complex analysis to solve flow problems.

Exercise 3.16. Find the component functions on a suitably chosen domain of each irrotational and incompressible vector field corresponding to the following analytic functions. Verify directly that the resulting vector fields have both curl and divergence equal to zero.

(a) $V(z) = z^2$

(b) $V(z) = e^z$

(c) $V(z) = \frac{1}{z}$

Try it out!

Exercise 3.17. Verify that the following vector field is irrotational and incompressible and find the corresponding analytic function.

$$\overline{V(z)} = (x^2 - y^2 - 2x) + i\,(-2xy + 2y)$$

Try it out!

The third example in Exercise 3.16 deserves some additional comments. The function $V(z) = 1/z$ is not defined at $z = 0$, but is analytic at every point in any neighborhood of $z = 0$. Such a point is called a *singular point* or a *singularity* of the function. The corresponding Polya vector field $\overline{V(z)}$ is also undefined at $z = 0$. Graphing the vector field reveals that the field lines emanate radially from the origin; see Figure 3.7. We call such a point a *source*. Conversely, if the field lines went towards the origin, we would refer to it as a *sink*. This leads to the following definition.

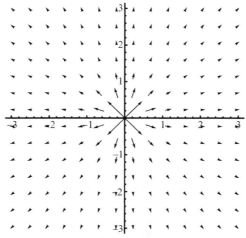

FIGURE 3.7. Polya vector field of the analytic function $f(z) = \frac{1}{z}$.

Definition 3.18. Let $V(z)$ be a function that is analytic except at $z = 0$. If $V(z)$ has a pole of order 1 at $z = 0$, then zero is either a source or sink of the Polya field $\overline{V(z)}$.

Sources and sinks have physical interpretations as a quantity being created or destroyed. For flow problems a source is a location where fluid is pumped into the region, whereas a sink is a drain, removing fluid from the region. For an electric field, sources and sinks correspond to positive and negative charges respectively. Allowing vector fields to have sources and sinks extends the examples to which we may apply our methods.

Exercise 3.19. Let $V(z) = 1/z$. Show that the Polya field $\overline{V(z)}$ emanates radially from the origin and that the magnitude of the field at a point is inversely proportional to the distance from the point to the origin. ***Try it out!***

To have a vector field with a source at $z = z_1$ rather than $z = 0$ we need only to translate the analytic function to obtain $V(z) = 1/(z - z_1)$. A sink is obtained by changing the sign of the function. The strength of a source or sink is changed by multiplying $V(z)$ by a real scalar. To account for more than one source or sink we use the *principle of superposition*. This states that if several components act to generate a vector field, then the field is the sum of the fields generated from each component separately. This is nothing more than the linearity of vector addition, but its consequences are far-reaching: complicated systems may be analyzed by studying the simpler components from which they are generated. For example, Figure 3.8 shows the field generated by equal sources at $z = 0$ and $z = 1$ and sink of twice the strength located at $z = 1 + 2i$. The analytic function corresponding to the field is

$$V(z) = \frac{1}{z} + \frac{1}{z - 1} - \frac{2}{z - (1 + 2i)}. \tag{3.8}$$

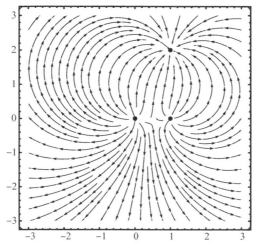

FIGURE 3.8. Integral curves of a vector field with two point sources and one sink.

Qualitatively, only the relative strengths of sources or sinks matters in the analysis of the situation. For completeness we compute the actual strength of a source or sink. We define the strength of a source to be the flux of the vector field over any simple closed loop C that encloses it, but no other singularities of the function.

Definition 3.20. Let $\vec{F}(x, y) = \langle P(x, y), Q(x, y) \rangle$ be an irrotational, incompressible vector field with a source at $z = z_0$. The strength of the source is defined to be

$$\int_C P(x, y)\,dy - Q(x, y)\,dx, \tag{3.9}$$

where C is a loop traversed in a counter clockwise direction containing z_0 and no other singularities of the field. If the strength is negative, then the singularity is a sink.

There is a relationship between the flux and circulation of a vector field and the complex integral of the associated Polya field. The strength of the source can be determined directly from the complex integral $\int_C \overline{V(z)}\,dz$ as shown in the following exercise.

Exercise 3.21. Let $V(z) = u(x, y) + i\,v(x, y)$ be a vector field. Let C be a simple loop not passing through any singular points of the field. Expand the integral $\int_C \overline{V(z)}\,dz$ in real and imaginary parts. (Hint: let $dz = dx + i\,dy$.) Show that

$$\int_C \overline{V(z)}\,dz = \int_C u\,dx + v\,dy + i \int_C u\,dy - v\,dx.$$

This shows that the flux of $V(z)$ outward across C is obtained from the imaginary part of the integral $\int_C \overline{V(z)}\,dz$. As a consequence, the circulation of $V(z)$ is given by the real part of the integral. ***Try it out!***

This exercise gives a useful characterization and interpretation of complex integration in terms of the familiar concepts of work and flux. An immediate consequence is that if $\overline{V(z)}$ is irrotational and incompressible inside and on C, then the work along C and the flux across C must be zero. This implies that $V(z)$ is analytic, as was already established. See Needham [2] for a thorough discussion of Polya vector fields and their relationship to complex integration.

If $V(z)$ is analytic everywhere except for a finite number of simple poles located at $z_i, i = 1, 2, \ldots, n$, then the strength of the source or sink of $\overline{V(z)}$ at z_k is given by the imaginary part of $\int_C V(z)\, dz$, where C is a simple closed loop traversed in the counterclockwise direction such that z_k is contained in the interior of C and none of the other z_i are inside C.

Let C be a simple closed curve containing z_k but no other z_i for $i \neq k$. The function can be represented as

$$V(z) = \frac{g(z)}{z - z_k},$$

where $g(z)$ is analytic inside and on C. By Cauchy's integral formula,

$$\int_C V(z)\, dz = 2\pi i\, g(z_k).$$

Exercise 3.22. In constructing vector fields with sources or sinks, the most common case is where $V(z)$ has the form

$$V(z) = \sum_{j=1}^{n} \frac{a_j}{z - z_j},$$

where $a_j \in \mathbb{R}$. Show that for the Polya field that the strength of the source or sink located at $z = z_k$ is $2\pi a_k$. ***Try it out!***

Exercise 3.23. Let $a \in \mathbb{R}$. Show that the vector field $\overline{V(z)} = a/(\bar{z} - \overline{z_0})$ has a source of strength $2\pi a$ at $z = z_0$ by explicitly computing the imaginary part of $\int_C V(z)\, dz$, where C is a circle centered at z_0. Do this by using a parameterization of C instead of Cauchy's integral formula. ***Try it out!***

Exercise 3.24. The Polya field of the function

$$V(z) = \frac{\sin z}{z^2}$$

has a simple pole at the origin. What is the strength of the source? ***Try it out!***

3.4   Complex Potential Functions

We have established a one-to-one correspondence using complex conjugation between irrotational, incompressible vector fields in the plane with a finite collections of point sources and sinks and complex functions that are analytic except at finitely many points where they have poles of order one.

We now construct the complex potential function of the underlying vector field. Let $V(z)$ be analytic on a simply connected domain so that $\overline{V(z)}$ is irrotational and incompressible, implying $\overline{V(z)}$ has a potential function ϕ that is harmonic. By Theorem 3.10

we know there exists a harmonic function ψ such that the complex function $\Omega(z) = \phi(x, y) + i\,\psi(x, y)$ is analytic. ψ is determined only up to an additive constant. The function Ω is called a *complex potential function* of $\overline{V(z)}$.

Definition 3.25. Given an irrotational, incompressible vector field $\overline{V(z)}$ on a simply connected domain, a *complex potential function* of $\overline{V(z)}$ is the analytic function $\Omega(z) = \phi(x, y) + i\,\psi(x, y)$, where ϕ is the real potential function of $\overline{V(z)}$ and ψ is a harmonic conjugate of ϕ.

A globally defined complex potential function may not exist. For example, $\overline{V(z)} = 1/\bar{z}$ is irrotational and incompressible on the punctured plane, but a complex potential for this field involves choosing a branch of the complex logarithm function that is not analytic on any neighborhood containing the origin.

As we already know, the level curves of ϕ, the real potential function, are orthogonal to the direction of $\overline{V(z)}$. Those curves form the equipotential lines of the field. For a velocity field of an ideal fluid, they represent points where the velocity is constant. In the case of an electric field, the level curves of ϕ represent curves of constant electrostatic potential. What do the level curves of ψ represent? The answer is found in the following standard result from complex variables.

Theorem 3.26. *Let $\Omega(z) = \phi(x, y) + i\,\psi(x, y)$ be analytic at $z_0 = x_0 + i\,y_0$ and suppose that $\Omega'(z_0) \neq 0$. Then the tangents to the level curves of ϕ and ψ are orthogonal at the point (x_0, y_0).*

For a proof see Zill and Shanahan [5].

Theorem 3.26 implies that the level curves of ψ are parallel to the underlying vector field $\overline{V(z)}$. Thus, they are the integral curves of the field and are exactly the curves that were sketched by the *FlowTool* applet in the first example in this chapter. The function ψ is often referred to as the *stream function* for a flow problem. The level curves of ψ are called the *streamlines*.

The next exercise shows the vector field can be obtained from the complex potential function.

Exercise 3.27. Let $\overline{V(z)}$ be an irrotational and incompressible vector field with analytic complex potential function $\Omega(z)$. Show that $\Omega'(z) = V(z)$. (Hint: Write $\Omega(z)$ in terms of its real and imaginary parts and differentiate using the Cauchy-Riemann equations.) ***Try it out!***

An equivalent statement to the last exercise is that

$$\Omega(z) = \int V(z)\,dz. \tag{3.10}$$

This formulation is useful in the next exercise.

Exercise 3.28. $\overline{V(z)} = 3\bar{z}^2 - 4\bar{z}$ is an irrotational, incompressible vector field.

(a) Determine a complex potential function on $\mathbb{C}$ using (3.10).

(b) Find the real and imaginary parts of the complex potential function to easily find the real potential function and the stream function.

Try it out!

The last exercise of this section illustrates a mapping property of the complex potential function.

Exercise 3.29. Let Ω be the complex potential for an irrotational, incompressible vector field. Show that when viewed as a mapping of the z-plane to the w-plane, Ω maps equipotential curves to vertical lines and flow lines to horizontal lines. ***Try it out!***

3.5 Uniform Flows in the Plane and other Regions

We now give examples that allow us to solve a wide variety of problems in different regions of the plane. We start by considering simple flows in the entire plane. Consider the function $\Omega(z) = z$. Thinking of Ω as a complex potential function, a natural problem is to determine the underlying vector field. The following exercise asks you to do this.

Exercise 3.30. Show that $\Omega(z) = z$ is the complex potential function for a uniform flow to the right. Explicitly compute the vector field $V(z)$. See Exercise 3.27. Show directly that the level curves of the imaginary part of Ω are stream lines for the flow. ***Try it out!***

Exercise 3.31. How does the answer to Exercise 3.30 change if the complex potential function is $\Omega(z) = (2 + 3i)z$? Show that the stream lines are given by the family of linear equations $3x + 2y = c$. See Figure 3.9. ***Try it out!***

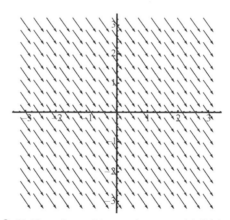

FIGURE 3.9. Uniform flow with complex potential $\Omega(z) = (2 + 3i)z$.

Our goal is to solve flow problems in regions such as sectors, strips, and disks. However, so far we have studied flows only in the entire complex plane. The extension to other regions lies in the theory of conformal mappings. A conformal mapping is a one-to-one complex function mapping a region in the z-plane to a region in the w-plane so that angles and orientation are preserved. A key result is that an analytic function is conformal at all points where its derivative is not zero. The critical property of conformal mappings for our development is the theorem:

Theorem 3.32. *Let $\Omega(w)$ be analytic on a domain $D' \in \mathbb{C}$, and let $f : D \to D'$ be conformal for some domain $D \in \mathbb{C}$. Then the composition $\widetilde{\Omega}(z) = (\Omega \circ f)(z)$ is analytic on D.*

The proof is found in many texts, such as [5].

The significance of Theorem 3.32 is that it allows a problem to be translated to a simpler domain, solved, and then translated back to the original domain. Unfortunately, it is usually impossible to map a given region conformally to the entire complex plane. Instead, we make use of a deep and powerful theorem from complex analysis.

Theorem 3.33 (Riemann Mapping Theorem). *Every simply connected domain in the complex plane, with the exception of the entire plane, can be conformally mapped to the upper half plane,* $\mathbb{H} = \{z \mid \operatorname{Im} z > 0\}$.

The Riemann mapping theorem gives no indication of how to find the conformal mapping. There are tables of conformal mappings that give specific instances of the mappings for a large collection of regions, as in [5]. Despite being an existence result, Theorem 3.33 indicates that it is valuable to know how to solve problems on the upper half plane. We will translate different problems to equivalent problems in $\mathbb{H}$ in order to solve them.

Suppose an ideal fluid is flowing from left to right in $\mathbb{H}$, so the real axis acts as a boundary for the flow. That is, the real axis acts as a streamline for the flow. It is clear that a constant vector field such as $\overline{V(z)} = 1$ is such a flow. It is easy to see the complex potential of this field is $\Omega(z) = z$. Thus, the identity function may be regarded as the complex potential for a uniform flow to the right in $\mathbb{H}$. Our goal is to use this and the theory of conformal mappings to find the complex potential for flows in other regions and in regions that have sources and sinks.

Our first example deals with a uniform flow in a quadrant, often called flow around a corner.

Example 3.34. Suppose a vertical barrier is inserted along the imaginary axis in a uniform flow in $\mathbb{H}$. The anticipated flow is shown in Figure 3.10. We want to find its complex potential function for the underlying vector field. The key is Theorem 3.32. We seek a conformal mapping that maps the region in Figure 3.10 to $\mathbb{H}$ with the property that the boundaries of the region are mapped to the real axis. This is important because the boundaries of the region are always streamlines for the flow (that is the flow is parallel to the

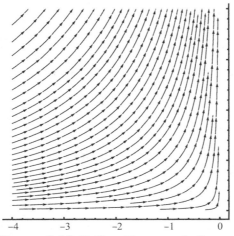

FIGURE 3.10. Uniform Flow around a Corner.

boundaries and there is no friction). If we let $h(z) = -z^2$ then the function will map the second quadrant to $\mathbb{H}$. The negative real axis is mapped to the negative real axis and the imaginary axis is mapped to the positive real axis. We know that the complex potential for the uniform flow in $\mathbb{H}$ is given by $\Omega(z) = z$, so we set $\widetilde{\Omega}(z) = \Omega(h(z)) = h(z) = -z^2 = -x^2 + y^2 - 2xyi$, and that the underlying vector field is $\overline{\widetilde{\Omega}'(z)} = -2\bar{z}$. Plotting the streamlines gives Figure 3.10. The streamlines are the level curves of the imaginary part of $\widetilde{\Omega}$, namely $-2xy = c$, where c is a constant.

Exercise 3.35. Let R be the region in polar coordinates

$$R = \{(r, \theta) \mid r \geq 0 \text{ and } 0 \leq \theta \leq \frac{\pi}{4}\}.$$

Find the complex potential function for a uniform, irrotational, incompressible flow with no source or sinks in R. **Try it out!**

The discussion above along with Exercise 3.35 demonstrates the technique of using conformal mappings to solve uniform flow problems. Any conformal mapping from a region D onto $\mathbb{H}$ is the complex potential for the velocity field of an ideal fluid flowing in D.

Another example demonstrating these techniques is flow around a cylinder.

Example 3.36. The function $h(z) = z + 1/z$ is a conformal mapping from $\mathbb{H}$ minus the upper half of the unit disk onto $\mathbb{H}$. That is, h maps $\{z \mid |z| \geq 1 \text{ and } \operatorname{Im} z > 0\}$ onto $\mathbb{H}$. The streamlines shown in Figure 3.11 are the family of curves

$$y - \frac{y}{x^2 + y^2} = c.$$

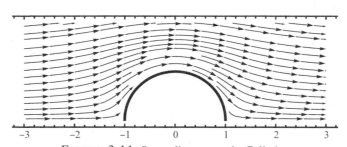

FIGURE 3.11. Streamlines around a Cylinder.

3.6   Sources and Sinks

Having seen how to solve for uniform flows in various regions, we now look at regions that have sources or sinks on the boundary of the region, as the example of the channel in the introduction. First, we will solve the problem in $\mathbb{H}$.

Let $a \in \mathbb{R}$. We have already seen that $\overline{V(z)} = 1/(\bar{z} - a)$ is an irrotational, incompressible vector field with a source at $z = a$ of strength 2π. We are interested only in the part of $V(z)$ that lies in $\mathbb{H}$. Since this is half of the field, we define the *effective strength* of the source on the boundary to be half of the strength, so the effective strength of the source is π. Only the relative strengths of the sources or sinks matter for the qualitative behavior of the flow. However, if we wish to balance a source on the boundary with a sink in the interior of the region, the notion of effective strength comes into play.

Before proceeding, we revisit the idea of a source in a vector field. The following exploration shows how physical principles lead to our complex function representation of a source or sink.

Exploration 3.37. Suppose that we have an irrotational and incompressible vector field defined on the punctured plane with a source of strength S at $z = 0$. Assume that that the flow lines are directed radially away from the origin. We wish to determine a formula for the field. Let $\vec{F}(x, y)$ denote the vector field at points other than the origin. We assumed the direction of $\vec{F}$ is directly away from the origin, so we need only determine the magnitude of $\vec{F}$. Since $\vec{F}$ is incompressible, the amount of fluid crossing into a closed region must equal the amount leaving it. We apply this principle to circles centered at the origin. The source having strength S means that there are S units of fluid entering the region per unit time, so S units of fluid cross the circle $x^2 + y^2 = R^2$ per unit time. But the amount crossing the circle is equal to the magnitude of the vector field multiplied by the length of the circle. Setting these two quantities equal gives

$$S = |\vec{F}(x, y)| \cdot 2\pi R. \tag{3.11}$$

Solve this equation for the magnitude of the field and combine the result with the fact that the direction is radial to find an explicit formula for the field. Show that its complex representation is $\overline{V(z)} = S/2\pi\bar{z}$. Hint: $1/\bar{z} = z/|z|^2$.

Immediate consequences of Exploration (3.37) are the results that we have already established:

(a) If the source is at $z = a$ instead of $z = 0$, then the complex representation of the vector field is $\overline{V(z)} = 1/(\bar{z} - \bar{a})$.

(b) If there is a sink at $z = a$, then the function is $\overline{V(z)} = -1/(\bar{z} - \bar{a})$

(c) The strength of a source or sink is changed by multiplying $V(z)$ by a real scalar.

Exercise 3.38. Find the complex representation of the irrotational, incompressible vector field in $\mathbb{H}$ if there is a source of effective strength π at $z = 3$ and a sink of effective strength 2π at $z = -2$. Hint: The vector field at any point is the sum of the vector fields determined from each source or sink. Use *FlowTool* to view the solution. ***Try it out!***

If $\overline{V(z)} = 1/(\bar{z} - \bar{a})$, then the complex potential function on $\mathbb{H}$ can be obtained from $V(z)$ by using (3.10):

$$\Omega(z) = \int V(z)\,dz = \text{Log}(z - a). \tag{3.12}$$

Thus, we see that the complex logarithm is involved when there are sources and sinks along the boundary. For the upper half plane, we use the principal value of the logairthm defined by $-\pi < \text{Arg} z \le \pi$. The complex logarithm extends the real logarithm function, and $\Omega(z)$ is analytic for $\text{Im}\, z > 0$, but is not analytic in a domain containing $z = a$.

Writing the logarithm in terms of real and imaginary parts

$$\text{Log}(z - a) = \ln|z - a| + i \, \text{Arg}(z - a),$$

the streamlines are given by the family of equations

$$\arctan\left(\frac{y}{x - a}\right) = c.$$

This is easily manipulated into the family of rays emanating from $z = a$ given by $y = (\tan c)(x - a)$. The values $c = 0$ and $c = \pi$ correspond to the two streamlines that run along the boundary of $\mathbb{H}$ in the positive and negative directions from the source at $z = a$.

This preceding development allows us to write the complex potential for an irrotational, incompressible vector field in $\mathbb{H}$ with any combination of sources and sinks along the boundary. The complete result is

Theorem 3.39. *Let $\overline{V(z)}$ be the irrotational, incompressible vector field in $\mathbb{H}$ generated by a finite collection of simple sources or sinks. Assume there are sources located along the boundary of $\mathbb{H}$ at $z = a_i$, $i = 1, \ldots, n$, $a_i \in \mathbb{R}$, with effective strengths, $S_i \pi$, $i = 1, \ldots, n$ (if $S_j < 0$, then there is a sink at a_j). Then the complex potential of $\overline{V(z)}$ is given by*

$$\Omega(z) = \sum_{j=1}^{n} S_j \, \text{Log}(z - a_j).$$

The next exercise asks you to experiment with *FlowTool* to develop some intuition regarding flows in $\mathbb{H}$. You may go beyond the suggestions in the exercise and experiment with other situations.

Exercise 3.40. Use *FlowTool* to view the flow in $\mathbb{H}$ for

(a) A source of strength 2π at $z = -1$ and a sink of strength 2π at $z = 1$.

(b) A source of strength 2π at $z = -1$ and a source of strength 2π at $z = 1$.

(c) A source of strength 4π at $z = -1$ and a sink of strength 2π at $z = 1$.

(d) Sources of strength 2π at $z = -3$ and $z = 0$ and a sink of strength 4π at $z = 2$.

Try it out!

In Exercise 3.40 sometimes all the fluid emanating from the sources is taken in by the sinks, but at other times some of the fluid either escapes to infinity or seems to emanate from infinity. If we think of infinity as a point on the boundary of $\mathbb{H}$, then we are led to the notion that there is a sink or source at infinity.

To examine the behavior at infinity, we revisit a flow in the entire plane with a finite collection of sources or sinks. Consider the vector field

$$\overline{V(z)} = \sum_{j=1}^{n} \frac{S_j}{2\pi} \frac{1}{\bar{z} - \bar{z}_j}.$$

Let C be a simple closed contour that encloses the singularities of $V(z)$ and consider $\int_C V(z)\,dz$. By the residue theorem,

$$\text{Im} \int_C V(z)\,dz = \sum_{j=1}^{n} S_j,$$

which gives the total flux across C. The integral $\int_C V(z)\,dz$ can also be viewed as a line integral around infinity traversed in the clockwise direction. When viewed in this way, its value has the opposite sign of the total flux outward across C. For example, if $V(z)$ has a source of strength 3 and a sink of strength 1 at points in the plane, then there must a sink of strength 2 at infinity. If the net flux in the finite plane is zero, then the sum of the strengths of the sinks and sources at infinity must be zero as well. That infinity can be both a source and a sink will be explored further. Section 3.7 investigates flow in an infinite channel, where infinity is a simultaneous source and sink.

3.7 Flow in a Channel

We return to a uniform flow in an infinitely long channel. Before dealing with any sources or sinks along the boundary, we investigate the case of uniform flow to the right in the channel. Assume the channel has one edge along the real axis and the other along the line $\operatorname{Im} z = \pi$. Since the edges of the channel are horizontal, the flow is the restriction of the uniform flow to the right in $\mathbb{H}$ restricted to the channel. Hence the complex potential is $\Omega(z) = z$.

On the other hand, the function $f(z) = e^z$ is a conformal mapping of the strip onto $\mathbb{H}$ such that boundary edges of the strip are mapped to the real axis. Using the example of the quadrant as motivation and Theorem 3.32, we see that $\widetilde{\Omega}(z) = \Omega(f(z)) = e^z$ should be the complex potential function for the uniform flow in the strip. However, $\operatorname{Im} \widetilde{\Omega}(z) = e^x \sin y$. Plotting the level curves gives the streamlines shown in Figure 3.12. They should simply go from left to right parallel to the edges of the strip. What went wrong?

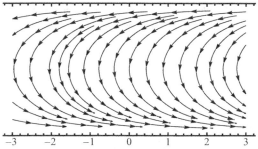

FIGURE 3.12. Incorrect uniform flow in a channel.

The incorrect flow occurred from a failure to account for the behavior of infinity under the conformal mapping. We may think of the flow in the channel as having a source at the left end of the strip and a sink of equal strength at the right end. Points in the strip with large negative real parts are mapped to points near zero. For all $z = x + iy$ with $0 \le y \le \pi$, we have

$$\lim_{x \to -\infty} e^{x+iy} = 0.$$

Thus, under the conformal mapping e^z, we expect to see a source at $z = 0$ in $\mathbb{H}$. On the other hand, as $x \to \infty$, the values of e^z approach ∞ and the sink is mapped to infinity in $\mathbb{H}$. This means that the complex potential in $\mathbb{H}$ has a source at 0. Its complex potential in

$\mathbb{H}$ is $\Omega(z) = \mathrm{Log}(z)$. Thus, the complex potential for the uniform flow in the strip as given by Theorem 3.32 is $\widetilde{\Omega}(z) = \Omega(e^z) = \mathrm{Log}(e^z) = z$, as expected.

With an understanding of how sources and sinks at infinity must be accounted for, it is easy to incorporate sources and sinks along the boundary. The following exercise asks you to find the complex potential for various flows in a channel. In each case the sum of all sources and sinks must be zero. Use *FlowTool* to see the streamlines for each flow and to note the behavior at infinity.

Exercise 3.41. An infinitely long channel of width π with its lower edge along the real axis has a source of strength 6π at $z = 0$ and a sink of strength 6π at $-3 + \pi i$. Find the complex potential for the flow. Use *FlowTool* to visualize the streamlines of the flow. Does any of the flow escape to infinity? Why or why not? **Try it out!**

Exercise 3.42. Use the same scenario as in Exercise 3.41, but suppose that the sink at $-3 + \pi i$ has strength 2π. Describe the behavior at infinity. Find the complex potential and use *FlowTool* to plot the streamlines. Does the result make physical sense? **Try it out!**

3.8   Flows in Other Regions

The methods used to solve flow problems in a strip apply to any region with sources or sinks along the boundary. All that we need is a conformal mapping from the region to the upper half-plane $\mathbb{H}$ such that the boundary of the region is mapped to the real axis. It is then a matter of determining where the sources and sinks are mapped and accounting for behavior at infinity. We can find the complex potential for the transformed problem in $\mathbb{H}$ and compose it with the conformal mapping to obtain the solution.

Example 3.43. For the region $R = \mathbb{H} - \{z \mid |z| \leq 1\}$ suppose a flow in R is generated by a source of strength 2π at $z = -1$ and a sink of strength 2π at $z = 1$. We find the complex potential for this flow as follows:

(a) The mapping $f(z) = z + 1/z$ maps R conformally to $\mathbb{H}$.

(b) $f(-1) = -2$ and $f(1) = 2$.

(c) The complex potential in $\mathbb{H}$ is $\Omega(z) = \mathrm{Log}(z + 2) - \mathrm{Log}(z - 2)$.

(d) Composing Ω with f gives $\widetilde{\Omega}(z) = \mathrm{Log}(z + 1/z + 2) - \mathrm{Log}(z + 1/z - 2)$.

(e) The flow lines are given by the level curves of $\mathrm{Im}\,\widetilde{\Omega}$.

The flow lines are shown in Figure 3.13.

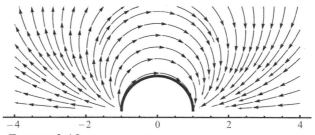

FIGURE 3.13. Flow around cylinder with sources and sinks.

Apply the methods outlined in the example in the next exercise.

Exercise 3.44. Let $R = \{z \mid 0 \le \arg(z) \le \pi/3\}$. Assume there is a source of strength 4π at $z = 0$ and a sink of strength 2π at $z = 2$. Find the complex potential for the flow in R. The streamlines are shown in Figure 3.14.

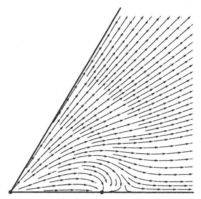

FIGURE 3.14. Streamlines for Exercise 3.44.

Even with a sophisticated graphing program it may not be easy to obtain a sketch of the flow lines. The Mathematica code used to generate Figure 3.14 is

```
H[z_] = 4*Log[z] - 2*Log[z - 8];
F[z_] = H[z^3] // Simplify;
G[z_] = F'[z] // Simplify;
a = StreamPlot[{Re[G[x + I y]], -Im[G[x + I y]]},
    {x, 0.01, 4}, {y, 0.01, 4},
    AspectRatio -> Automatic, StreamPoints -> Fine,
    FrameStyle -> Thick, Frame -> None,
    RegionFunction -> Function[{x, y, z}, y < Sqrt[3]*x]];
b = Plot[Sqrt[3]*x, {x, 0, 4}, PlotStyle -> {Thick, Black}];
c = Graphics[{PointSize[.02], Point[{0, 0}]}];
d = Graphics[{PointSize[.02], Point[{2, 0}]}];
e = Plot[0, {x, 0, 4}, PlotStyle -> {Thick, Black}];
Show[a, b, c, d, e]
```

In the Mathematica code, the RegionFunction statement is used to plot streamlines only in the region of interest. We have chosen to use the StreamPlot command rather than the ContourPlot command. Both are useful, but StreamPlot uses arrows that indicate the direction of the flow. Observe that StreamPlot requires a vector field as the input which we find by differentiating the complex potential function and then taking the conjugate, which gives us the Polya field. When the potential function contains the complex logarithm function, the plot may have some breaks in the flow lines that do not look right. Sometimes it is possible to fix this by combining the logarithm terms. For example, in the code fragment, the function could have been defined as

```
H[z_]=Log[z^4/(z-8)^2]
```

The *FlowTool* applet provides the first quadrant as a domain available for study. In the absence of any sources or sinks on the boundary, the uniform flow in this region is the flow around a corner that we have already mentioned. By including sources or sinks some some interesting flows can be seen.

Exploration 3.45. Develop a procedure for finding the complex potential for a flow in the first quadrant with sources or sinks on the boundary. Use your method to find the complex potentials for (a) - (c). Use the *FlowTool* applet to investigate the flow.

(a) Sources of equal strength at $z = 1$ and $z = i$.

(b) A source and a sink of equal strengths at $z = 1$ and $z = i$.

(c) Sources of equal strength at $z = 1$ and $z = i$, and a sink of double strength at the origin.

3.9   Flows inside the Disk

The methods can be applied to fluid flow inside a disk. The next set of exercises explores several flows inside a disk. Obtaining the graphical output for these flows can be challenging. We present the Mathematica code to generate one of the plots.

Exercise 3.46. Let $\mathbb{D} = \{z \in \mathbb{C} \mid |z| < 1\}$. Show that

$$f(z) = i\,\frac{1-z}{1+z}$$

is a conformal mapping from the unit disk $\mathbb{D}$ to the upper half plane $\mathbb{H}$. Determine the images of the points ± 1 and $\pm i$. Use the conformal mapping to find a uniform flow inside $\mathbb{D}$. The flow lines are shown in Figure 3.15. Explain why there appears to be both a source and a sink at $z = -1$. *Try it out!*

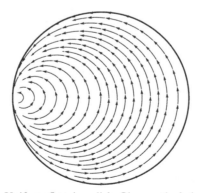

FIGURE 3.15. Uniform flow in a disk. Observe the behavior at $z = -1$.

The Mathematica code for Figure 3.15 is

```
H[z_] = z;
F[z_] = H[I*(1 - z)/(1 + z)] // Simplify;
G[z_] = F'[z] // Simplify;
```

```
a = StreamPlot[{Re[G[x + I y]], -Im[G[x + I y]]},
    {x, -1.2, 1.2}, {y, -1.2, 1.2}, AspectRatio -> Automatic,
    StreamPoints -> Automatic, Frame -> None, FrameStyle -> Thick,
    RegionFunction -> Function[{x, y, z}, x^2 + y^2 < 1]];
b = ParametricPlot[{Cos[t], Sin[t]}, {t, 0, 2*Pi},
    PlotStyle -> {Thick, Black}];
Show[a, b]
```

Exploration 3.47. This exploration considers flows inside a disk.

(a) Find the velocity field of the ideal fluid flow inside $\mathbb{D}$ if there is a source of strength 2π located at $z = 1$. The total strength of the sources in a problem must be balanced by a sink or sinks of equal strength. Where is the sink in this problem? Can you control its location?

(b) Find the complex potential for the flow in $\mathbb{D}$ if there is a sink of strength 2π at $z = i$.

(c) Suppose there are equal strength sources on the boundary of $\mathbb{D}$ located at $z = 1$ and $z = -1$. Consider the location of the resulting sink. Why does this occur?

3.10   Interval Sources and Sinks

In this section we extend some of our methods. Instead of looking at point sources or sinks, we consider interval sources or sinks. This will allow us to model phenomena such as flow through a levee or the electric field generated by a line of charges. We present the material as a series of exercises which lead to the key results.

We begin by examining the flow lines generated by a uniformly distributed source of total strength 2π located along the interval $a \leq x \leq b$. We expect to see a vector field as in Figure 3.16.

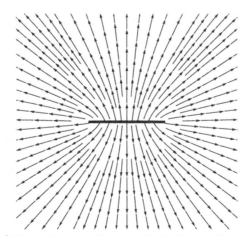

FIGURE 3.16. Integral curves of a vector field with a uniform interval source.

The next set of exercises shows that the complex potential for an interval source can be derived as a limiting process of a collection of point sources as the number of points approaches infinity.

Exercise 3.48. Let $a = x_0, x_1, \ldots, x_n = b$ be $n + 1$ equally spaced points in the interval $[a, b]$. Assume there is a source of strength $2\pi/(n + 1)$ at each x_j. Find the complex potential $\Omega(z)$ for the flow in $\mathbb{H}$. ***Try it out!***

Now we let $n \to \infty$ and recognize the limit as a definite integral.

Exercise 3.49. Let $\Delta x_j = x_{j+1} - x_j$, $j = 0, 1, \ldots, n$. Express the result of the last exercise as a Riemann sum on $[a, b]$. Show that its limit is the definite integral

$$\frac{1}{b - a} \int_a^b \text{Log}(z - x) \, dx. \tag{3.13}$$

Try it out!

Exercise 3.50. Obtain the definition of the complex potential for a uniform interval source by using integration by parts to show the that the integral (3.13) is

$$\Omega_a^b(z) = \frac{b - z}{b - a} \text{Log}(z - b) + \frac{z - a}{b - a} \text{Log}(z - a) - 1. \tag{3.14}$$

Try it out!

Exercise 3.51. Use (3.14) to find the complex potential in $\mathbb{H}$ for a flow that has a uniform source of strength 2π along the interval $[0, 3]$ and a uniform sink of strength 2π along the interval $[-2, -1]$. The graph appears in Figure 3.17. ***Try it out!***

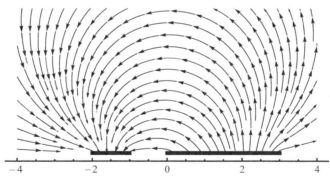

FIGURE 3.17. Flow lines with interval source and sink.

Next we extend interval sources or sinks to the boundaries of other regions. We must be sure the interval sources or sinks are still uniform, as we now show. Suppose in the first quadrant R in $\mathbb{C}$ we have interval sources of equal strength located along the intervals $[1, 3]$ and $[i, 3i]$. We proceed as in previous sections. The conformal mapping $f(z) = z^2$ maps R to $\mathbb{H}$. We determine the behavior of the key intervals under the conformal mapping. The interval $[1, 3]$ is mapped to $[1, 9]$ and $[i, 3i]$ is mapped to $[-9, -1]$. We know how to find the complex potential in $\mathbb{H}$ from (3.14). We see that $\Omega(z) = \Omega_1^9(z) + \Omega_{-9}^{-1}(z)$. We then find the complex potential in R by composition: $\widetilde{\Omega}(z) = \Omega(z^2)$. Unfortunately this result is not quite correct because sources should be uniformly distributed across the intervals in R. When we apply the conformal mapping the intervals in $\mathbb{H}$ are no longer uniformly distributed.

To understand what is happening, focus on the interval $[1, 3]$ under the mapping $f(z) = z^2$. It is mapped to $[1, 9]$. The first half, $[1, 2]$, is mapped to $[1, 4]$ and the second half, $[2, 3]$, is mapped to $[4, 9]$. Hence the density of points is less in the second half of the interval. Thus, when solving for the complex potential in $\mathbb{H}$ we cannot treat the intervals as though the source is uniformly distributed. Recall that (3.14) was derived under the assumption that the source was uniformly distributed along the interval. We must instead use (3.13) and take into account the non-uniform distribution. Thus instead of computing the integral $\int_1^9 \text{Log}(z - x)\, dx$ we compute $\int_1^3 \text{Log}(z - x^2)\, dx$.

Exercise 3.52. Find a complex potential for the flow in the first quadrant generated by uniform interval sources along the intervals $[1, 3]$ and $[i, 3i]$. Plot the streamlines.

Exercise 3.53. Find the complex potential for an ideal flow in the region

$$R = \{re^{i\theta} \mid r \geq 0, \quad 0 \leq \theta \leq \frac{\pi}{3}\}$$

with a uniform source of strength 2π located along the interval $[2, 4]$. *Try it out!*

Exercise 3.54. Find the complex potential for an ideal flow in the infinite channel

$$R = \{z \mid 0 \leq \text{Im}\, z \leq 2\}$$

with a uniform source of strength 2π located on the boundary along the interval $[1+2i, 4+2i]$. *Try it out!*

A natural extension of the previous material on interval sources is to consider intervals with a non-uniform density. This small project was inspired by Potter [3].

Small Project 3.55. Suppose a function $\lambda : [a, b] \subset \mathbb{R} \to \mathbb{R}$ has the property that

$$\int_a^b \lambda(x)\, dx = S,$$

where S is the total strength of the generalized source on $[a, b]$. We think of $\lambda(x)$ as giving the source density at $x \in [a, b]$. The goal is to find the complex potential for the interval source with variable density. If $\lambda(x) = S/(b - a)$, for all x in $[a, b]$, then we obtain a uniformly distributed source along the interval.

Rather than attempting to find the complex potential directly, it is better to first find the underlying vector field $\overline{V(z)}$. Subdivide the interval into n equal subintervals, where the n th subinterval is $[x_i, x_{i+1}]$. Now, consider the vector field $\overline{V_i(z)}$ having a source at x_i with strength

$$S_i = \frac{b - a}{n}\lambda(x_i).$$

Show by summing the individual vector fields and taking a limit as $n \to \infty$ that the vector field is given by

$$\overline{V(z)} = \int_a^b \frac{\lambda(x)}{\bar{z} - x}\, dx. \tag{3.15}$$

From (3.15) it follows that the complex potential is

$$\Omega(z) = \int \left(\int_a^b \frac{\lambda(x)}{z - x}\, dx \right) dz. \tag{3.16}$$

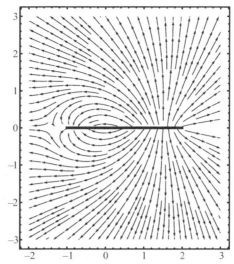

FIGURE 3.18. Flow lines for variable density interval source.

For $\lambda(x) = x$ on the interval $[-1, 2]$ the source density takes on both positive and negative values, with the positive density increasing towards the right end of the interval. The strength of the variable density source is

$$\int_{-1}^{2} x \, dx = \frac{3}{2}.$$

Compute $V(z)$ by using the substitution $w = z - x$ in the integral. Use the result to find the complex potential and plot the flow lines. The result is shown in Figure 3.18.

If λ is a polynomial, then a formula for $V(z)$ can be found explicitly using the same substitution, $w = z - x$. The reader is encouraged to find and plot the complex potential for $\lambda(x) = x^2$ on $[0, 1]$.

3.11   Steady State Temperature Problems

A common problem in applied mathematics is to find the steady state temperature in a region of the plane given the temperature distribution along the boundary of the region. For the upper half plane $\mathbb{H}$, the problem takes the form: Given a piecewise continuous, bounded function f defined on the real axis (i.e., the temperature), find a function T in $\mathbb{H}$ such that $\Delta T = 0$ and T agrees with f along the boundary. The solution involves finding a harmonic function in $\mathbb{H}$. We are seeking solutions that are physically meaningful. For example, if the boundary values are identically zero, then $T_1(x, y) = 0$ and $T_2(x, y) = e^x \sin y$ are both solutions in $\mathbb{H}$. However, T_2 is unbounded, which does not correspond to a real temperature distribution.

Every time we construct a complex potential function in $\mathbb{H}$, both the real and imaginary parts are harmonic. To see if our methods are useful, we discuss how to manipulate the boundary values in an example.

We begin with the complex potential for an ideal flow in $\mathbb{H}$ with a single point source of strength 2π at $z = 0$. We know that the complex potential is $\Omega(z) = \text{Log } z$, so

$$\Omega(z) = \ln|z| + i \text{ Arg}(z). \tag{3.17}$$

Along the real axis, when $x > 0$, $\text{Arg}(x) = 0$, and when $x < 0$, $\text{Arg}(z) = \pi$. Thus the argument function is piecewise constant along the real axis.

Suppose the steady state temperature distribution along the boundary is

$$f(x) = \begin{cases} 0 & \text{if } x < 0 \\ 100 & \text{if } x \geq 0. \end{cases} \tag{3.18}$$

From (3.17), we see that multiplying $\text{Im } \Omega(z)$ by $100/\pi$ gives a harmonic function in $\mathbb{H}$ satisfying the boundary condition

$$\frac{100}{\pi} \text{Arg}(x) = \begin{cases} 100 & \text{if } x < 0 \\ 0 & \text{if } x \geq 0. \end{cases} \tag{3.19}$$

This is almost correct. To get the boundary values switched to the correct intervals we need some properties of the argument function. The next exercise leads to the correct result.

Exercise 3.56. If $\text{Im } z > 0$, then show that reflecting $\text{Arg}(z)$ across the y-axis gives $\text{Arg}(-\bar{z})$. Show that

$$\text{Arg}(-\bar{z}) = -\text{Arg}(-z).$$

Try it out!

From Exercise 3.56, we see that

$$g(x, y) = -\frac{100}{\pi} \text{Arg}(-z)$$

is harmonic and satisfies the correct boundary conditions. The graph of $g(x, y)$ is shown in Figure 3.19. The solution $g(x, y)$ is the imaginary part of the complex potential

$$\Omega(z) = -\frac{100}{\pi} \text{Log}(-z).$$

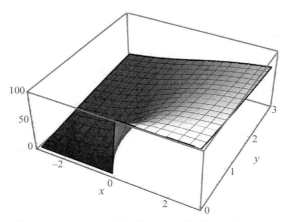

FIGURE 3.19. Steady state temperature distribution with piecewise constant boundary values.

The streamlines correspond to isotherms—curves of constant temperature.

The next exercises explore slight variations on this problem. We then look at problems where segments of the boundary are not insulated (i.e., there are segments that have non-constant temperatures) and see how our methods can be applied.

Exercise 3.57. Solve the steady state temperature problem in $\mathbb{H}$ where the boundary temperature is given below. Hint: This is similar to the above example: start with a slightly different complex potential function.

$$f(x) = \begin{cases} 100 & \text{if } x < 2 \\ 0 & \text{if } x \geq 2. \end{cases} \tag{3.20}$$

Try it out!

Exercise 3.58. Solve the steady state temperature problem in $\mathbb{H}$ where the boundary temperature is given below. Hint: Add a constant to the function, seeing the effect it has on the other boundary value.

$$f(x) = \begin{cases} 100 & \text{if } x < 2 \\ 50 & \text{if } x \geq 2. \end{cases} \tag{3.21}$$

Try it out!

The boundary conditions we have been considering can be extended to more segments. The next exercise leads to the algorithm for solving the general problem.

Exercise 3.59. Solve the steady state temperature problem in $\mathbb{H}$ where the boundary temperature is given by

$$f(x) = \begin{cases} 100 & \text{if } x < -3 \\ 50 & \text{if } -3 < x < 2 \\ 25 & \text{if } x \geq 2. \end{cases} \tag{3.22}$$

Make a contour plot of your solution and observe that the contours represent curves of constant temperature. *Try it out!*

The general problem of finding the temperature distribution in the half-plane given piecewise constant boundary conditions is a standard application in complex analysis texts, as in [5]. The problem usually is presented as a boundary value problem for Laplace's equation.

Exploration 3.60. Solve:
$$T_{xx} + T_{yy} = 0$$

subject to

$$T(x,0) = \begin{cases} k_0 & \text{if } -\infty < x < x_1 \\ k_1 & \text{if } x_1 < x < x_2 \\ \vdots & \cdots \\ k_n & \text{if } x_n < x < \infty. \end{cases}$$

Using the techniques developed in the exercises, derive the general solution.

Next we consider a more general problem where segments of the boundary are not kept at constant temperature. Suppose we have the boundary temperature distribution

$$f(x) = \begin{cases} 100 & \text{if } x < 0 \\ 0 & \text{if } x > 1. \end{cases} \tag{3.23}$$

No temperature is specified on the interval $(0, 1)$. If it is not insulated then we expect the temperature to change linearly from 100 degrees to 0 degrees (which can be seen by solving the one-dimensional heat flow problem on the interval $[0, 1]$ with the endpoints held at 100 and 0 degrees).

We now investigate how our techniques can be applied to this problem. We find the steady state temperature distribution in $\mathbb{H}$ for the boundary condition given in (3.23). The complex potential for an ideal flow with an interval source uniformly distributed along the interval $[0, 1]$ is, from (3.14),

$$\Omega_0^1(z) = (1 - z)\text{Log}(z - 1) + z\text{Log}(z) - 1. \tag{3.24}$$

We know that $\text{Im}\,\Omega_0^1(z)$ must be constant along the intervals $x < 0$ and $x > 1$. We determine the constants by choosing a test value from each interval:

$$\text{Im}\,\Omega_0^1(-1) = \text{Im}(2\text{Log}(2) - \text{Log}(-1) - 1) = -\pi,$$

and

$$\text{Im}\,\Omega_0^1(2) = \text{Im}(-\text{Log}(1) + 2\text{Log}(2) - 1) = 0.$$

Hence multiplying our complex potential by $-100/\pi$ gives the correct temperatures on the two intervals that are held constant. But what about the interval $(0, 1)$? In Figure 3.20, we show the graph of $\text{Im}\,\Omega_0^1(x)$ for $-3 < x < 3$ and observe the linear behavior on the interval $(0, 1)$.

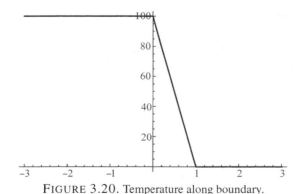

FIGURE 3.20. Temperature along boundary.

The graph of the solution on $\mathbb{H}$ is shown in Figure 3.21. The surface representing the temperature matches the boundary values.

The next small project asks you to generalize the example to a general problem of the same type.

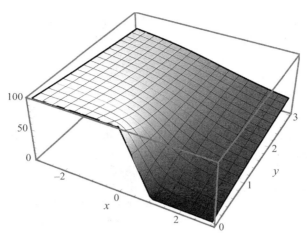

FIGURE 3.21. Solution to heat equation.

Small Project 3.61. Find a general formula for the steady-state temperature distribution $T(x, y)$ in the half-plane with the boundary data

$$T(x,0) = \begin{cases} k_1 & \text{if } -\infty < x < x_1 \\ k_2 & \text{if } x_2 < x < x_3 \\ \vdots & \dots \\ k_n & \text{if } x_n < x < \infty \end{cases}$$

We assume $x_1 < x_2 < \dots < x_n$ and that $T(x, 0)$ is linear on the intervals between the x_i locations.

3.12   Flows with Source and Sinks not on the Boundary

In this section we extend the ideas developed thus far to a wider array of applications. Consider the upper half plane $\mathbb{H}$, with a source located at $z = i$, not on the boundary. We expect the flow to look like an ordinary source near $z = i$, but since it is constrained to stay in $\mathbb{H}$ the real axis deflects it so that it runs parallel to the boundary. Figure 3.22 shows the flow lines we expect.

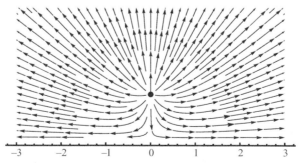

FIGURE 3.22. Flow in $\mathbb{H}$ with a single source at $z = i$.

To obtain the complex potential that gives Figure 3.22, we balance the source at $z = i$ with a source of equal strength at $z = -i$. By the principle of superposition, the vertical components of the underlying vector fields will sum to zero along the real axis. The resulting complex potential is $\Omega(z) = \text{Log}(z - i) + \text{Log}(z + i)$ or, equivalently, $\Omega(z) = \text{Log}(z^2 + 1)$. It is not analytic in on any punctured neighborhood of i. Thus, Ω is not formally a complex potential in all of $\mathbb{H}$. However, by choosing different branches of the complex logarithm, it always possible to construct an analytic complex potential in a simply connected domain not containing i.

Balancing sources or sinks across the boundaries, combined with our earlier work, can be used to deal with sources or sinks in the interior of a region. Needham [2] refers to this approach as the *method of images*. The next exercise finds the complex potential for a flow in the first quadrant with a source located at the interior point $z = 1 + 2i$.

Exercise 3.62. Find the flow of an ideal fluid in the first quadrant with a single source located at $z = 1 + 2i$. Hint: Use the conformal mapping $f(z) = z^2$ to map the first quadrant to $\mathbb{H}$. Determine the location of source by computing $f(1+2i)$. Find the complex potential in $\mathbb{H}$ by balancing the source with another source symmetrically located across the real axis. Compose the result with $f(z)$ to obtain the complex potential of the flow in the first quadrant. The result gives the flow shown in Figure 3.23. ***Try it out!***

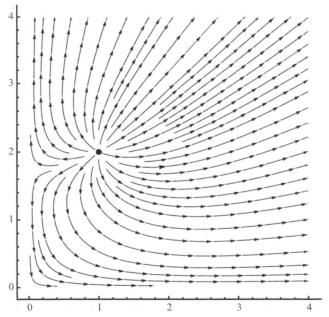

FIGURE 3.23. Flow in the first quadrant with a single source at $z = 1 + 2i$

When dealing with sources and sinks on the boundary and in the interior the notion of *effective strength* comes into play. A source on the boundary must have twice the strength as one in the interior to have the same effective strength.

Exercise 3.63. Find the complex potential for a uniform flow in $\mathbb{H}$ with a source at $z = 0$

and a sink of equal magnitude at $z = i$. Answer: $\Omega(z) = \text{Log}(z^2/(z^2+1))$ The streamlines are shown in Figure 3.24 **Try it out!**

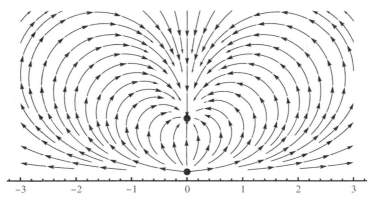

FIGURE 3.24. Balancing a source on the boundary with a sink in the interior.

The techniques developed thus far allow us to combine sources and sinks of various relative strengths both on the boundary and in the interior of a region. The next exercise shows that the complex potential for a complicated flow can be built up from simpler pieces.

Exercise 3.64. Let R be the infinite strip $0 \le \text{Im}\, z \le \pi$. Suppose there is a source of strength 2π at $z = 1$, a sink of strength 2π at $z = 4$, and a source of strength 2π in the middle of the strip at $z = \pi i/2$. Construct the complex potential of the velocity field. The streamlines are shown in Figure 3.25. **Try it out!**

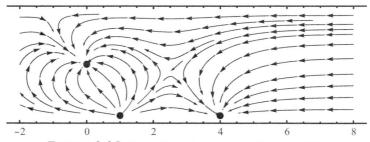

FIGURE 3.25. Streamlines for the flow in Exercise 3.64.

3.13   Vector Fields with Other Types of Singularities

Another application arises when we interpret ideal flows as electric field lines generated by uniform density current in a wire running perpendicular to the plane. We imagine the plane being a copper plate with wires cutting through the plane at different locations. It is convenient to think of these locations as positive or negative charges on the plane. A *dipole* is obtained when a positive and negative charge of equal strength are separated by a small distance. Suppose we have a source of strength 1 located at $z = \epsilon$ and a sink of strength

-1 located at $z = -\epsilon$. The Polya vector field for the electric field is

$$\overline{f(z)} = \frac{1}{\bar{z} - \epsilon} - \frac{1}{\bar{z} + \epsilon}.$$

As ϵ approaches zero, the field vanishes. To prevent the field from vanishing, we need to increase the strengths of the source and sink inversely to the distance between them, so as to keep the field strength constant while the distance between the charges approaches zero.

Exercise 3.65. For the Polya vector field

$$\overline{f(z)} = \frac{1}{2\epsilon} \left(\frac{1}{\bar{z} - \epsilon} - \frac{1}{\bar{z} + \epsilon} \right)$$

compute the limiting field as ϵ goes to zero and the complex potential for the field. Hint: Recall that $\Omega'(z) = f(z)$. **Try it out!**

As shown in Exercise 3.65, the vector field generated by the dipole is $\overline{f(z)} = 1/\bar{z}^2$ and the complex potential is $\Omega(z) = -1/z$. The electric field lines are the level curves of Im $\Omega(z)$ and are shown in Figure 3.26.

Exercise 3.66. Show that the electric field lines for the dipole are circles with centers on the imaginary axis. **Try it out!**

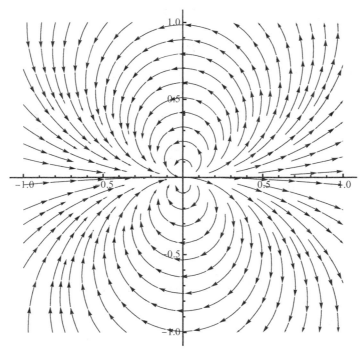

FIGURE 3.26. Electric field lines for a dipole.

Figure 3.26 shows one possible orientation of the dipole. In general, the orientation of the dipole depends on the direction from which the source and the sink approach each other. In this case, the Polya field was $1/\bar{z}^2$ and the *dipole moment* was 1. Other dipole moments are obtained by considering the field $d/\bar{z}^2$, where d is a complex number. For example if $d = 1 + i$, then we obtain the dipole field shown in Figure 3.27.

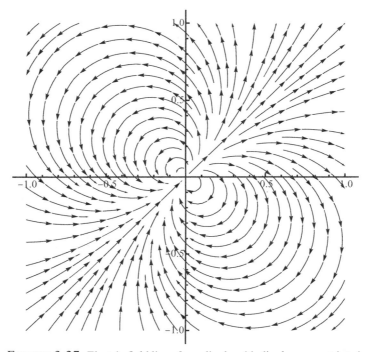

FIGURE 3.27. Electric field lines for a dipole with dipole moment $1 + i$.

Exploration 3.67. Investigate the behavior of the dipole whose field is given by $\overline{f(z)} = (a + bi)/\bar{z}^2$.

It is also possible to consider the problem of multiple charges approaching each other to obtain *multipoles*. As an example, consider the complex potential function $\Omega(z) = 1/z^2$. The electric field lines are shown in Figure 3.28.

Exercise 3.68. What charges are converging to give the electric field lines shown in Figure 3.28? *Try it out!*

An interesting problem involves looking at sources or sinks and multipoles at the same location. For example, consider the Polya field $\overline{V(z)} = 1/\bar{z}^2 + 1/\bar{z}$. The complex potential function is $\Omega(z) = -1/z + \text{Log } z$. As an exercise, plot the electric field lines near zero and on a larger scale. The dipole dominates the behavior near zero and the source dominates the behavior far from zero.

We close this chapter with a discussion of sources and sinks at infinity. Since we now understand the idea of a multipole, we can classify the behavior of a flow at infinity.

Consider the example of uniform flow to the right in the entire plane $\mathbb{C}$. Since fluid appears from the left and disappears to the right it seems reasonable to say that there is both a source and a sink at infinity. We now recognize this as a dipole at infinity.

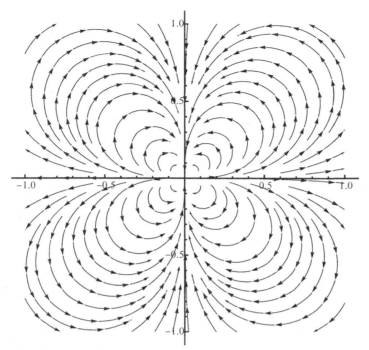

FIGURE 3.28. Electric field lines for a multipole.

To study the behavior of a complex function at infinity, it is standard to replace z with $1/z$ and study the behavior of the resulting function at $z = 0$. For the uniform flow, the complex potential is $\Omega(z) = z$. The behavior at infinity is determined by examining $\Omega(1/z) = 1/z$ at $z = 0$. Thus we have a potential function of $1/z$. Differentiating and taking the conjugate to obtain the underlying vector field, we get $\overline{1/z^2}$, which we recognize as a dipole.

More generally, for a vector field where the net sum of the sources and sinks is N, the strength of the source or sink at infinity is $-N$. If $N = 0$, then there could be a multipole at infinity.

3.14 Bibliography

[1] Ruel V. Churchill and J. W. Brown, *Complex Variables and Applications, 4th ed.,* McGraw-Hill Book Company, New York, 1984.

[2] Tristan Needham, *Visual Complex Analysis,* Oxford University Press, Oxford, UK, 1997.

[3] Harrison Potter, *On Conformal Mappings and Vector Fields*, Senior Thesis, Marietta College, Marietta, OH, 2008.

[4] James Stewart, *Multivariable Calculus, 6th ed.,* Thompson Brooks/Cole, 2008.

[5] Dennis G. Zill and P. D. Shanahan, *A First Course in Complex Analysis with Applications, 2nd ed.,* Jones and Bartlett, Boston-Toronto-London-Singapore, 2009.

4

Anamorphosis, Mapping Problems, and Harmonic Univalent Functions

Michael J. Dorff (text),
James S. Rolf (software)

4.1 Introduction

Complex-valued analytic functions have many nice properties that are not possessed by real-valued functions. For example, we say a complex-valued function is analytic if we can differentiate it one time. It is true that if a complex-valued function f is analytic, then we can differentiate it infinitely many times. Complex-valued analytic functions can always be represented as a Taylor series, and they are conformal (that is, they preserve angles when $f' \neq 0$), properties that are not true for real-valued functions that can be differentiated one time. Why does an analytic function have these properties? If $f = u + iv$ is an analytic function, then its real part, $u(x, y)$, and its imaginary part, $v(x, y)$, satisfy Laplace's equation and are both harmonic. Also, u and v satisfy the Cauchy-Riemann equations and are harmonic conjugates of each other. In this chapter we discuss some ideas and problems related to a collection of univalent (i.e., one-to-one) complex-valued functions, $f = u + iv$, where u and v satisfy Laplace's equation but not necessarily the Cauchy-Riemann equations. These functions are known as *harmonic univalent functions* or mappings and contain analytic univalent functions as a subset. Analytic univalent functions have been studied since the early 1900s, and thousands of research papers have been written about them. The study of harmonic univalent mappings is a fairly recent area of research. So, it is natural to consider the properties of analytic univalent functions as a starting point for our study of harmonic univalent mappings. A general question will be "What properties of analytic univalent functions are true for the larger class of harmonic univalent functions?"

Section 4.2 discusses how to determine the image of domains in $\mathbb{C}$ under a collection of complex-valued functions known as Möbius maps and introduces the applet *ComplexTool* as an aid to visualize them. Section 4.3 presents some background about the family of univalent analytic functions. Section 4.4 introduces the fundamentals of harmonic univalent functions. The study of harmonic univalent functions from the perspective of univalent complex-valued analytic functions is a new area of research. Finding examples of them is not easy, but a useful method of doing so is discussed in Section 4.5. Sections 4.2–4.5

should be read first. After that, the remaining sections can be read in any order and are independent of each other. Three applets are used in this chapter:

1. *ComplexTool* is used to plot the image of domains in $\mathbb{C}$ under complex-valued functions.

2. *ShearTool* is used to plot the image of domains in $\mathbb{C}$ under a complex-valued harmonic function that is formed by shearing an analytic function and a dilatation; the user enters the analytic function and dilatation without having to solve explicitly for the harmonic function.

3. *LinComboTool* is used to plot and explore the convex combination of complex-valued harmonic polygonal maps.

They can be accessed online at `www.maa.org/ebooks/EXCA/applets.html`. Each section contains examples, exercises, and explorations that involve using the applets. You should do all of the exercises and explorations, many of which present functions and concepts that will be used in the chapter. There are additional exercises at the end of the chapter. In the study of harmonic univalent functions, there are many unsolved problems, some of which are mentioned. There are also short and long projects that are suitable as research problems for undergraduate students.

The goal of the chapter is not to give a comprehensive introduction to this topic, but to engage readers with its general notions, questions, and techniques, and to encourage readers to actively pose and answer their own questions. To understand the nature and purpose of this text better, the reader should read the Introduction. The study of harmonic univalent functions has many interesting problems that undergraduates can investigate with computers and the applets. I hope that students will explore the ideas in this chapter and will prove some new results in the field.

4.2   Anamorphosis and Möbius Maps

Numb3rs is a U.S. television show that ran from 2005 to 2010 that dealt with a group of FBI special agents trying to solve crimes. One of the FBI agents is Don Eppes who has a mathematical genius brother, Charlie, who helps the FBI solve their cases. In the episode "Jack of All Trades" (season 5, episode 4), the FBI is trying to catch a thief who has eluded them for two years. They do not even have an accurate image of his face. At one point in this episode, the thief escapes by stealing a bus at night and driving away. Before he escapes, one agent takes his picture with a flash camera. Unfortunately, the flash from the camera reflects off the bus window obscuring the face of the thief. However, there is a metal cylindrical thermos in the photo that displays a distorted image of the thief. Using mathematics, Charlie recreates the image of the thief's face, giving the FBI information that leads to his capture.

The method of producing a distorted image that appears normal by viewing it through a curved mirror or from an unusual angle is known as anamorphosis. Anamorphosis was studied by painters in the 15–16th centuries as they were trying to understand perspective. Two famous paintings that display anamorphosis are Jan van Eyck's "The Arnolfini Marriage" (see Figure 4.1) and Hans Holbein's "The Ambassadors" (see Figure 4.2). Van Eyck's painting is thought to depict the Italian merchant Giovanni Arnolfini and his wife.

FIGURE 4.1. Van Eyck's "The Arnolfini Marriage."

On the wall behind them is a curved mirror reflecting a distorted image of the scene. In Holbein's painting there is a distorted shape lying diagonally at the bottom. That it is a skull can be seen more clearly when the viewer is looking at the painting from a certain angle.

FIGURE 4.2. Holbein's "The Ambassadors."

In modern times anamorphosis has continued to be used. In the sketch "Mysterious Island" by the Hungarian artist István Orosz (see Figure 4.3) the viewer sees the image of a seashore with hills and the sun in the background, two men walking, and a ship being tossed in the sea. If a cylindrical mirror is placed on top of the sun, the image of the author Jules Verne appears on the cylinder.

In 2010 the group Preventable (also known as the Community against Preventable Injuries) caused drivers to slow down while driving near a school in British Columbia,

FIGURE 4.3. Orosz's "Mysterious Island." Images copyright István Orosz. Used with permission.

Canada by putting an anamorphosis image on the highway (see Figure 4.4). The image is not clear until the driver reached a point on the highway when it appeared that a young girl had darted into the road chasing after a pink ball. In the same way that the anamorphosis image could appear to the driver, so could a child suddenly run in front of the driver.

FIGURE 4.4. Anamorphosis image of a child chasing a ball into the street. Used with permission.

This idea of how an image is distorted is used in complex analysis when trying to visualize how a domain is mapped by a complex-valued function. We begin by looking at maps known as Möbius transformations. If we have a one-dimensional real-valued function, such as $f(x) = x^2$, the graph of the function tells us about some of the properties of f (i.e., zeros, one-to-one, increasing, etc.). We want to do the same thing for complex-valued functions, but we would need a 4-dimensional graph (two dimensions for the domain and two for the range). We can represent some of the properties of a complex-valued function by looking at specific sets in the domain and seeing where the complex-valued function $w = f(z)$ takes them.

We can use the accompanying applet *ComplexTool* to graph the image of domains under complex-valued functions or maps. Open *ComplexTool* (see Figure 4.5). Suppose that we want to find the image of the unit disk under the map $f(z) = z+2-i$. In the middle section near the top there is a box that has **f(z) =** before it. In it, enter $z + 2 - i$. Below it, there

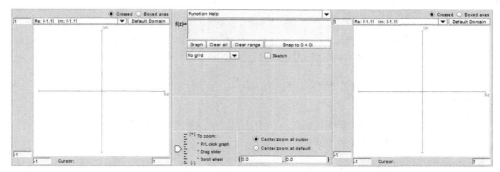

FIGURE 4.5. The applet *ComplexTool*.

is a window that states **No grid**. Click on the down arrow ▼ and choose **Circular grid**. An image of a circular grid will appear on the left, which we call the z-plane. Click on the button **Graph** in the middle section below the function you entered. The image of the circular grid will appear on the right, which we call the w-plane. To reduce the size of the image, click on the down arrow ▼ above the image and chose a different size, such as **Re: [−3,3] Im: [−3,3]**. You can move the axes so that the image is centered by positioning the cursor over the image, clicking on the left mouse button, and dragging the image to the left (see Figure 4.6).

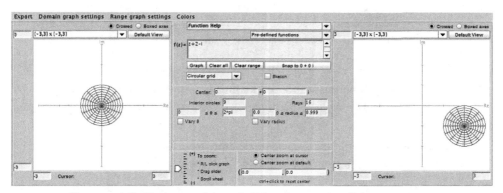

FIGURE 4.6. The image of the unit disk under the map $f(z) = z + 2 - i$.

Exploration 4.1. You can graph the halfdisk, $S = \{z \mid |z| < 1, \operatorname{Im} z > 0\}$, by using a circular disk and restricting the the the θ value to be between 0 and π in the middle panel of the applet. For the following functions determine the image of S. Make a conjecture about how a domain is transformed under the maps $f(z) = z + A$, $f(z) = Bz$, and $f(z) = e^{i\theta}z$, where $A = a_1 + ia_2 \in \mathbb{C}$, $B > 0$, and $\theta \in \mathbb{R}$.

(a) $f(z) = z - 1$ (b) $f(z) = z + i$ (c) $f(z) = z - 1 + i$

(d) $f(z) = 0.5z$ (e) $f(z) = 2z$ (f) $f(z) = 2.5z$

(g) $f(z) = e^{i\pi/4}z$ (h) $f(z) = e^{i\pi/2}z$ (i) $f(z) = e^{-i\pi/4}z$.

Try it out!

The previous exercise helps us see characteristics of the maps:

(a) if $A = a_1 + ia_2 \in \mathbb{C}$, then the map $f(z) = z + A$ moves the image domain a_1 units horizontally and a_2 units vertically

(b) if $B > 0$, then the map $f(z) = Bz$ scales (i.e., expands or shrinks) the domain by a factor of B

(c) if $\theta \in \mathbb{R}$, then the map $f(z) = e^{i\theta}z$ rotates the domain about the origin in a counter-clockwise direction by an angle of θ.

Exercise 4.2. Justify that $f(z) = z + A$, $f(z) = Bz$, and $f(z) = e^{i\theta}z$ map circles onto circles and lines onto lines. ***Try it out!***

If we think of a line as a circle with infinite radius, then we can say that the functions preserve circles. That is, they map circles onto circles.

The inversion map is $f(z) = \frac{1}{z}$ and if we write $z = re^{i\theta}$,

$$f(z) = \frac{1}{z} = \frac{1}{re^{i\theta}} = \frac{1}{r}e^{-i\theta}.$$

Thus, the function $f(z) = \frac{1}{z}$ scales the domain by a factor of $\frac{1}{r}$ and the factor $e^{-i\theta}$ reflects it across the real axis.

Exploration 4.3. Open *ComplexTool*. We want to explore the image of the unit circle under $f(z) = \frac{1}{z}$. To do this, choose the option **Circular grid**. In the middle panel below this, select **Interior circles:** to be 1 and **Rays:** to be 0. This should give the domain in the z-plane as the unit circle. Click on the down arrow ▼ above the image and chose **Re: [−3,3] Im: [−3,3]** for the z-plane. Do this also for the w-plane. Left click on the **Graph** button to produce the image of the unit circle in the right box. Left click on the domain circle and drag it around. See what happens to the image as you do this.

(a) As you move the center of the domain circle away from the origin in the z-plane, what happens to the size of the image circle in the w-plane?

(b) What shape does the image take when the domain circle intersects the origin?

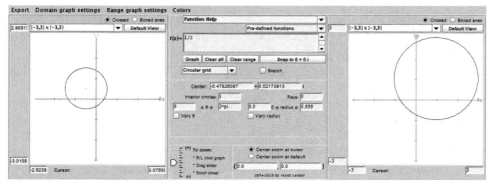

FIGURE 4.7. The image of a circle not intersecting the origin under the map $f(z) = \frac{1}{z}$.

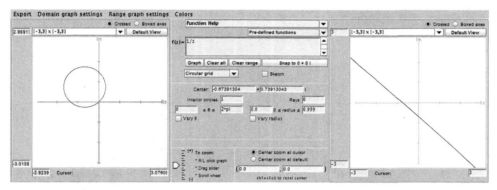

FIGURE 4.8. The image of a circle intersecting the origin under the map $f(z) = \frac{1}{z}$.

(c) What happens to the size of the image circle as the domain circle moves off the screen?

(d) Can you move the domain circle to a position so that the image is neither a circle nor a line?

(e) Explain your observation in parts (a)–(d) in terms of the fact that the function $f(z) = \frac{1}{z}$ scales the domain by a factor of $\frac{1}{r}$ and reflects it across the real axis

Try it out!

The inversion map $f(z) = \frac{1}{z}$ maps circles to either circles or lines. Let's look at the image of vertical lines under $f(z) = \frac{1}{z}$. If the line L_1 starts at the bottom of the imaginary axis and travels upward, it can be described as $z = 0 + iy$, where y varies from $-\infty$ to ∞. The image of L_1 is

$$\frac{1}{z} = \frac{1}{iy} = i\frac{-1}{y}.$$

Because the real part of the image is zero, L_1 is mapped into the imaginary axis. As y starts at $-\infty$ and increases, the image starts at 0 and moves upward. When y passes through 0, the image will wrap around from the top of the imaginary axis and go to the bottom of the imaginary axis. As y continues up along the positive imaginary axis, the image will continue up from the bottom of the imaginary axis. You can use the applet to visualize this. Open *ComplexTool* and enter the function $f(z) = 1/z$. Chose the window size of **Re: [-3,3] Im: [-3,3]** for the z-plane and the w-plane. Click on the box next to the **Sketch** feature so that a checkmark appears. Then click on the **Graph** button. Place the cursor arrow on the imaginary axis in the z-plane, hold down the left mouse button, and drag the cursor upward. As you do so, the image in the w-plane should appear (see Figure 4.9).

Let's see what the image of the vertical line with $\mathrm{Re}\, z = \frac{1}{2}$ is. In the z-plane, let $z = x + iy$. Because

$$\frac{1}{z} = \frac{\bar{z}}{z\bar{z}} = \frac{1}{|z|^2}\bar{z} = \frac{x}{x^2 + y^2} - i\frac{y}{x^2 + y^2},$$

if we let $w = u + iv$ in the image domain, then we have

$$u = \frac{x}{x^2 + y^2} \quad \text{and} \quad v = \frac{-y}{x^2 + y^2}.$$

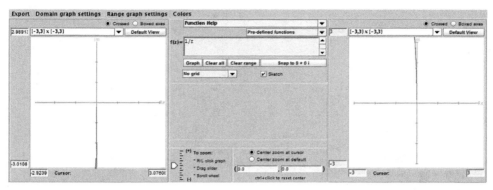

FIGURE 4.9. The image of a vertical line on the imaginary axis under the map $f(z) = \frac{1}{z}$.

In the z-plane, the vertical line with $\operatorname{Re} z = \frac{1}{2}$ can be described by $z = \frac{1}{2} + iy$. So the image of the vertical line under the inversion map is

$$\frac{1}{z} = u + iv = \frac{\frac{1}{2}}{\frac{1}{4} + y^2} + i\frac{-y}{\frac{1}{4} + y^2} = \frac{2}{4y^2 + 1} + i\frac{-4y}{4y^2 + 1}.$$

What is the shape described by u and v? It is a circle centered at $u = 1$ of radius 1 (see Figure 4.10). To see this, notice

$$(u - 1)^2 + v^2 = \left(\frac{2}{4y^2 + 1} - 1\right)^2 + \left(\frac{-4y}{4y^2 + 1}\right)^2$$
$$= \frac{(-4y^2 + 1)^2 + (-4y)^2}{(4y^2 + 1)^2}$$
$$= \frac{16y^4 + 8y^2 + 1}{(4y^2 + 1)^2}$$
$$= 1$$

which is the equation of a circle centered at $u = 1$ of radius 1.

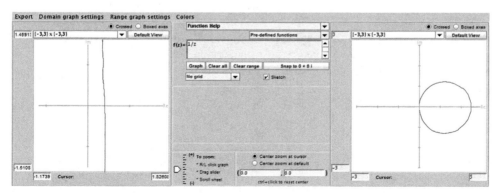

FIGURE 4.10. The image of a vertical line with $\operatorname{Re} z = \frac{1}{2}$ under the map $f(z) = \frac{1}{z}$.

Exercise 4.4. Using this approach, show that $f(z) = \frac{1}{z}$ maps the vertical line $\operatorname{Re} z = c$ onto the circle $\left(u - \frac{1}{2c}\right)^2 + v^2 = \left(\frac{1}{2c}\right)^2$. ***Try it out!***

The inversion map $f(z) = \frac{1}{z}$ takes circles and lines onto circles and lines.

(a) A circle not intersecting the origin is mapped onto a circle not intersecting the origin.

(b) A line intersecting the origin is mapped onto a line intersecting the origin.

(c) A circle intersecting the origin is mapped onto a line not intersecting the origin, and vice versa.

The maps we have discussed so far are special cases of *Möbius transformations*

$$M(z) = \frac{Az + B}{Cz + D},$$

where $A, B, C, D \in \mathbb{C}$ and $AD \neq BC$.

Exercise 4.5. What happens to $M(z) = \frac{Az+B}{Cz+D}$ if $AD = BC$? ***Try it out!***

If $A = 1$, $C = 0$, and $D = 1$, we have $M(z) = z + B$, a translation map. If $B = 0$, $C = 0$, and $D = 1$, we have $M(z) = Az$, which is a scaling, a rotation, or both. If $A = 0$, $B = 1$, $C = 1$, and $D = 0$, we have $M(z) = \frac{1}{z}$, the inversion map. Because

$$\frac{Az + B}{Cz + D} = \frac{\frac{A}{C}(Cz + D) - \frac{AD}{C} + B}{Cz + D} = \frac{A}{C} + \frac{B - \frac{AD}{C}}{Cz + D}$$

if we let

$$f_1(z) = Cz + D, \quad f_2(z) = \frac{1}{z}, \quad \text{and} \quad f_3(z) = \left(B - \frac{AD}{C}\right)z + \frac{A}{C},$$

then the Möbius transformation $M(z) = \frac{Az+B}{Cz+D}$ can be expressed as $(f_3 \circ f_2 \circ f_1)(z)$. We know how the maps f_1, f_2, and f_3 affect a given domain, and since they map circles (lines) to circles (lines), a Möbius transformation will also. By including the point at ∞ in our domain and range, the Möbius transformation is a one-to-one and onto function. Thus, its inverse function exists and is

$$M^{-1}(z) = \frac{Dz - B}{-Cz + A},$$

so the inverse of a Möbius transformation is a Möbius transformation.

Exercise 4.6. Starting with a Möbius transformation M, compute that M^{-1} is $\frac{Dz-B}{-Cz+A}$. ***Try it out!***

We can determine the image of a circle under a Möbius transformation M since a circle is determined by three distinct points. To do this, take three points on the circle, find their images under M, and then determine the circle on which they lie.

Example 4.7. Let's find the image of the unit circle $\{z \mid |z| = 1\}$ under the map

$$M(z) = \frac{z}{1 - z}.$$

Choose three points on the unit circle. We will choose i, -1, and $-i$. Then

$$M(i) = -\frac{1}{2} + \frac{1}{2}i, \quad M(-1) = -\frac{1}{2}, \quad \text{and} \quad M(-i) = -\frac{1}{2} - \frac{1}{2}i,$$

and $-\frac{1}{2} + \frac{1}{2}i$, $-\frac{1}{2}$, and $-\frac{1}{2} - \frac{1}{2}i$ lie on the line, $\{z \mid \text{Re}\{z\} = -\frac{1}{2}\}$, which is the image of the unit circle.

Question: What is the image of the unit disk $\{z \mid |z| < 1\}$ under the map $\frac{z}{1-z}$? Möbius maps will send the interior of the domain into the interior of the image. Since $M(0) = 0$, we know that the unit disk is mapped onto the right half-plane $\{w \mid \text{Re}\{w\} > -\frac{1}{2}\}$.

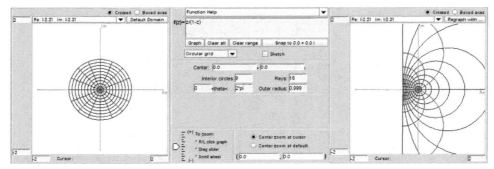

FIGURE 4.11. The image of the unit disk under the Möbius map $M(z) = \frac{z}{1-z}$.

Exercise 4.8. Using the approach in Example 4.7 determine the image of the unit disk $\{z \mid |z| < 1\}$ under the transformation

$$M(z) = \frac{z - i}{z}.$$

Use *ComplexTool* to check your answer. ***Try it out!***

Suppose that we want to map a given circle or line C_1 to another circle or line C_2. How do we construct a function that will do this? Three points uniquely determine a circle and Möbius transformations send circles (lines) to circles (lines). Choose three points z_1, z_2, and z_3 on C_1 and three points w_1, w_2, and w_3 on C_2. Then a Möbius transformation $M(z)$ with

$$M(z_1) = w_1, \quad M(z_2) = w_2, \quad \text{and} \quad M(z_3) = w_3$$

will map C_1 onto C_2.

We can construct $M(z)$ by using the cross-ratio formula

$$\frac{(w - w_1)(w_2 - w_3)}{(w - w_3)(w_2 - w_1)} = \frac{(z - z_1)(z_2 - z_3)}{(z - z_3)(z_2 - z_1)}. \tag{4.1}$$

If one of the terms z_i or w_i is ∞ (i.e., one of the circles C_1 or C_2 is a line), then we omit the expressions containing that term. For example, if $w_1 = \infty$, then (4.1) becomes

$$\frac{w_2 - w_3}{w - w_3} = \frac{(z - z_1)(z_2 - z_3)}{(z - z_3)(z_2 - z_1)}.$$

Let's use (4.1) to find a Möbius transformation that maps z_1 to w_1, z_2 to w_2, and z_3 to w_3.

Example 4.9. Suppose that we want to determine a Möbius transformation that maps the vertical line $l = \{z \mid \text{Re}\{z\} = -\frac{1}{2}\}$ onto the unit circle, C. Pick three points on l. We choose

$$z_1 = -\frac{1}{2}, \quad z_2 = -\frac{1}{2} + i\frac{1}{2}, \quad \text{and} \quad z_3 = \infty.$$

Next, we pick three points on C. Let's choose

$$w_1 = -1, \quad w_2 = i, \quad \text{and} \quad w_3 = 1.$$

Using (4.1), we have

$$\frac{z + \frac{1}{2}}{(-\frac{1}{2} + i\frac{1}{2}) + \frac{1}{2}} = \frac{(w+1)(i-1)}{(w-1)(i+1)}.$$

Solving for w yields

$$w = \frac{z}{1+z}.$$

Thus, the Möbius transformation $M(z) = \frac{z}{1+z}$ maps l onto C. It is the inverse of the transformation used in Example 4.7.

Exercise 4.10. In Example 4.9, work out the details to show that

$$\frac{z + \frac{1}{2}}{(-\frac{1}{2} + i\frac{1}{2}) + \frac{1}{2}} = \frac{(w+1)(i-1)}{(w-1)(i+1)} \quad \text{simplifies to} \quad w = \frac{z}{1+z}.$$

Try it out!

In Example 4.9, we found a Möbius transformation M that maps the line l onto the circle C. We see that M maps the right half-plane $D = \{z \mid \text{Re}\{z\} > -\frac{1}{2}\}$ to the unit disk because $M(0) = 0$. It could have happened that we had constructed a Möbius transformation that maps the right half-plane to the exterior part of the unit circle instead of the unit disk. What can we do at the beginning to guarantee that the Möbius transformation we construct will map the domain onto the region we want? We could have switched the values of w_2 and w_3 so that $w_1 = -1$, $w_2 = 1$, and $w_3 = i$, and constructed a different Möbius transformation that maps l to C. Then, the right half-plane D would have been mapped onto the exterior of the unit circle. To guarantee that the Möbius transformation we construct will map our initial domain onto the region we want, we need to choose the order of the points w_1, w_2, and w_3 correctly. How do we do this? As we travel along l from $z_1 = -\frac{1}{2}$ to $z_2 = -\frac{1}{2} + i\frac{1}{2}$ to $z_3 = \infty$, the right half-plane D is on the right side of the line l. To guarantee that M maps D onto the unit disk, we need to choose the order of the points w_1, w_2, and w_3 so that the unit disk is also on the right side of the unit circle C. This happens in Example 4.9, because we go from $w_1 = -1$ to $w_2 = i$ to $w_3 = 1$.

Exercise 4.11. Determine a Möbius transformation that maps the unit disk $\{z \mid |z| < 1\}$ onto the upper half-plane $\{z \mid \text{Im}\{z\} > 0\}$. Use *ComplexTool* to check your answer. *Try it out!*

There is a nice short video called "Moebius Transformations Revealed" by Douglas Arnold and Jonathan Rogness that illustrates some of the connections between Möbius transformations and motions of the sphere. It can be found by searching the internet.

4.3   The Family S of Analytic, Normalized, Univalent Functions

We will be discussing mapping problems in complex analysis. These problems deal with the properties of a collection of functions that map one domain onto certain image domains. We need some background material. Let $G \subset \mathbb{C}$ be a simply-connected domain and let $\mathbb{D} = \{z : |z| < 1\}$, the unit disk.

Definition 4.12. A function f is *univalent* in G if f is one-to-one in G.

Univalent analytic functions have inverse functions that are analytic.

Example 4.13. Suppose we want to prove that $f(z) = (1 + z)^2$ is univalent in $\mathbb{D}$. A standard argument for that is to let $z_1, z_2 \in \mathbb{D}$ and suppose $f(z_1) = f(z_2)$. Then

$$f(z_1) = f(z_2) \Rightarrow (1 + z_1)^2 = (1 + z_2)^2$$
$$\Rightarrow 1 + 2z_1 + z_1^2 = 1 + 2z_2 + z_2^2$$
$$\Rightarrow z_1^2 - z_2^2 + 2(z_1 - z_2) = 0$$
$$\Rightarrow (z_1 - z_2)(z_1 + z_2 + 2) = 0.$$

Since $|z_1|, |z_2| < 1$, we know that $z_1 + z_2 + 2 \neq 0$. Hence, we must have $z_1 - z_2 = 0$. So, $z_1 = z_2$. Thus, f is one-to-one.

The image of $\mathbb{D}$ under the map $f(z) = (1 + z)^2$ is shown in Figure 4.12.

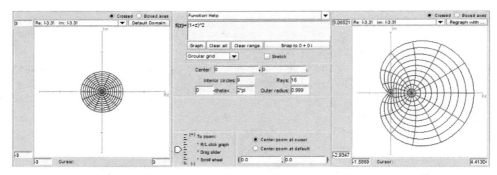

FIGURE 4.12. The image of the unit disk under the map $f(z) = (1 + z)^2$.

Exploration 4.14. From Example 4.13, we know that $f(z) = (1 + z)^2$ is univalent on $\mathbb{D}$. It can be shown that $f(z) = (1 + z)^4$ is not univalent on $\mathbb{D}$ (see Exploration 4.15). Use *ComplexTool* to graph the image of $\mathbb{D}$ under the analytic function $f(z) = (1 + z)^2$ and under the analytic function $f(z) = (1 + z)^4$. What aspects of the images suggest that a function is univalent or is not univalent? Explore by plotting the following further functions in *ComplexTool* and conjecture which are univalent:

(a) $g_1(z) = z - z^2$ (b) $g_2(z) = z - \frac{1}{2}z^2$

(c) $g_3(z) = 2z - z^2$ (d) $g_4(z) = z + \frac{3}{4}z^2$

(e) $g_5(z) = \frac{z}{1-z}$ (f) $g_6(z) = \frac{z^2}{1-z}$

(g) $g_7(z) = \frac{z}{(1-z)^2}$.

Try it out!

Exploration 4.15. Prove that $f(z) = (1 + z)^4$ is not univalent in $\mathbb{D}$.

One way to do this is to find two distinct points $z_1, z_2 \in \mathbb{D}$ such that $f(z_1) = f(z_2)$. You can use *ComplexTool* to help you find z_1 and z_2. Plot the image of $\mathbb{D}$ under $f(z) = (1+z)^4$. You can increase the size of the image by clicking on the left button on the mouse and decrease it by clicking on the right mouse button. Check the **Sketch** box in the top middle section. The **Sketch** command allows you to draw a shape in the domain on the left and see its image under the function on the right. For example, draw a line along the imaginary axis from 0 to i and a line along the imaginary axis from 0 to $-i$; the image curves meet at $f(i) = f(-i)$. Compute $f(i)$ and $f(-i)$ to prove that this is true. This does not prove that f is not univalent in $\mathbb{D}$ since $i, -i \notin \mathbb{D}$. Use the **Sketch** feature of *ComplexTool* to help you find two points $z_1, z_2 \in \mathbb{D}$ such that $f(z_1) = f(z_2)$. To delete the shapes you have drawn, use the **Clear all** button and then regraph your image. Hint: Using *ComplexTool* find two lines in the original domain that are mapped to the line on the real axis from -4 to 0. Parametrize the lines so that for each t, the image of the parametrized lines under f give the same image point. ***Try it out!***

In Exploration 4.14, you may have noticed that the image of $\mathbb{D}$ under $g_2(z) = z - \frac{1}{2}z^2$ is similar to its image under $g_3(z) = 2z - z^2$. This is because $g_3 = 2g_2$. We want to avoid such repetitions. To do so, we will *normalize* all functions in the family of analytic univalent functions defined on $\mathbb{D}$. To do this suppose f_1 is univalent and analytic in the simply connected domain $G \neq \mathbb{C}$. The Riemann mapping theorem can be stated in the following form:

Theorem 4.16 (Riemann Mapping Theorem). *Let $a \in G$. Then there exists a unique function $f_2 : G \to \mathbb{C}$ such that*

(a) $f_2(a) = 0$ and $f_2'(a) > 0$

(b) f_2 is univalent

(c) $f_2(G) = \mathbb{D}$.

Thus $f_3 = f_1 \circ f_2^{-1}$ maps $\mathbb{D}$ to $f_1(G)$ with f_3 being univalent and analytic. So when studying mappings of simply-connected domains, we can just let $\mathbb{D}$ be our domain. Let $f_3 : \mathbb{D} \to \mathbb{C}$ be univalent and analytic. Since f_3 is analytic, it has a power series about the origin,

$$f_3(z) = \alpha_0 + \alpha_1 z + \alpha_2 z^2 + \alpha_3 z^3 + \cdots,$$

that converges in $\mathbb{D}$. Adding a constant translates the image domain and does not affect the univalency. Hence

$$f_4(z) = f_3(z) - \alpha_0 = \alpha_1 z + \alpha_2 z^2 + \alpha_3 z^3 + \cdots$$

is also univalent and analytic in $\mathbb{D}$. Next, $\alpha_1 \neq 0$ because f_4 is univalent, so $f_4'(z) \neq 0$ (for all $z \in \mathbb{D}$); but $f_4'(0) = \alpha_1$. Let

$$f_5(z) = \frac{1}{\alpha_1} f_4(z) = z + \frac{\alpha_2}{\alpha_1} z^2 + \frac{\alpha_3}{\alpha_1} z^3 + \cdots.$$

Multiplying f_4 by $\frac{1}{\alpha_1}$ rotates or stretches (or shrinks) the image domain. Hence f_5 is still univalent and analytic in $\mathbb{D}$. We have normalized our original function f_3 so that $f'(0) = 1$, and $f(0) = 0$.

Definition 4.17. The family of analytic, normalized, univalent functions is denoted by S (from the German word "schlicht" that means "simple" or "plain"); that is,

$$S = \{f : \mathbb{D} \to \mathbb{C} \mid f \text{ is analytic and univalent with } f(0) = 0, f'(0) = 1\}.$$

Thus $f \in S$ implies $f(z) = z + a_2 z^2 + a_3 z^3 + \cdots$.

Exercise 4.18. Show that $f(z) = z + a_2 z^2$ is univalent in $\mathbb{D} \iff |a_2| \leq \frac{1}{2}$. *Try it out!*

Example 4.19. In Exploration 4.14, you graphed the image of $\mathbb{D}$ under $g_2(z) = z - \frac{1}{2}z^2$. While computer images are helpful, they can be misleading and inaccurate. So it is important for us to determine them analytically. How can we determine $g_2(\mathbb{D})$ analytically? We find the image of the boundary of $\mathbb{D}$:

$$
\begin{aligned}
w = g_2(e^{i\theta}) &= e^{i\theta} - \frac{1}{2}e^{2i\theta} \\
&= (\cos\theta + i\sin\theta) - \frac{1}{2}(\cos 2\theta + i\sin 2\theta) \\
&= \left(\cos\theta - \frac{1}{2}\cos 2\theta\right) + i\left(\sin\theta - \frac{1}{2}\sin 2\theta\right) \\
&= u + iv.
\end{aligned}
$$

Thus, $g_2(\partial\mathbb{D})$ is parametrized by

$$
\begin{aligned}
u(\theta) &= \cos\theta - \frac{1}{2}\cos 2\theta \\
v(\theta) &= \sin\theta - \frac{1}{2}\sin 2\theta.
\end{aligned}
$$

This is a cardiod, which is also known as an epicycloid with one cusp (see Figure 4.13).

Definition 4.20. An *epicycloid* is the path traced out by a point on a circle of radius b rolling on the outside of a circle of radius a:

$$
\begin{aligned}
x(\theta) &= (a + b)\cos\theta - b\cos\left(\left(\frac{a}{b} + 1\right)\theta\right) \\
y(\theta) &= (a + b)\sin\theta - b\sin\left(\left(\frac{a}{b} + 1\right)\theta\right).
\end{aligned}
$$

Exploration 4.21. In Exercise 4.18, you showed that $f(z) = z + a_2 z^2$ is univalent in $\mathbb{D} \iff |a_2| \leq \frac{1}{2}$. We want to make a conjecture about the generalization of this result. Use *ComplexTool* to graph $f(z) = z + a_3 z^3$ for various values of a_3. What do you conjecture is the bound on a_3 for which f is univalent on $\mathbb{D}$? Do the same for $f(z) = z + a_4 z^4$, $f(z) = z + a_5 z^5$, etc. What do you conjecture is the bound on a_n for which $f(z) = z + a_n z^n$ is univalent on $\mathbb{D}$? What do you conjecture $f(\mathbb{D})$ is when $a_n = -\frac{1}{n}$? *Try it out!*

Let's determine $f(\mathbb{D})$ analytically for a few examples that were included in Exploration 4.14.

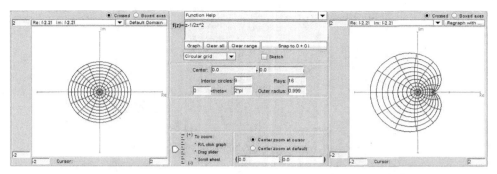

FIGURE 4.13. The image of the unit disk under the map $z - \frac{1}{2}z^2$.

Example 4.22. If

$$f_r(z) = \frac{z}{1 - z} \in S,$$

because $\dfrac{1}{1 - z} = \displaystyle\sum_{n=0}^{\infty} z^n$, we can multiply by z to get

$$f(z) = \frac{z}{1 - z} = \sum_{n=1}^{\infty} z^n = z + z^2 + z^3 + \cdots.$$

This is the Möbius transformation that maps $\mathbb{D}$ onto the right half-plane whose boundary is the line $\mathrm{Re}\{w\} = -\frac{1}{2}$ (see Figure 4.14).

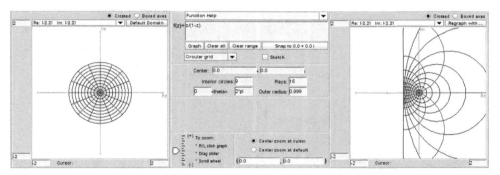

FIGURE 4.14. The image of the unit disk under the analytic right half-plane map in S.

Example 4.23. The power series for

$$f_k(z) = \frac{z}{(1 - z)^2} \in S$$

can be found by differentiating the series for $\frac{1}{1-z}$ and then multiplying by z:

$$\frac{z}{(1 - z)^2} = \sum_{n=1}^{\infty} n z^n = z + 2z^2 + 3z^3 + \cdots.$$

For this function $a_n = n$ for all n. We now show that the image of $\mathbb{D}$ under f_k is a slit domain. That is, it is a domain consisting of the entire complex plane except that a ray (or a slit) is cut out of it. To determine $f_k(\mathbb{D})$, let

$$u_1(z) = \frac{1+z}{1-z}, \quad u_2(z) = z^2, \quad u_3(z) = \frac{1}{4}[z-1].$$

Then,

$$u_3 \circ u_2 \circ u_1(z) = \frac{1}{4}\left[\left(\frac{1+z}{1-z}\right)^2 - 1\right] = \frac{z}{(1-z)^2}.$$

The Möbius transformation u_1 maps $\mathbb{D}$ onto the right half-plane whose boundary is the imaginary axis, u_2 is the squaring function, and u_3 translates the image one space to the left and multiplies it by a factor of $\frac{1}{4}$.

Thus the image $\mathbb{D}$ is the entire complex plane except for a slit along the negative real axis from $w = \infty$ to $w = -\frac{1}{4}$ (see Figure 4.15). The function $f_k(z) = \frac{z}{(1-z)^2}$ is known as the Koebe function.

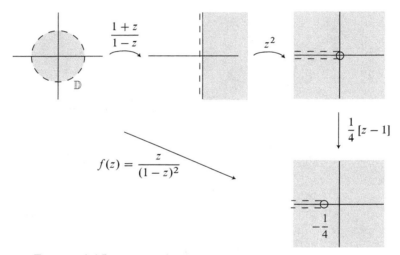

FIGURE 4.15. The image of the unit disk under the Koebe function.

Exploration 4.24. It is difficult to interpret the image of $\mathbb{D}$ under the Koebe function using *ComplexTool*. One way is to use increasing values that approach 1 for the radius of circles in $\mathbb{D}$. We can do this by using the left box in the "$\leq$ **radius** $\leq$" feature in the center panel of the applet. Graph $\mathbb{D}$ under the map $\frac{z}{(1-z)^2}$ using the values of 0.8, 0.85, 0.9, 0.95, and 0.999 in the left box in the "$\leq$ **radius** $\leq$" feature. *Try it out!*

Suppose we have an analytic function f with a Taylor series representation $f(z) = z + a_2 z^2 + a_3 z^3 + \cdots$. One question to ask is for what values of a_n is f in the family of schlicht functions? If all $a_n = 0$ except possibly a_2, then $f(z) = z + a_2 z^2$. By Exercise 4.18, we know that $|a_2| \leq \frac{1}{2}$ if and only if $f \in S$. The function $f(z) = z - \frac{1}{2}z^2$ from Example 4.19 is an extremal function. An extremal function is on the boundary between those that satisfy a condition and those that do not. Here, $f(z)$ is extremal, because if we increase $|a_2| = |-\frac{1}{2}|$ even just a little bit, then $f(z) = a + a_2 z^2$ is no longer schlicht. In general, how large can $|a_n|$ be and $f(z) = z + a_2 z^2 + a_3 z^3 + \cdots$ still be schlicht? For the Koebe function,

$$\frac{z}{(1-z)^2} = \sum_{n=1}^{\infty} n z^n,$$

and so we have $a_n = n$. This led Bieberbach in 1916 to make his famous conjecture.

Bieberbach Conjecture. *For $f \in S$, $|a_n| \le n$, for all n. In particular, $|a_2| \le 2$.*

Because the image of $\mathbb{D}$ under the Koebe function covers all of $\mathbb{C}$ except a slit along the real axis, it seems plausible that the the Bieberbach conjecture is true with the Koebe function being extremal. The conjecture is true but was not proved until 1984 by de Branges [7].

We say an inequality involving a function sharp if it is impossible to improve it (that is, we cannot decrease an upper bound or increase a lower bound). We can show that an inequality is sharp by finding a function for which the inequality becomes equality, an extremal function. For the Bieberbach conjecture or de Branges' theorem, the Koebe function is extremal.

There is another case in which the Koebe function is extremal. We know that if $f \in S$, then $f(\mathbb{D})$ is not the entire complex plane, so there is a point $a \in \mathbb{C}$ such that $a \notin f(\mathbb{D})$. This leads to the question of how small can $|a|$ be. For example, if $f(z) = z$, then $|a| = 1$; if $f(z) = \frac{z}{1-z}$, the right half-plane mapping in Example 4.22, then $|a| = \frac{1}{2}$. The answer is that for all $f \in S$, $|a| \ge \frac{1}{4}$. This is known as the Koebe $\frac{1}{4}$-Theorem. The Koebe function is extremal, because $|a| = \frac{1}{4}$ for the Koebe function. For these reasons and others, the Koebe function is important in the study of schlicht functions.

Exploration 4.25. If $f(z) = \dfrac{z - tz^2}{(1-z)^2}$, where $0 \le t \le 1$, what is $f(\mathbb{D})$ when $t = 0$? What is $f(\mathbb{D})$ when $t = 1$? Using your answers to these two questions and not *ComplexTool*, make a conjecture of what $f(\mathbb{D})$ is, when $0 < t < 1$. Now, use *ComplexTool* to modify or strengthen your conjecture. Using *ComplexTool*, what happens to $f(\mathbb{D})$ for $t > 1$? For $t = ik, 0 \le k \le 1$? **Try it out!**

The family S has been studied extensively. Here are a few facts about normalized, analytic univalent functions that will be used later:

(a) (uniqueness in the Riemann mapping theorem) Let $G \ne \mathbb{C}$ be a simply-connected domain with $a \in G$. Because of the Riemann mapping theorem, the map $f \in S$ that maps $\mathbb{D}$ onto G with $f(0) = a$ and $f'(0) > 0$ is unique.

(b) (de Branges' Theorem) For $f \in S$, $|a_n| \le n$, for all n.

(c) (Koebe $\frac{1}{4}$-Theorem) The range of every function in class S contains the disk $G = \{w : |w| < \frac{1}{4}\}$. This consequence of the fact that $|a_2| \le 2$ was proved by Bieberbach in 1916.

(d) Let $f \in S$. Then $f(\mathbb{D})$ omits a value on each circle $\{w : |w| = R\}$ where $R \ge 1$. There is no function $f \in S$ for which $f(\mathbb{D})$ contains $\partial \mathbb{D}$, the unit circle.

4.4   The Family S_H of Normalized, Harmonic, Univalent Functions

About the same time that de Branges proved the Bierbach conjecture, Clunie and Sheil-Small studied a family, S_H, of complex-valued harmonic functions that contained S as a proper subset and considered some of the properties of S_H that had been investigated in S.

A function $\phi(x, y)$ is harmonic if and only if $\phi_{xx} + \phi_{yy} = 0$.

Definition 4.26. A continuous function $f = u + iv$ defined in G is a *complex-valued harmonic function* in G if u and v are real harmonic (but not necessarily harmonic conjugates) in G.

Example 4.27. The function

$$f(x, y) = u(x, y) + iv(x, y) = (x^2 - y^2) + i2xy$$

is complex-valued harmonic because

$$u_{xx} + u_{yy} = 2 - 2 = 0$$
$$v_{xx} + v_{yy} = 0 + 0 = 0.$$

Exercise 4.28. Show that

$$f(x, y) = u(x, y) + iv(x, y) = \left(x + \frac{1}{2}x^2 - \frac{1}{2}y^2\right) + i(y - xy)$$

is complex-valued harmonic. ***Try it out!***

Although harmonic functions are more general than analytic functions, some theorems about analytic functions have equivalent forms for harmonic functions. These include the mean-value theorem, the maximum-modulus theorem, Liouville's Theorem, and the Argument Principle. However, by considering all harmonic functions instead of the subclass of analytic functions we can sometimes get more information. For example, we can use harmonic functions to study of minimal surfaces (see Chapter 2 on Minimal Surfaces).

One way of thinking of a function $f(x, y) = u(x, y) + iv(x, y)$ as being analytic is that f can be expressed in terms of $z = x + iy$ only without using $\bar{z} = x - iy$. Hence, the function $f = z^2$ is analytic while $f = z\bar{z}$ is not. To explore this idea, let $\zeta = z = x + iy$ and $\xi = \bar{z} = x - iy$. Then, we can formally write $x = \frac{1}{2}(\zeta + \xi)$ and $y = \frac{1}{2i}(\zeta - \xi)$. Applying the chain rule on $f(x(\zeta, \xi), y(\zeta, \xi))$ with $\zeta = z$ and $\xi = \bar{z}$, we can show that

$$\frac{\partial f}{\partial z} = \frac{1}{2}\left(\frac{\partial u}{\partial x} + \frac{\partial v}{\partial y}\right) + \frac{i}{2}\left(\frac{\partial v}{\partial x} - \frac{\partial u}{\partial y}\right) \tag{4.2}$$

$$\frac{\partial f}{\partial \bar{z}} = \frac{1}{2}\left(\frac{\partial u}{\partial x} - \frac{\partial v}{\partial y}\right) + \frac{i}{2}\left(\frac{\partial u}{\partial y} + \frac{\partial v}{\partial x}\right). \tag{4.3}$$

Exercise 4.29.

(a) Derive (4.2) and (4.3).

(b) Use these equations and the Cauchy-Riemann equations to prove that $f(x, y) = u(x, y) + iv(x, y)$ is analytic $\Longleftrightarrow \dfrac{\partial f}{\partial \overline{z}} = 0$.

Try it out!

Exercise 4.30.

(a) Using $x = \frac{1}{2}(z + \overline{z})$ and $y = \frac{1}{2i}(z - \overline{z})$, rewrite

$$f(x, y) = u(x, y) + iv(x, y) = \left(x + \frac{1}{2}x^2 - \frac{1}{2}y^2\right) + i(y - xy)$$

in terms of z and $\overline{z}$.

(b) Use Exercise 4.29 to determine if f is analytic.

(c) Show that all analytic functions are complex-valued harmonic, but not all complex-valued harmonic functions are analytic.

Try it out!

The next theorem tells us that a complex-valued harmonic function defined on $\mathbb{D}$ is related to analytic functions, and in fact can be expressed in terms of a *canonical decomposition*.

Theorem 4.31. *If $f = u + iv$ is harmonic in a simply-connected domain G, then $f = h + \overline{g}$, where h and g are analytic.*

Proof. If u and v are real harmonic on a simply-connected domain, then there exist analytic functions K and L such that $u = \operatorname{Re} K$ and $v = \operatorname{Im} L$. Hence,

$$f = u + iv = \operatorname{Re} K + i \operatorname{Im} L = \frac{K + \overline{K}}{2} + i\frac{L - \overline{L}}{2i} = \frac{K + L}{2} + \frac{\overline{K - L}}{2} = h + \overline{g}.$$

$\square$

Exercise 4.32. Let

$$f(x, y) = u(x, y) + iv(x, y) = \left(x + \frac{1}{2}x^2 - \frac{1}{2}y^2\right) + i(y - xy)$$

on $\mathbb{D}$. In Exercise 4.28 you showed that f is harmonic. Find analytic functions h and g such that $f = h + \overline{g}$. ***Try it out!***

We can use the applet *ComplexTool* to graph complex-valued harmonic functions. For example, to graph the image of $\mathbb{D}$ under the harmonic function $f(z) = z + \frac{1}{2}\overline{z}^2$, enter the function in *ComplexTool* in the form $z + 1/2\,\mathbf{conj}\,(z \wedge 2)$ (see Figure 4.16).

The harmonic function $f(z) = h(z) + \overline{g(z)}$ can be written as

$$f(z) = \operatorname{Re}\{h(z) + g(z)\} + i \operatorname{Im}\{h(z) - g(z)\}. \tag{4.4}$$

Hence, in Example 4.32, $f(z) = z + \frac{1}{2}\overline{z}^2$ can be written as $f(z) = \operatorname{Re}\{z + \frac{1}{2}z^2\} + i \operatorname{Im}\{z - \frac{1}{2}z^2\}$. In *ComplexTool* you can also enter the harmonic function in this form by typing $\mathbf{re}(z + 1/2z \wedge 2) + \mathbf{i*im}\,(z - 1/2z \wedge 2)$.

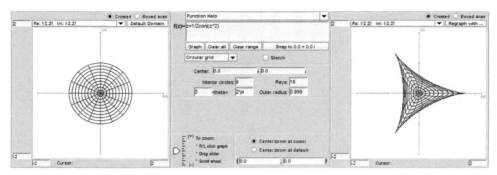

FIGURE 4.16. Image of $\mathbb{D}$ under the harmonic function $f(z) = z + \frac{1}{2}\bar{z}^2$.

Exercise 4.33. Prove that the representations $f(z) = h(z) + \overline{g(z)}$ and $f(z) = \mathrm{Re}\,\{h(z) + g(z)\} + i\,\mathrm{Im}\,\{h(z) - g(z)\}$ are equivalent. ***Try it out!***

Exploration 4.34. Graph the image of $\mathbb{D}$ under the following harmonic maps. Describe characteristics that appear to be different for harmonic mappings as compared to analytic mappings.

(a) $f_1(z) = z + \frac{1}{3}\bar{z}^3$

(b) $f_2(z) = \mathrm{Re}\left(\frac{z}{1-z}\right) + i\,\mathrm{Im}\left(\frac{z}{(1-z)^2}\right)$

(c) $f_3(z) = \frac{z}{1-z} - \frac{1}{2}e^{\frac{\bar{z}+1}{\bar{z}-1}}$

(d) $f_4(z) = \mathrm{Re}\left(\frac{i}{\sqrt{3}}\ln\left(\frac{1+e^{-i\frac{\pi}{3}}z}{1+e^{i\frac{\pi}{3}}z}\right)\right) + i\,\mathrm{Im}\left(\frac{1}{3}\ln\left(\frac{1+z+z^2}{1-2z+z^2}\right)\right)$

(e) $f_5(z) = z + 2\ln(z+1) + (\bar{z}+1)e^{\frac{\bar{z}-1}{\bar{z}+1}}$.

Try it out!

Since $f = h + \bar{g}$ where h and g are analytic, f has the series representation

$$f(z) = \sum_{n=0}^{\infty} a_n z^n + \sum_{n=1}^{\infty} b_n \bar{z}^n.$$

Hence, we may normalize harmonic univalent functions as we normalized analytic univalent functions.

Definition 4.35. Let S_H be the family of complex-valued harmonic, univalent mappings that are normalized on the unit disk. That is,

$S_H = \{f : \mathbb{D} \to \mathbb{C} \mid f$ is harmonic, univalent

with $f(0) = a_0 = 0,\ f_z(0) = a_1 = 1\}$.

If we restrict this family by requiring $b_1 = 0$, we have the family

$S_H^O = \{f \in S_H \mid f_{\bar{z}}(0) = b_1 = 0\}.$

Notice $S \subset S_H^O \subset S_H$.

Let's look at some examples.

Example 4.36. In the next section we will prove that

$$f(z) = h(z) + \overline{g(z)} = z + \frac{1}{2}\overline{z}^2$$

is univalent and hence in S_H^O. For now we will assume it and look at the image of $\mathbb{D}$ under f (see Figure 4.16). The function f maps $\mathbb{D}$ onto the interior of the region bounded by a hypocycloid with three cusps.

Definition 4.37. A *hypocycloid* is the curve produced by a fixed point on a small circle of radius b rolling the inside of a larger circle of radius a. It has the parametric equations

$$x(\theta) = (a - b)\cos\theta + b\cos\left(\left(\frac{a}{b} - 1\right)\theta\right)$$

$$y(\theta) = (a - b)\sin\theta + b\sin\left(\left(\frac{a}{b} - 1\right)\theta\right).$$

In Example 4.53, we will see that f is a shearing of $F(z) = z - 1/2z^2$ which maps $\mathbb{D}$ to an epicycloid (see Example 4.19).

Exploration 4.38.

(a) Use *ComplexTool* to plot the image of $\mathbb{D}$ under the analytic polynomial map $F(z) = z - \frac{1}{2}e^{it\frac{\pi}{6}}z^2$ for $t = 0, 1, 2, 3, 4, 5, 6$. Describe what happens to the image as t varies.

(b) Use *ComplexTool* to plot the image of $\mathbb{D}$ under the harmonic polynomial map $f(z) = z + \frac{1}{2}e^{it\frac{\pi}{6}}\overline{z}^2$ for $t = 0, 1, 2, 3, 4, 5, 6$. Describe what happens to the image as t varies.

(c) What differences do you notice between the images in (a) and (b) as t increases? Explain why it is reasonable for these differences to occur.

Try it out!

Small Project 4.39.

(a) Use *ComplexTool* to plot the images of the following polynomials

Harmonic Functions	Analytic Functions
(i) $f_1(z) = z + \frac{1}{3}\overline{z}^3$	(ii) $F_1(z) = z + \frac{1}{3}z^3$
(iii) $f_2(z) = z + \frac{1}{4}z^2 + \frac{1}{4}\overline{z}^2 + \frac{1}{3}\overline{z}^3$	(iv) $F_2(z) = z + \frac{1}{2}z^2 + \frac{1}{3}z^3$
(v) $f_3(z) = z + \frac{1}{6}\overline{z}^2 + \frac{1}{6}\overline{z}^4$	(vi) $F_3(z) = z + \frac{1}{6}z^6$
(vii) $f_4(z) = z + \frac{1}{6}z^2 + \frac{1}{6}\overline{z}^4$.	

(b) List the similarities and differences between the images of $\mathbb{D}$ under harmonic and the analytic functions.

(c) Questions to consider are: If a polynomial has three or more terms then how large can its coefficients be in modulus to guarantee univalency? If a polynomial has three terms, what difference does it make if the last two terms are $\overline{z}^2$ and $\overline{z}^3$ instead of z^2 and $\overline{z}^3$?

(d) Plot your own examples of harmonic and analytic polynomials and see if the properties from (b) are still valid.

Optional

Example 4.40. Let

$$f(z) = h(z) + \overline{g(z)} = \frac{z - \frac{1}{2}z^2}{(1-z)^2} - \frac{\frac{1}{2}\overline{z}^2}{(1-\overline{z})^2}$$

$$= \operatorname{Re}(h(z) - g(z)) + i \operatorname{Im}(h(z) - g(z)) = \operatorname{Re}\left(\frac{z}{1-z}\right) + i \operatorname{Im}\left(\frac{z}{(1-z)^2}\right).$$

We will prove that f is univalent in the next section. The image of $\mathbb{D}$ under f using *ComplexTool* is shown in Figure 4.17.

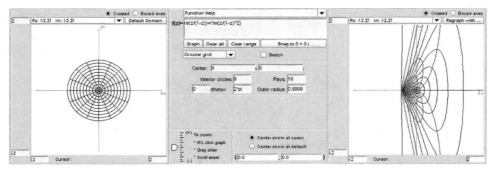

FIGURE 4.17. Image of $\mathbb{D}$ under $f(z) = \operatorname{Re}\left(\frac{z}{1-z}\right) + i \operatorname{Im}\left(\frac{z}{(1-z)^2}\right)$.

The image of $\mathbb{D}$ under the harmonic map f is the right half-plane $\{w \in \mathbb{C} \,|\, \operatorname{Re}\{w\} \geq -\frac{1}{2}\}$, the same region as the image of $\mathbb{D}$ under the analytic map $\frac{z}{1-z}$ although the boundary behavior is different.

Exploration 4.41.

(a) Use *ComplexTool* to plot $\mathbb{D}$ under the analytic right half-plane map $\frac{z}{1-z}$. Use the **Sketch** box to draw radial lines from the origin to the boundary of $\mathbb{D}$ in the domain. What are the images of points on the unit circle under this analytic map?

(b) Use *ComplexTool* to plot $\mathbb{D}$ under the harmonic right half-plane map $\frac{z - \frac{1}{2}z^2}{(1-z)^2} - \frac{\frac{1}{2}\overline{z}^2}{(1-\overline{z})^2}$. Use the **Sketch** box to draw radial lines from the origin to the boundary of $\mathbb{D}$ in the domain. What is the image of points on the unit circle under this analytic map?

(c) Using (a) and (b), describe how the boundary behavior is different between the analytic right half-plane map and this harmonic right half-plane map.

Try it out!

We are interested in harmonic univalent functions. However, results about univalent functions can be difficult to obtain. So it is sometimes useful to consider locally univalent functions instead of globally univalent. Locally univalent functions are functions that are locally one-to-one. This means that there is a small neighborhood around a point $z_0 \in \mathbb{D}$ such as a small disk in $\mathbb{D}$ centered at z_0, and the function is one-to-one for points z in that neighborhood. A locally univalent function may not be univalent in $\mathbb{D}$.

Exploration 4.42. In Exploration 4.15, you proved that $f(z) = (1+z)^4$ is not univalent in $\mathbb{D}$. This can be shown by seeing that the points $-\frac{1}{2} + \frac{1}{2}i$ and $-\frac{1}{2} - \frac{1}{2}i$ in $\mathbb{D}$ both map to $-\frac{1}{4}$ under $f(z) = (1+z)^4$. However, $f(z) = (1+z)^4$ is locally univalent in $\mathbb{D}$. We won't prove this. To get a sense of what local univalence means, use *ComplexTool* to graph $f(z) = (1+z)^4$ on the sector $\{z = re^{i\theta} \in \mathbb{D} \mid \frac{\pi}{2} < \theta < \frac{3\pi}{2}\}$. You can do this by entering the values for θ into the left and right boxes for the "$\leq \theta \leq$" feature in the center panel. The point $z_0 = -\frac{1}{2} + \frac{1}{2}i$ is in the sector. Also, the image indicates that $f(z) = (1+z)^4$ is not univalent in it. This does not mean that f is not locally univalent. Local univalence means that at z_0 there is some neighborhood in which the function is one-to-one for points z in the neighborhood. Use *ComplexTool* to determine a neighborhood of the points z_0 in which f is one-to-one. You can increase the size of the image by clicking on the left button on the mouse and decrease the size by clicking on the right mouse button.

(a) $z_0 = -\frac{1}{2} + \frac{1}{2}i$ (b) $z_0 = -\frac{1}{2} - \frac{1}{2}i$

(c) $z_0 = -\frac{5}{6} + \frac{1}{6}i$ (d) $z_0 = -0.9 + 0.1i$

Try it out!

Let's look at the idea of local univalence.

Definition 4.43. A function $f = h + \overline{g}$ is *locally univalent* on G if $J_f \neq 0$ on G, where J_f is the Jacobian of $f = u + iv$:

$$J_f = \det \begin{vmatrix} u_x & u_y \\ v_x & v_y \end{vmatrix}$$

$$= \det \begin{vmatrix} (\text{Re } h)_x + (\text{Re } g)_x & (\text{Re } h)_y + (\text{Re } g)_y \\ (\text{Im } h)_x - (\text{Im } g)_x & (\text{Im } h)_y - (\text{Im } g)_y \end{vmatrix}.$$

For analytic functions F, the Cauchy-Riemann equations yield $(\text{Re } F)_y = -(\text{Im } F)_x$ and $(\text{Im } F)_y = (\text{Re } F)_x$. Hence we have

$$J_f = \det \begin{vmatrix} (\text{Re } h)_x + (\text{Re } g)_x & -(\text{Im } h)_x - (\text{Im } g)_x \\ (\text{Im } h)_x - (\text{Im } g)_x & (\text{Re } h)_x - (\text{Re } g)_x \end{vmatrix}$$

$$= (\text{Re } h)_x^2 - (\text{Re } g)_x^2 + (\text{Im } h)_x^2 - (\text{Im } g)_x^2$$

$$= |h'|^2 - |g'|^2.$$

Thus, we want $|h'|^2 - |g'|^2 \neq 0$.

Besides local univalence, another important property of these functions is being sense-preserving. A continuous function f is *sense-preserving* (or *orientation-preserving*) if it preserves orientation. Let f_1, f_2 be defined on the punctured disk $\mathbb{D} - \{0\}$ by

$$f_1(z) = \frac{1}{z} \qquad \text{and} \qquad f_2(z) = \overline{z}.$$

Both map the unit circle, $\partial\mathbb{D}$, onto itself, and both map the points $A = 1$, $B = e^{i\frac{\pi}{4}}$, and $C = i$ to the points $A' = 1$, $B' = e^{-i\frac{\pi}{4}}$, and $C' = -i$, respectively (see Figure 4.18). In

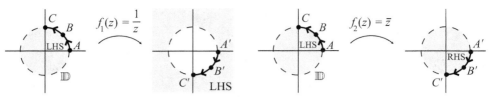

FIGURE 4.18. f_1 is sense-preserving and f_2 is sense-reversing.

the domain, as we travel along the unit circle in a counterclockwise direction (i.e., going from A to B to C), the region to the left of this path is $\mathbb{D}$. We will call this region the left-hand side domain (LHS) and the region to the right of the path we will call the right-hand side domain (RHS). So, $\mathbb{C} - \mathbb{D}$ is RHS. Where does $\mathbb{D}$, a LHS, get mapped by the functions? As we travel along $\partial\mathbb{D}$ in a counterclockwise direction in the domain set, the image curve under both functions will be $\partial\mathbb{D}$ traversed in a clockwise direction. So, in the image, $\mathbb{D}$ is now RHS while $\mathbb{C} - \mathbb{D}$ is LHS. The function f_1 maps the point $\frac{1}{2} \in \mathbb{D}$ to $2 \in \mathbb{C} - \mathbb{D}$ and so f_1 maps the LHS onto the LHS. Functions that map the LHS onto the LHS are sense-preserving. On the other hand, f_2 maps $\frac{1}{2} \in \mathbb{D}$ to $\frac{1}{2} \in \mathbb{D}$ and so f_2 maps the LHS onto the RHS. Functions that map the LHS onto the RHS are sense-reversing.

When we travel counterclockwise along a simple closed contour $\gamma \in G$, there exists a left-hand side domain (LHS) and a right-hand side domain (RHS). Let the image curve be $f(\gamma)$. The function f is sense-preserving if the original LHS domain with regard to γ is mapped to the LHS domain with regard to $f(\gamma)$. The function f is *sense-reversing* if the LHS domain with regard to γ is mapped to the RHS domain with regard to $f(\gamma)$. All analytic functions are sense-preserving, while some complex-valued harmonic functions are sense-preserving and some are sense-reversing.

Exploration 4.44. For the following harmonic functions, use *ComplexTool* to conjecture if the function is (a) locally univalent and (b) sense-preserving. Hint: for sense-preserving, you can use the Sketch feature to draw a counterclockwise curve on the unit circle and see its image under the function.

    (a) $z + 2\overline{z}$ (b) $z + \frac{1}{2}\overline{z}$

    (c) $z + 2\overline{z}^2$ (d) $z + \frac{1}{2}\overline{z}^2$

    (e) $2z^2 + \overline{z}$ (f) $\frac{1}{2}z^2 + \overline{z}$

Try it out!

Now we will need the following definition.

Definition 4.45. $\omega(z) = g'(z)/h'(z)$ is the *dilatation* of $f = h + \overline{g}$.

There is a connection between the dilatation of a harmonic function and its locally univalent and sense-preserving nature.

Theorem 4.46 (Lewy). *The function $f = h + \overline{g}$ is locally univalent and sense-preserving* $\iff |\omega(z)| < 1$ *for all $z \in G$.*

Exercise 4.47. Compute $\omega(z)$ for the functions in Exploration 4.44. Use your results from Exploration 4.44 to verify Lewy's Theorem for the functions in that Exploration. *Try it out!*

Exercise 4.48. Show that for $z \in \mathbb{D}$, $|\omega_k(z)| < 1$ for

(a) $\omega_1(z) = e^{i\theta} z$, where $\theta \in \mathbb{R}$

(b) $\omega_2(z) = z^n$, where $n = 1, 2, 3, \ldots$

(c) $\omega_3(z) = \dfrac{z + a}{1 + \overline{a}z}$, where $|a| < 1$

(d) $\omega_4(z)$, the composition of any of ω_1, ω_2, and ω_3.

Try it out!

Remark 4.49. Let $f = h + \overline{g}$ be a sense-preserving harmonic map with dilatation $\omega = g'/h'$. If $|\omega(z)| = 1$ for all z in an arc γ of $\partial\mathbb{D}$, then the image of γ under f is either

(a) a concave arc (i.e., there are two points in $f(\mathbb{D})$ such that the line connecting them goes outside of $f(\mathbb{D})$) (see Example 4.36); or

(b) a stationary point (see Example 4.40).

We discuss the dilatation more in Section 4.6.

4.5   The Shearing Technique

Finding examples of univalent harmonic mappings that are not analytic is not easy. The *shearing technique* by Clunie and Sheil-Small is a useful way to construct new examples of univalent harmonic mappings. Shearing starts with an analytic function F and a dilatation ω that have certain properties. Then, we can find a univalent harmonic function $f = h + \overline{g}$ by writing F as $F = h - g$ and ω as $\omega = g'/h'$, and solving for h and g. Before we proceed, we need to discuss a property of F.

Definition 4.50. A domain Ω is *convex in the direction* $e^{i\varphi}$ if for every $a \in \mathbb{C}$ the set $\Omega \cap \{a + te^{i\varphi} : t \in \mathbb{R}\}$ is either connected or empty. A domain is convex in the horizontal direction (CHD) if every line parallel to the real axis has a connected intersection with Ω.

CHD   not CHD

Exercise 4.51. For which values of $n = 1, 2, 3, \ldots$ do the following functions map $\mathbb{D}$ onto a CHD domain?

(a) $f(z) = z^n$

(b) $f(z) = z - \frac{1}{n}z^n$ (see Example 4.19 and Definition 4.20)

(c) $f(z) = \frac{z}{(1-z)^n}$ (see Examples 4.22 and 4.23 to get you started).

Try it out!

We have a theorem by Clunie and Sheil-Small that forms the basis of the shearing technique.

Theorem 4.52. *Let $f = h + \overline{g}$ be a harmonic function that is locally univalent in $\mathbb{D}$ (i.e., $|\omega(z)| < 1$ for all $z \in \mathbb{D}$). The function $F = h - g$ is an analytic univalent mapping of $\mathbb{D}$ onto a CHD domain $\Longleftrightarrow$ $f = h + \overline{g}$ is a univalent mapping of $\mathbb{D}$ onto a CHD domain.*

Before we prove Theorem 4.52, let's look at an example.

Example 4.53. The shearing technique starts with a univalent analytic function F that maps $\mathbb{D}$ onto a CHD domain and a dilatation ω where $|\omega(z)| < 1$ for all $z \in \mathbb{D}$. Then, we construct a univalent harmonic function $f = h + \overline{g}$ by writing F as $F = h - g$ and ω as $\omega = g'/h'$, and solving for h and g.

In Example 4.36, we claimed that the harmonic polynomial $f(z) = z + \frac{1}{2}\overline{z}^2$ (see Figure 4.19) is univalent and is related to the analytic function $F(z) = z - \frac{1}{2}z^2$ (see Figure 4.20). We can use Theorem 4.52 to show this.

Let's start with the analytic univalent function

$$F(z) = h(z) - g(z) = z - \frac{1}{2}z^2$$

that maps $\partial\mathbb{D}$ to an epicycloid with one cusp (see Example 4.19) resulting in a CHD do-

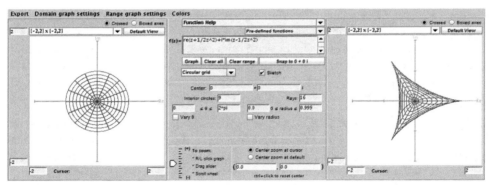

FIGURE 4.19. Image of $\mathbb{D}$ under the harmonic map $f(z) = z + \frac{1}{2}\overline{z}^2$.

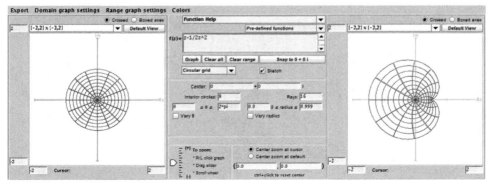

FIGURE 4.20. Image of $\mathbb{D}$ under the analytic map $F(z) = z - \frac{1}{2}z^2$.

main. Choose $\omega(z) = g'(z)/h'(z) = z$ and apply the shearing technique:

$$h'(z) - g'(z) = 1 - z \Rightarrow h'(z) - zh'(z) = 1 - z$$
$$\Rightarrow h'(z) = 1$$
$$\Rightarrow h(z) = z.$$

Since $g'(z) = zh'(z) = z$, we have $g(z) = \frac{1}{2}z^2$. Both h and g are normalized; that is, $h(0) = 0$ and $g(0) = 0$. So the corresponding harmonic univalent function is

$$f(z) = h(z) + \overline{g(z)} = z + \frac{1}{2}\overline{z}^2 \in S_H^O,$$

which was derived from shearing the analytic univalent function

$$F(z) = z - \frac{1}{2}z^2 \in S$$

with the dilatation $\omega(z) = z$.

Remark 4.54. This technique is known as shearing. The word shear means to cut (as in shearing sheep). As in Theorem 4.52, suppose $F = h - g$ is an analytic univalent function convex in the horizontal direction. The harmonic shear is

$$f = h + \overline{g} = h - g + g + \overline{g} = h - g + 2\,\mathrm{Re}\{g\},$$

which differs from the analytic function by the adding of a real function. Geometrically, this takes the image $F(\mathbb{D})$, that is convex in the horizontal direction and is formed from the analytic univalent function F, and cuts it into horizontal slices. These slices are then translated or scaled continuously to form the image $f(\mathbb{D})$ that is also convex in the horizontal direction and is formed from the harmonic univalent function f (see Figure 4.21). This is why the method is called shearing.

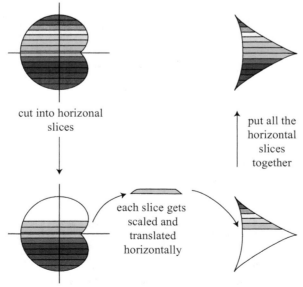

cut into horizonal slices

put all the horizontal slices together

each slice gets scaled and translated horizontally

FIGURE 4.21. Shearing an analytic function to preserve univalency.

Exercise 4.55. Let $f = h + \overline{g}$ with $h(z) - g(z) = z$ and $\omega(z) = z$. Compute h and g so that $f \in S_H^O$ and use *ComplexTool* to sketch $f(\mathbb{D})$.

Exercise 4.56. Let $f = h + \overline{g}$ with $h(z) - g(z) = z - \frac{1}{3}z^3$ and $\omega(z) = z^2$. Compute h and g so that $f \in S_H^O$ and use *ComplexTool* to sketch $f(\mathbb{D})$.

For exploring shears of functions, the applet *ShearTool* (see Figure 4.22) is better than *ComplexTool*, because it allows you to see the image of $\mathbb{D}$ under a shear of $h(z) - g(z)$ without having to compute the harmonic function $f = h + \overline{g}$. When using *ShearTool*, enter an analytic function that is convex in the horizontal direction in **h−g=** box in the upper left section and enter the dilatation in the ω box below it. The click on the **Graph** button.

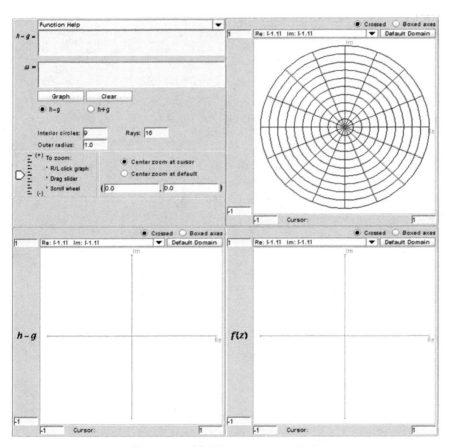

FIGURE 4.22. The applet *ShearTool*.

In Example 4.53, we sheared $h(z) - g(z) = z - \frac{1}{2}z^2$ with $\omega(z) = z$. Entering them into *ShearTool* we get the image of a hypocycloid with three cusps (see Figure 4.23).

Exploration 4.57.

(a) Use *ShearTool* to graph the image of $\mathbb{D}$ under $f = h + \overline{g}$ where $h(z) - g(z) = z$ and $\omega(z) = -z$. Note the difference between this and Exercise 4.55.

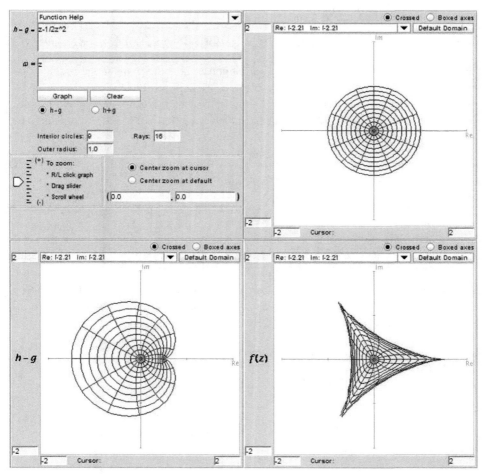

FIGURE 4.23. The image of $\mathbb{D}$ when shearing $h(z) - g(z) = z - \frac{1}{2}z^2$ with $\omega(z) = z$.

(b) Use *ShearTool* to graph the image of $\mathbb{D}$ under $f = h + \overline{g}$ where $h(z) - g(z) = z - \frac{1}{3}z^3$ and $\omega(z) = -z^2$. Note the difference between this and Exercise 4.56.

Try it out!

To prove Theorem 4.52, we will use the lemma:

Lemma 4.58. *Let $\Omega \subset \mathbb{C}$ be a CHD domain and let ρ be a real-valued continuous function in Ω. Then the map $\Psi(w) = w + \rho(w)$ is one-to-one in $\Omega \iff \Psi$ is locally one-to-one. If Ψ is one-to-one, then its range is a CHD domain.*

Proof. ($\Rightarrow$) Trivial.

($\Leftarrow$) Suppose the map $\Psi(w) = w + \rho(w)$ is not one-to-one. Then, there are distinct points $w_1 = u_1 + iv_1$, $w_2 = u_2 + iv_2 \in \Omega$ such that $\Psi(w_1) = \Psi(w_2)$. Since ρ is real valued, we have that $\text{Im}(\Psi(w_1)) = \text{Im}(w_1 + \rho(w_1)) = \text{Im}(w_1) = v_1$. Similarly, $\text{Im}(\Psi(w_2)) = v_2$. Since $\Psi(w_1) = \Psi(w_2)$, we have $v_1 = v_2$. Let's say $k = v_1 = v_2$ for a constant $k \in \mathbb{R}$. Define $\Phi : \widetilde{\Omega} \subset \mathbb{R} \to \mathbb{R}$ by $\Phi(u) = u + \rho(u + ik)$. Because Ψ is not one-to-one, we have

$$\Phi(u_1) = u_1 + \rho(u_1 + k) = \operatorname{Re}\{\Psi(w_1)\} = \operatorname{Re}\{\Psi(w_2)\} = u_2 + \rho(u_2 + k) = \Phi(u_2).$$

Thus, Φ is not a strictly monotonic function. Suppose u_0 is the point at which the monotonicity of Φ changes. Because Φ maps into $\mathbb{R}$, it cannot be locally one-to-one in any neighborhood of u_0. Hence, Φ is not locally one-to-one, and Ψ is not locally one-to-one.

Geometrically, Ψ acts as a shear in the horizontal direction and hence its range is CHD.

$\square$

Proof of Theorem 4.52. ($\Rightarrow$) Assume $f = h + \overline{g}$ is one-to-one and $\Omega = f(\mathbb{D})$ is CHD (see Figure 4.24). Then $f = h - g + g + \overline{g} = h - g + 2\operatorname{Re}\{g\}$. So

$$(h - g) \circ f^{-1}(w) = (f - 2\operatorname{Re}\{g\}) \circ f^{-1}(w) = w - 2\operatorname{Re}\{g(f^{-1}(w))\} = w + p(w)$$

may be defined in Ω, where p is real valued and continuous. Since f is locally $1-1$, $|g'| < |h'| \iff g'(z) \neq h'(z), \forall z \in \mathbb{D}$. Hence $h - g$ is locally $1-1$ in $\mathbb{D}$, and thus $w \to w + p(w)$ is also locally $1-1$ on Ω since it is the composition of locally $1-1$ functions. By Lemma 4.58, $w \to w + p(w)$ is univalent and its range is CHD. Hence, $(h - g)(z) = [w + p(w)] \circ f(z)$ is univalent, being the composition of univalent functions, and its image is CHD.

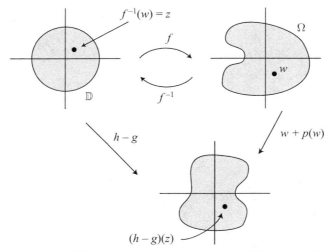

FIGURE 4.24. Proof in the forward direction of Theorem 4.52.

($\Leftarrow$) Assume that $F = h - g$ is univalent on $\mathbb{D}$ and that $\Omega = F(\mathbb{D})$ is CHD (see Figure 4.25). Then $f = F + 2\operatorname{Re}\{g\}$ and

$$f(F^{-1}(w)) = w + 2\operatorname{Re}\{g(F^{-1}(w))\}$$
$$= w + q(w).$$

is locally one-to-one (being the composition of locally one-to-one functions) in Ω. By Lemma 4.58, $f \circ F^{-1}$ is univalent in Ω and has a range that is CHD.

$\square$

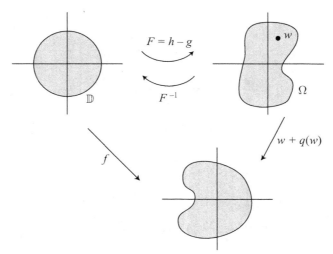

FIGURE 4.25. Proof in the reverse direction of Theorem 4.52.

Example 4.59. Let's shear the Koebe function with a standard dilatation. Let

$$h(z) - g(z) = \frac{z}{(1 - z)^2} \tag{4.5}$$

and

$$\omega(z) = g'(z)/h'(z) = z.$$

Applying the shearing technique, we have

$$h'(z) - g'(z) = \frac{1 + z}{(1 - z)^3} \Rightarrow h'(z) - z h'(z) = \frac{1 + z}{(1 - z)^3}$$

$$\Rightarrow h'(z) = \frac{1 + z}{(1 - z)^4}.$$

Integrating $h'(z)$ and normalizing so that $h(0) = 0$ yields

$$h(z) = \frac{z - \frac{1}{2}z^2 + \frac{1}{6}z^3}{(1 - z)^3}. \tag{4.6}$$

We can use this method to solve for normalized $g(z)$, where $g(0) = 0$. Instead we will find $g(z)$ by using (4.5) and (4.6):

$$g(z) = h(z) - \frac{z}{(1 - z)^2} = \frac{\frac{1}{2}z^2 + \frac{1}{6}z^3}{(1 - z)^3}.$$

So

$$f(z) = h(z) + \overline{g(z)} = \mathrm{Re}\left(\frac{z + \frac{1}{3}z^3}{(1 - z)^3}\right) + i\,\mathrm{Im}\left(\frac{z}{(1 - z)^2}\right) \in S_H^O.$$

The image of $\mathbb{D}$ under f is similar to the image of $\mathbb{D}$ under the analytic Koebe function (see Figure 4.15) with the slit on the negative real axis except that the tip of the slit is at $-\frac{1}{6}$ instead of $-\frac{1}{4}$. In Example 4.23 we used the transformation $\frac{1+z}{1-z} = w = u + iv$. Let's use this transformation again (it is often useful in horizontal slit domains). Since $z \in \mathbb{D}$, $w = \frac{1+z}{1-z}$ is the right half-plane $\{w = u + iv \in \mathbb{C} \mid \operatorname{Re} w = u > 0, -\infty < v < \infty\}$. Then $z = \frac{w-1}{w+1}$. Substituting into $h(z)$ and $g(z)$ and simplifying, we get

$$h\left(\frac{w-1}{w+1}\right) = \frac{1}{8}\left[\frac{2}{3}w^3 + w^2 - \frac{5}{3}\right]$$

$$g\left(\frac{w-1}{w+1}\right) = \frac{1}{8}\left[\frac{2}{3}w^3 - w^2 + \frac{1}{3}\right].$$

Because $f = \operatorname{Re}(h + g) + i \operatorname{Im}(h - g)$,

$$f\left(\frac{w-1}{w+1}\right) = \frac{1}{6}\operatorname{Re}\left\{w^3 - 1\right\} + \frac{1}{4}i \operatorname{Im}\left\{w^2 - 1\right\}.$$

Using $w = u + iv$ and taking the real and imaginary parts gives

$$f\left(\frac{w-1}{w+1}\right) = \frac{1}{6}\left(u^3 - 3uv^2 - 1\right) + i\frac{1}{2}uv.$$

If we let $uv = 0$ (so $v = 0$), then the imaginary part vanishes and, because $u > 0$, the real part varies from $-\frac{1}{6}$ to $+\infty$. Thus, for $uv = 0$, $f(\mathbb{D})$ contains the line segment on the real axis from $-\frac{1}{6}$ to $+\infty$. If we let $uv = c \neq 0$, then the imaginary part is constant and the real part is $\frac{u^3}{6} - \frac{c^2}{2u}$, which varies between $-\infty$ and $+\infty$. Thus, for $c \neq 0$, $f(\mathbb{D})$ contains the entire line parallel to the real axis that goes through ic. Therefore, $f(\mathbb{D})$ is the entire complex plane except the slit on the negative real axis from $-\frac{1}{6}$ to $-\infty$.

Exercise 4.60. Let $f = h + \overline{g}$ with $h(z) - g(z) = \frac{z}{(1-z)^2}$ and $\omega(z) = z^2$. Compute h and g so that $f \in S_H^O$ and determine $f(\mathbb{D})$. *Try it out!*

Exploration 4.61. If we shear $h(z) - g(z) = \frac{z}{(1-z)^2}$ with $\omega(z) = \frac{z^2 + az}{1+az}$, where $-1 < a \leq 1$, then the image of $\mathbb{D}$ under $f = h + \overline{g}$ is a slit domain (like the image of $\mathbb{D}$ under the analytic and harmonic Koebe maps) with the tip of the slit varying as a varies. Use *ShearTool* to graph the domains decreasing a from 1 to -1 (You might find it beneficial to set the **Outer radius** value to less than 1.0. Try to place the cursor on the tip of the slit in the $f(z)$–domain box and look at the coordinates to estimate the distance of the tip from the origin. How close to the origin can you get the tip of the slit? How far from the origin can you get the tip of the slit? Try finding other ω expressions that result in slit domains when shearing $h(z) - g(z) = \frac{z}{(1-z)^2}$). *Try it out!*

Because of the importance and extremal nature of the analytic Koebe function (see Example 4.23), it is conjectured that the harmonic Koebe function in Example 4.59 is extremal. The coefficients of the harmonic function

$$f(z) = \sum_{n=0}^{\infty} a_n z^n + \sum_{n=1}^{\infty} b_n \overline{z}^n$$

$$= \operatorname{Re}\left(\frac{z + \frac{1}{3}z^3}{(1-z)^3}\right) + i \operatorname{Im}\left(\frac{z}{(1-z)^2}\right) \tag{4.7}$$

satisfy $|a_n| = \frac{1}{6}(n+1)(2n+1)$, $|b_n| = \frac{1}{6}(n-1)(2n-1)$, and $\left||a_n| - |b_n|\right| = n$.

Conjecture 1 (Harmonic Bieberbach Conjecture). *Let*

$$f(z) = \sum_{n=0}^{\infty} a_n z^n + \sum_{n=1}^{\infty} b_n \overline{z}^n \in S_H^O.$$

Then

$$|a_n| \le \frac{1}{6}(n+1)(2n+1),$$
$$|b_n| \le \frac{1}{6}(n-1)(2n-1), \tag{4.8}$$
$$\big||a_n| - |b_n|\big| \le n.$$

In particular,

$$|a_2| \le \frac{5}{2} \tag{4.9}$$

Exercise 4.62. Verify that the harmonic Koebe function given in (4.7) satisfies equality in (4.8) and (4.9). **Try it out!**

Currently, the best bound is that for all functions $f \in S_H^O$, $|a_2| < 49$ (see [12]). There is room for improvement.

Large Project 4.63. Read and understand the proof that $|a_2| \le 2$ for analytic functions in S (for example, see [1, Section 5.1]). Read and understand two proofs that give bounds on $|a_2|$ for harmonic functions in S_H^O (see [6, Theorem 4.1] and [12, p. 96]). Investigate ways to modify them, or other proofs, to establish that for $f \in S_H^O$, $|a_2| \le K$ for some K, where $\frac{5}{2} \le K < 49$. **Optional**

Open Problem 4.64. Prove the Conjecture 1.

Because the tip of the the harmonic Koebe function is at $-\frac{1}{6}$, we have a conjecture that is the analogue of the analytic Koebe $\frac{1}{4}$-Theorem:

Conjecture 2 (Harmonic Koebe $\frac{1}{6}$-Conjecture). *The range of a function in class S_H^O contains the disk $G = \{w : |w| < \frac{1}{6}\}$.*

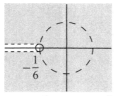

disk $= \{w : |w| < \frac{1}{6}\}$

FIGURE 4.26. The disk of radius $\frac{1}{6}$ is contained in the image of $\mathbb{D}$ under the harmonic Koebe function.

Open Problem 4.65. Prove Conjecture 2. The best result so far is that the range of $f \in S_H^O$ contains the disk $\{w : |w| < \frac{1}{16}\}$ (see [6]), so it would be interesting to increase the radius to some K with $\frac{1}{16} < K \le \frac{1}{6}$.

A domain convex in the general direction φ was defined in Definition 4.50. The shearing theorem by Clunie and Sheil-Small can be generalized to apply to such domains.

Corollary 4.66. *A harmonic function* $f = h + \overline{g}$ *locally univalent in* $\mathbb{D}$ *is a univalent mapping of* $\mathbb{D}$ *onto a domain convex in the direction* φ $\Longleftrightarrow$ $h - e^{2i\varphi}g$ *is an analytic univalent mapping of* $\mathbb{D}$ *onto a domain convex in the direction* φ.

Example 4.67. The analytic right half-plane function $\frac{z}{1-z}$ maps $\mathbb{D}$ onto a convex domain, so it is convex in all directions φ. Hence, it is convex in the direction of the imaginary axis ($\varphi = \frac{\pi}{2}$). Let's apply Corollary 4.66 with $\varphi = \frac{\pi}{2}$ to $\frac{z}{1-z}$ and use a dilatation that simplifies calculations. With $\varphi = \frac{\pi}{2}$, we set $F(z) = \frac{z}{1-z}$ equal to $h(z) - e^{2i\varphi}g(z) = h(z) + g(z)$ instead of $h(z) - g(z)$. So we start with

$$h(z) + g(z) = \frac{z}{1-z}$$

and choose

$$\omega(z) = g'(z)/h'(z) = -z.$$

Computing h and g as in the previous two examples yields

$$h(z) = \frac{z - \frac{1}{2}z^2}{(1-z)^2}$$
$$g(z) = -\frac{\frac{1}{2}z^2}{(1-z)^2}.$$ (4.10)

Hence, the harmonic function is

$$f(z) = h(z) + \overline{g(z)} = \operatorname{Re}\left(\frac{z}{1-z}\right) + i\operatorname{Im}\left(\frac{z}{(1-z)^2}\right) \in S_H^O.$$

This function is a harmonic right half-plane map.

Exercise 4.68. Verify that shearing $h(z) + g(z) = \frac{z}{1-z}$ with $\omega(z) = -z$, yields (4.10). *Try it out!*

The harmonic right half-plane map in Example 4.67 is the same harmonic map we discussed in Example 4.40.

Exploration 4.69. Shear $h + g = \frac{z}{1-z}$ using $\omega = -z^n$, where $n = 1, 2, 3, \ldots$ and sketch $f(\mathbb{D})$ using *ShearTool*. Describe what happens to $f(\mathbb{D})$ as n varies. Pay attention to the number of points the green lines go to as n increases. *Try it out!*

One significance of $\mathbb{D}$ being mapped to the same domains under these two functions is that the uniqueness of the Riemann mapping theorem for analytic functions does not hold for harmonic functions. This leaves the open question:

Open Problem 4.70. What is the analogue of the Riemann mapping theorem for harmonic functions?

Small Project 4.71. Let $f = h + \overline{g}$ with $h(z) - g(z) = \dfrac{z}{(1-z)^2}$ and $\omega(z) = \dfrac{z^2 + az}{1 + az}$.

(a) Show that for $-1 \le a \le 1$, $|\omega(z)| < 1$, $\forall z \in \mathbb{D}$.

(b) Compute h and g so that $f \in S_H^O$.

(c) Show that for $a = -1$, $f(\mathbb{D})$ is a right half-plane.

(d) Show that for $-1 < a \le 1$, $f(\mathbb{D})$ is a slit domain like the Koebe domain. For each a, determine where the tip of the slit is located.

Optional

Example 4.72. Here is a shearing example that we will use later in our discussion of minimal surfaces. Let

$$h(z) - g(z) = \frac{1}{2} \log \left(\frac{1+z}{1-z} \right), \tag{4.11}$$

which is an analytic function that maps $\mathbb{D}$ onto a horizontal strip convex in the direction of the real axis. Let

$$\omega(z) = g'(z)/h'(z) = -z^2.$$

Using *ShearTool* we see that the shear of $h - g$ with $-z^2$ results in a univalent harmonic function that maps onto the interior of the region bounded by a square (see Figure 4.27).

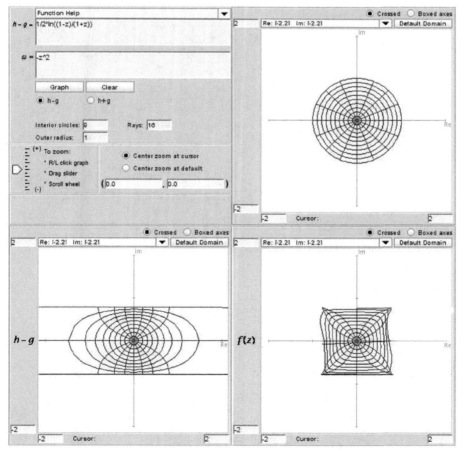

FIGURE 4.27. The image of $\mathbb{D}$ when shearing $h(z) - g(z) = \frac{1}{2} \log \left(\frac{1+z}{1-z} \right)$ with $\omega(z) = -z^2$.

Let's compute $h(z)$ and $g(z)$ and prove that the image is a square region. Applying the shearing method, we have

$$h'(z) - g'(z) = \frac{1}{1-z^2} \Rightarrow h'(z) + z^2 h'(z) = \frac{1}{1-z^2}$$

$$\Rightarrow h'(z) = \frac{1}{1-z^4} = \frac{1}{4}\left[\frac{1}{1+z} + \frac{1}{1-z} + \frac{i}{i+z} + \frac{i}{i-z} \right].$$

Integrating $h'(z)$ and normalizing so that $h(0) = 0$ yields

$$h(z) = \frac{1}{4}\log\left(\frac{1+z}{1-z}\right) + \frac{i}{4}\log\left(\frac{i+z}{i-z}\right). \tag{4.12}$$

We can use this method to solve for normalized $g(z)$, where $g(0) = 0$, or we can find $g(z)$ by using (4.11) and (4.12). Either way, we get

$$g(z) = -\frac{1}{4}\log\left(\frac{1+z}{1-z}\right) + \frac{i}{4}\log\left(\frac{i+z}{i-z}\right).$$

So

$$f(z) = h(z) + \overline{g(z)} = \mathrm{Re}\left[\frac{i}{2}\log\left(\frac{i+z}{i-z}\right) \right] + i\,\mathrm{Im}\left[\frac{1}{2}\log\left(\frac{1+z}{1-z}\right) \right] \in S_H^O.$$

What is $f(\mathbb{D})$? We have

$$f(z) = \left[-\frac{1}{2}\arg\left(\frac{i+z}{i-z}\right) \right] + i\left[\frac{1}{2}\arg\left(\frac{1+z}{1-z}\right) \right]$$

$$= u + iv.$$

Let $z = e^{i\theta} \in \partial\mathbb{D}$. Then

$$\frac{i+z}{i-z} = \frac{i+e^{i\theta}}{i-e^{i\theta}}\frac{-i-e^{-i\theta}}{-i-e^{-i\theta}} = \frac{1 - i(e^{i\theta} + e^{-i\theta}) - 1}{1 + i(e^{i\theta} - e^{-i\theta}) + 1} = -i\frac{\cos\theta}{1-\sin\theta}.$$

Thus,

$$u = -\frac{1}{2}\arg\left(\frac{i+z}{i-z}\right)\Bigg|_{z=e^{i\theta}} = \begin{cases} \frac{\pi}{4} & \text{if } \cos\theta > 0 \\ -\frac{\pi}{4} & \text{if } \cos\theta < 0. \end{cases}$$

Likewise, we can show that

$$v = \begin{cases} \frac{\pi}{4} & \text{if } \sin\theta > 0 \\ -\frac{\pi}{4} & \text{if } \sin\theta < 0. \end{cases}$$

So, we have that $z = e^{i\theta} \in \partial\mathbb{D}$ is mapped to

$$u + iv = \begin{cases} z_1 = \frac{\pi}{2\sqrt{2}}\,e^{i\frac{\pi}{4}} = \frac{\pi}{4} + i\frac{\pi}{4} & \text{if } \theta \in (0, \frac{\pi}{2}) \\ z_3 = \frac{\pi}{2\sqrt{2}}\,e^{i\frac{3\pi}{4}} = -\frac{\pi}{4} + i\frac{\pi}{4} & \text{if } \theta \in (\frac{\pi}{2}, \pi) \\ z_5 = \frac{\pi}{2\sqrt{2}}\,e^{i\frac{5\pi}{4}} = -\frac{\pi}{4} - i\frac{\pi}{4} & \text{if } \theta \in (\pi, \frac{3\pi}{2}) \\ z_7 = \frac{\pi}{2\sqrt{2}}\,e^{i\frac{7\pi}{4}} = \frac{\pi}{4} - i\frac{\pi}{4} & \text{if } \theta \in (\frac{3\pi}{2}, 2\pi). \end{cases}$$

Thus, this harmonic function maps $\mathbb{D}$ onto the interior of the region bounded by a square with vertices at z_1, z_3, z_5, and z_7.

This function can be obtained using the Poisson integral formula. Using the Poisson formula we can find a harmonic function that maps $\mathbb{D}$ onto a regular n-gon for any $n \geq 3$. Let

$$f(z) = \frac{1}{2\pi} \int_0^{2\pi} Re\left(\frac{1 + ze^{-it}}{1 - ze^{-it}}\right) e^{i\phi(t)} dt,$$

where $\phi(t) = \frac{\pi(2k+1)}{n}$ $(\frac{2\pi k}{n} \leq t < \frac{2\pi(k+1)}{n}, k = 0, ..., n-1)$. We can derive that

$$h(z) = \sum_{m=0}^{\infty} \frac{1}{nm + 1} z^{nm+1}$$

$$g(z) = \sum_{m=1}^{\infty} \frac{-1}{nm - 1} z^{nm-1}$$

$$h'(z) = \frac{1}{1 - z^n} \tag{4.13}$$

$$g'(z) = \frac{-z^{n-2}}{1 - z^n}, \qquad \text{and}$$

$$\omega(z) = -z^{n-2}.$$

Exercise 4.73. Let $f = h + \overline{g}$ with $h(z) - g(z) = \frac{1}{3} \log\left(\frac{1 + z + z^2}{1 - 2z + z^2}\right)$ and $\omega(z) = -z$.

(a) Show that $h'(z) = \frac{1}{1-z^3}$ and $g'(z) = \frac{-z}{1-z^3}$.

(b) According to the previous paragraph, what should be the image of $\mathbb{D}$ under $f = h + \overline{g}$?

(c) Use *ShearTool* to sketch the image of f.

(d) Compute h and g so $f \in S_H^O$.

Try it out!

Exercise 4.74. Let $f = h + \overline{g}$ with $h(z) + g(z) = \frac{1}{2} \log\left(\frac{1 + z}{1 - z}\right)$ and $\omega(z) = -z^2$. Compute h and g explicitly so that $f \in S_H^O$ and use *ComplexTool* to graph $f(\mathbb{D})$. Notice the differences in the dilatations and the images of $f(\mathbb{D})$ in this exercise and in Example 4.72. *Try it out!*

4.6   Properties of the Dilatation

Because of the importance of the dilatation, we will examine some of its properties. We know that an analytic function will map infinitesimal circles to infinitesimal circles at a point where its derivative is nonzero. For harmonic functions, this does not hold, as can be seen in the following exploration.

Exploration 4.75. In *ComplexTool*, enter values of 0.1, 0.4, 0.7, and 0.999 one at a time into the **Outer radius** box in the center panel to plot disks under the analytic right half-plane map

$$F(z) = \frac{z}{1-z}$$

and the harmonic right half-plane map

$$f(z) = \frac{z - \frac{1}{2}z^2}{(1-z)^2} - \frac{\frac{1}{2}\overline{z}^2}{(1-\overline{z})^2}.$$

Look at the images of small circles under F and under f. What appears to be the image of circles under the harmonic function f? ***Try it out!***

Now, let's explore the ideas of the geometric dilatation D_f and the analytic dilatation $\omega(z)$ that we discussed earlier.

Exercise 4.76.

(a) Prove that the formulas

$$\frac{\partial f}{\partial z} = \frac{1}{2}\left(\frac{\partial f}{\partial x} - i\frac{\partial f}{\partial y}\right) \qquad \text{and} \qquad \frac{\partial f}{\partial \overline{z}} = \frac{1}{2}\left(\frac{\partial f}{\partial x} + i\frac{\partial f}{\partial y}\right).$$

are equivalent to the formulas (4.2) and (4.3) given in Section 4.4.

(b) Prove that if $f = h + \overline{g}$ is harmonic, then

$$\frac{\partial f}{\partial z} = h'(z) \qquad \text{and} \qquad \frac{\partial f}{\partial \overline{z}} = \overline{g'(z)}.$$

Try it out!

Think of the differential of f as $df = f_z dz + f_{\overline{z}} d\overline{z}$. Then we have

$$|df| = |f_z dz + f_{\overline{z}} d\overline{z}| = |h'(z)dz + \overline{g'(z)}d\overline{z}|.$$

Bound the differential, and use the fact that if dz is small, then $d\overline{z}$ will be small and approximately equal to dz. When we examine upper and lower bounds, we have

$$(|h'(z)| - |g'(z)|)|dz| \leq |df| \leq (|h'(z)| + |g'(z)|)|dz|.$$

For sense-preserving harmonic functions, the left-hand side is always positive. The ratio of the upper and lower bounds gives the *geometric dilatation* D_f defined by

$$D_f = \frac{|h'| + |g'|}{|h'| - |g'|}.$$

Since this is a ratio of the maximum to minimum $|df|$, if we evaluate D_f at z_0, we will find a number that represents a ratio between the most and the least that an infinitesimal circle will be deformed by the function. Thus, if the function maps an infinitesimal circle to an infinitesimal ellipse, then the geometric dilatation gives the ratio of the major axis to the minor axis of the ellipse.

Exercise 4.77.

(a) Prove that for analytic functions, $D_f = 1$.

(b) Prove that for sense-preserving harmonic functions, $D_f \geq 1$.

(c) Find a formula for D_f for $z + \frac{\bar{z}^3}{3}$. What are the maximum and minimum of this function over the unit disk $\mathbb{D}$?

Try it out!

Exercise 4.78. Examine the geometric dilatation for $z + \frac{\bar{z}^3}{3}$ in greater detail. For the points $z = 0, 0.5, 0.9, 0.9e^{i\pi/4}$, and $0.9i$, find $D_f(z)$. Using *ComplexTool*, examine the images of circles of radius 0.05 centered at the points. Estimate the ratio of the major axis to the minor axis of the image ellipse. Does it match with your computation for D_f? ***Try it out!***

While the geometric dilatation provides useful information about the function, some information is lost when we take the moduli $|h'(z)|$ and $|g'(z)|$. It is often useful to examine the analytic part versus the anti-analytic part of f. We define what is sometimes called the *second complex dilatation* of f,

$$\omega_f(z) = \frac{\overline{f_{\bar{z}}(z)}}{f_z(z)} = \frac{g'(z)}{h'(z)},$$

where the representation in the last equality makes sense only for harmonic functions. When the function f is clear, we omit the subscript and refer to the dilatation as ω. Because $\omega(z)$ is analytic if and only if $f(z)$ is harmonic, the second complex dilatation is also called the *analytic dilatation* of f.

Exercise 4.79.

(a) Prove that $\omega(z) = \frac{\overline{f_{\bar{z}}(z)}}{f_z(z)}$ is analytic if and only if $f(z)$ is harmonic.

(b) Prove that $\omega(z)$ is identically 0 if and only if f is analytic.

(c) Prove that for sense-preserving non-analytic harmonic functions f, $0 < |\omega(z)| < 1$.

Try it out!

We can ask what the relationship is between the geometric dilatation and the analytic dilatation.

Exercise 4.80. Prove that $D_f(z) \leq K$ if and only if $|\omega_f(z)| \leq \frac{K-1}{K+1}$. ***Try it out!***

Exploration 4.81. Reexamine the function $z + \frac{1}{3}\bar{z}^3$, evaluating $\omega(z)$ at $z = 0, 0.5, 0.9$, $0.9e^{i\pi/4}$, and $0.9i$. What do you observe about the relationship between ω and images of a small circle centered at the points? ***Try it out!***

In Section 4.4, we remarked that if $f = h + \bar{g}$ is a sense-preserving harmonic map that has $|\omega(z)| = 1$ for all $z \in$ arc of $\partial\mathbb{D}$, then the image of the arc is either a concave arc or stationary. To further explore this, we will use *ShearTool* to graph the image of $\mathbb{D}$ under $f = h + \bar{g}$, when $h - g = z$ and varying ω.

Exploration 4.82.

(a) Shear $h(z) - g(z) = z$ using $\omega(z) = e^{i\pi n/6}z$, for $n = 0, \ldots, 6$ and sketch $f(\mathbb{D})$ using *ShearTool*. Describe what happens to $f(\mathbb{D})$ as n varies.

(b) Shear $h(z) - g(z) = z$ using $\omega(z) = z^n$, for $n = 1, 2, 3, 4$ and sketch $f(\mathbb{D})$ using *ShearTool*.

   (i) What patterns do you notice relating $f(\mathbb{D})$ and n?

   (ii) Make a sketch of $f(\mathbb{D})$ for $n = 5$. Then graph the shear using *ShearTool*.

   (iii) Make a sketch of $f(\mathbb{D})$ for $n = 6$. Then graph the shear using *ShearTool*.

(c) Shear $h(z) - g(z) = z$ using $\omega(z) = \frac{z+a}{1+\overline{a}z}$ for various values of $a \in \mathbb{D}$ and sketch $f(\mathbb{D})$ using *ShearTool*. Describe what happens to $f(\mathbb{D})$ as a varies.

Try it out!

Small Project 4.83. Investigate the shearing of $h(z) - g(z) = z - \frac{1}{n^2}z^n$ $(n = 2, 3, 4, \dots)$ with ω for various values of ω (the image of $\mathbb{D}$ under the analytic function $z - \frac{1}{n}z^n$ is not CHD for $n = 4, 5, 6, \dots$; however, it is if we use $z - \frac{1}{n^2}z^n$). Use the approach of Exploration 4.82 as a starting point and then explore new approaches. **Optional**

So far, we have used only dilatations that are finite Blaschke products. A *finite Blaschke product* $B(z)$ can be expressed in the form

$$B(z) = e^{i\theta} \prod_{j=1}^{n} \left(\frac{z - a_j}{1 - \overline{a}_j z} \right)^{m_j},$$

where $\theta \in \mathbb{R}$, $|a_j| < 1$, and m_j is the multiplicity of the zero a_j. The dilatations given in Exploration 4.82 and finite products of them are examples of finite Blaschke products. Harmonic univalent mappings whose dilatations are finite Blaschke products have been studied (see [17]). However, little is known about mappings whose dilatations are not finite Blaschke products. One important type of mappings that are not finite Blaschke products is a singular inner function, and we will now investigate harmonic univalent mappings with dilatations that are singular inner functions.

First, we need to know what singular inner functions are. Let $f : \mathbb{D} \to \mathbb{C}$ be an analytic function and denote its radial limit by

$$f^*(e^{i\theta}) = \lim_{r \to 1, \, r < 1} f(re^{i\theta}).$$

Definition 4.84. A bounded analytic function f is called an *inner function* if $|f^*(e^{i\theta})| = 1$ almost everywhere with respect to Lebesgue measure on $\partial\mathbb{D}$. If f has no zeros on $\mathbb{D}$, then f is called a *singular inner function*.

Every inner function can be written as

$$f(z) = e^{i\alpha} B(z) e^{\left(-\int \frac{e^{i\theta} + z}{e^{i\theta} - z} \, d\mu(e^{i\theta}) \right)},$$

where $\alpha, \theta \in R$, μ is a positive measure on $\partial\mathbb{D}$, and $B(z)$ is a Blaschke product. The function $f(z) = e^{\frac{z+1}{z-1}}$ is an example of a singular inner function.

Exercise 4.85. Show that if $\omega(z) = e^{\frac{z+1}{z-1}}$, then $|\omega(z)| < 1$, for $z \in \mathbb{D}$. *Try it out!*

It has been difficult to construct examples of harmonic mappings whose dilatations are singular inner functions. For a while there were no known examples [21] but Weitsman [30] provided two examples. We present them (see Example 4.88 and Example 4.92) giving a shorter proof of his second example. The proof provides a method to find more examples.

One way to find an example of a harmonic map with a singular inner function as its dilatation is to use the shearing technique. However, it is not often possible to find a closed form for $f = h + \overline{g}$. For example, let $h(z) - g(z) = z$ and the dilatation be $\omega(z) = e^{\frac{z+1}{z-1}}$. Then by the shearing technique

$$h(z) = \int \frac{1}{1 - e^{\frac{z+1}{z-1}}} \, dz.$$

The integral does not have a closed form and so we cannot find an explicit representation for $f = h + \overline{g}$.

Exercise 4.86. Using the shearing technique with $h(z) - g(z) = z - \frac{1}{n}z^n$ and $\omega(z) = e^{\frac{z+1}{z-1}}$, express h as an integral. It is not possible to integrate h to get a closed-form solution. *Try it out!*

Next we give an example by Weitsman in which the shearing technique allows us to solve the integrals for h and g. But first, we will need a result by Pommerenke [22].

Theorem 4.87. *Let f be an analytic function in $\mathbb{D}$ with $f(0) = 0$ and $f'(0) \neq 0$, and let*

$$\varphi(z) = \frac{z}{(1 + ze^{i\theta})(1 + ze^{-i\theta})},$$

where $\theta \in \mathbb{R}$. If

$$\mathrm{Re} \left\{ \frac{zf'(z)}{\varphi(z)} \right\} > 0, \text{ for all } z \in \mathbb{D},$$

then f is convex in the direction of the real axis.

Example 4.88. Consider shearing the analytic function

$$h(z) - g(z) = \frac{z}{1 - z} + \frac{1}{2}e^{\frac{z+1}{z-1}}$$

with

$$\omega(z) = e^{\frac{z+1}{z-1}}.$$

So

$$h'(z) - g'(z) = \frac{1}{(1 - z)^2}\left[1 - e^{\frac{z+1}{z-1}}\right].$$

Using $\theta = \pi$ in Theorem 4.87, we have

$$\mathrm{Re}\left[1 - e^{\frac{z+1}{z-1}}\right] > 0$$

because $\left|e^{\frac{z+1}{z-1}}\right| < 1$. Hence $h - g$ is convex in the direction of the real axis.

Shearing $h - g$ with $\omega(z) = e^{\frac{z+1}{z-1}}$ and normalizing yields

$$h(z) = \int \frac{1}{(1 - z)^2} \, dz = \frac{z}{1 - z},$$

and solving for g we get

$$g(z) = -\frac{1}{2}e^{\frac{z+1}{z-1}}.$$

The image given by the map is similar to the image given by the right half-plane map $\frac{z}{1-z}$ except that there are an infinite number of cusps (see Figure 4.28).

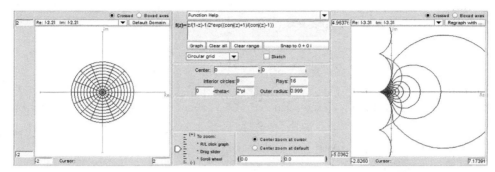

FIGURE 4.28. Image of $\mathbb{D}$ under $f(z) = \dfrac{z}{1-z} - \dfrac{1}{2}e^{\frac{z+1}{z-1}}$.

Exercise 4.89. Let $f = h + \overline{g}$ with $h(z) - g(z) = \frac{z}{1-z}$ and $\omega(z) = e^{\frac{z+1}{z-1}}$. Use the shearing method to compute h and g explicitly so $f \in S_H^O$ and use *ComplexTool* to sketch $f(\mathbb{D})$. Hint: In finding the h, use a u-substitution to evaluate the integral. ***Try it out!***

Another technique to find harmonic mappings whose dilatations are singular inner functions involves using the following theorem by Clunie and Sheil-Small [6].

Theorem 4.90. *Let $f = h + \overline{g}$ be locally univalent in $\mathbb{D}$ and suppose that $h + \epsilon g$ is convex for some $|\epsilon| \leq 1$. Then f is univalent.*

Theorem 4.90 gives us a way to show that a harmonic function is univalent. To develop the technique, we let $\epsilon = 0$ in the theorem. This means that if h is analytic convex and if ω is analytic with $|\omega(z)| < 1$, then $f = h + \overline{g}$ is a harmonic univalent mapping. To establish that a function f is convex, we will use the following theorem (see [13]).

Theorem 4.91. *Let f be analytic in $\mathbb{D}$ with $f(0) = 0$ and $f'(0) = 1$. Then f is univalent and maps onto a convex domain if and only if*

$$\mathrm{Re}\left[1 + \frac{zf''(z)}{f'(z)}\right] \geq 0, \text{ for all } z \in \mathbb{D}.$$

In the next example we will show how we can use these ideas to construct a harmonic univalent function whose dilatation is a singular inner function.

Example 4.92. Let

$$h(z) = z - \frac{1}{4}z^2 \quad \text{with} \quad \omega(z) = g'(z)/h'(z) = e^{\frac{z+1}{z-1}}.$$

We will use Theorem 4.91 to show that h is convex. Let

$$T(z) = 1 + \frac{zh''(z)}{h'(z)} = 1 + \frac{-\frac{1}{2}z}{1 - \frac{1}{2}z} = \frac{1 - z}{1 - \frac{1}{2}z}$$

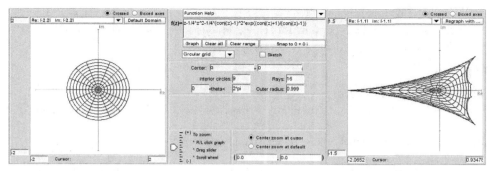

FIGURE 4.29. Image of $\mathbb{D}$ under $f_1(z) = z - \frac{1}{4}z^2 - \frac{1}{4}(\bar{z} - 1)^2 e^{\frac{\bar{z}+1}{\bar{z}-1}}$.

which is a Möbius transformation. By the mapping properties of Möbius transformations we can show that T maps $\mathbb{D}$ onto $|z - \frac{2}{3}| = \frac{2}{3}$, the circle centered at $\frac{2}{3}$ with radius $\frac{2}{3}$. Hence, $\text{Re}\{T(z)\} > 0$ and h is convex.

We can solve for g:

$$g(z) = \int h'(z)\omega(z)\, dz = \int \left(1 - \frac{1}{2}z\right)e^{\frac{z+1}{z-1}}\, dz = -\frac{1}{4}(z - 1)^2 e^{\frac{z+1}{z-1}}.$$

Hence,

$$f(z) = h(z) + \overline{g(z)} = z - \frac{1}{4}z^2 - \frac{1}{4}(\bar{z} - 1)^2 e^{\frac{\bar{z}+1}{\bar{z}-1}}.$$

By Theorem 4.90, $f = h + \bar{g}$ is univalent. The image of $\mathbb{D}$ under $f_1(z) = z - \frac{1}{4}z^2 - \frac{1}{4}(\bar{z} - 1)^2 e^{\frac{\bar{z}+1}{\bar{z}-1}}$ is similar to the map of a harmonic polynomial but with an infinite number of cusps in the middle section on the right side (see Figure 4.29).

This approach has been extended to find more examples of harmonic univalent functions with singular inner function dilatations [2].

Exercise 4.93. Let $h(z) = z + \frac{1}{11}z^3$ and $g(z) = -\frac{1}{11}(z - 3)(z + 1)^2 e^{\frac{z-1}{z+1}}$. Show that $f = h + \bar{g}$ is univalent and use *ComplexTool* to graph $f(\mathbb{D})$. ***Try it out!***

Exercise 4.94. Use this approach to show that $f = h + \bar{g}$ is harmonic univalent, where $h(z) = z + 2\log(z + 1)$ and $\omega(z) = e^{\frac{z-1}{z+1}}$. The graph of $\mathbb{D}$ under f is in Figure 4.30. ***Try it out!***

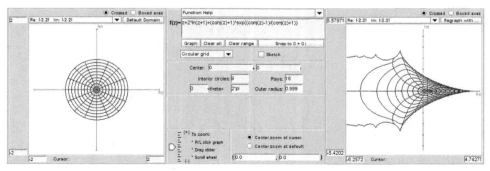

FIGURE 4.30. Image of $\mathbb{D}$ under $f(z) = h + \bar{g}$ in Exercise 4.94.

We mentioned that it is not often possible to find a closed form for $f = h + \overline{g}$ when the dilatation is a singular inner function. However, *ShearTool* can explore images of $f(\mathbb{D})$ when the dilatation is a singular inner product.

Exploration 4.95.

(a) If we shear $h(z) - g(z) = \frac{z}{(1-z)^2}$ with $\omega(z) = e^{\frac{z+1}{z-1}}$, then $f = h + \overline{g}$ will be univalent and convex in the direction of the real axis. Use *ShearTool* to sketch $f(\mathbb{D})$.

(b) Use *ShearTool* to sketch $f(\mathbb{D})$, where $h(z) - g(z) = z - \frac{1}{2}z^2$ with $\omega(z) = e^{\frac{z^2+1}{z^2-1}}$.

(c) Use *ShearTool* to sketch $f(\mathbb{D})$, where $h(z) - g(z) = \frac{1}{2} \log\left(\frac{1+z}{1-z}\right)$ with $\omega(z) = e^{\frac{z+1}{z-1}}$.

Try it out!

Open Problem 4.96. What are the properties of harmonic univalent mappings whose dilatation is a singular inner product?

4.7   Harmonic Linear Combinations

A common way to construct new functions with a given property is to take a linear combination of two functions with that property. This is done with derivatives and integrals in beginning calculus. In Exploration 4.25, this is done with the analytic Koebe mapping, f_k, and the right half-plane mapping, f_r, where

$$f_k(z) = \frac{z}{(1-z)^2} \qquad \text{and} \qquad f_r(z) = \frac{z}{1-z},$$

to derive the univalent analytic map

$$f_3(z) = t f_k(z) + (1-t) f_r(z) = \frac{z - tz^2}{(1-z)^2},$$

where $0 \le t \le 1$.

Is it true that a linear combination of two one-to-one functions is a one-to-one function? Let's look at real-valued functions. Suppose $f_1 : \mathbb{R} \to \mathbb{R}$ and $f_2 : \mathbb{R} \to \mathbb{R}$ are one-to-one functions. Will $f_3(x) = t f_1(x) + (1-t) f_2(x)$ also be one-to-one when $0 \le t \le 1$? Not necessarily. Let $f_1(x) = x^3$, $f_2(x) = -x^3$, and $t = \frac{1}{2}$. Because they satisfy the horizontal line test, f_1 and f_2 are one-to-one. But $f_3(x) = t f_1(x) + (1-t) f_2(x) = 0$, which is not one-to-one.

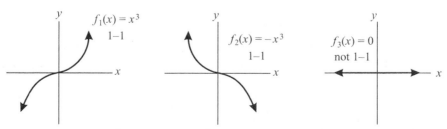

FIGURE 4.31. $f_1(x) = x^3$ and $f_2(x) = -x^3$ are one-to-one, but $f_3(x) = t f_1(x) + (1-t) f_2(x) = 0$ is not one-to-one when $t = \frac{1}{2}$.

The difficulty is that f_1 is an increasing one-to-one function, f_2 is a decreasing one-to-one function, and when $t = \frac{1}{2}$ the increase of f_1 cancels out the decrease of f_2. We can get around this difficulty by requiring that f_1, f_2 are either both increasing or both decreasing. This idea can be applied to complex-valued functions.

Condition A. Suppose f is complex-valued harmonic and non-constant in $\mathbb{D}$. There exists sequences $\{z_n'\}$, $\{z_n''\}$ converging to $z = 1$, $z = -1$, respectively, such that

$$\lim_{n \to \infty} \operatorname{Re}\{f(z_n')\} = \sup_{|z|<1} \operatorname{Re}\{f(z)\}$$
$$\lim_{n \to \infty} \operatorname{Re}\{f(z_n'')\} = \inf_{|z|<1} \operatorname{Re}\{f(z)\}. \tag{4.14}$$

The normalization in (4.14) can be thought of as if $f(1)$ and $f(-1)$ are the right and left extremes in the image domain in the extended complex plane.

Example 4.97. We will show that condition A is satisfied by $f(z) = z + \frac{1}{3}\bar{z}^3$ (see Figure 4.32).

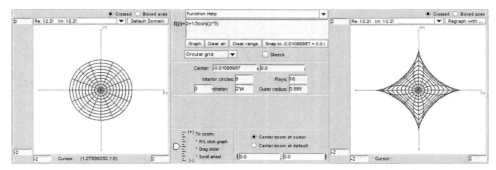

FIGURE 4.32. Image of $\mathbb{D}$ under $f(z) = z + \frac{1}{3}\bar{z}^3$.

One can see that f satisfies condition A by using the **Sketch** option in *ComplexTool* to draw several paths of sequences $\{z_n'\} \in \mathbb{D}$ that approach 1 and see that the images of the paths approach the right-side cusp of the image of $\mathbb{D}$, a hypocycloid of four cusps (see Figure 4.32). Likewise, paths $\{z_n''\} \in \mathbb{D}$ that approach -1 result in image paths that approach the left-side cusp of the hypocycloid.

To prove that f satisfies condition A, we have

$$f(e^{i\theta}) = e^{i\theta} + \frac{1}{3}e^{-3i\theta} = \left(\cos\theta + \frac{1}{3}\cos 3\theta\right) + i\left(\sin\theta - \frac{1}{3}\sin 3\theta\right).$$

So, $\operatorname{Re}\{f(e^{i\theta})\} = \cos\theta + \frac{1}{3}\cos 3\theta$. Hence $-\frac{4}{3} \le \operatorname{Re}\{f(e^{i\theta})\} \le \frac{4}{3}$ which implies $\sup_{|z|<1} \operatorname{Re}\{f(z)\} = \frac{4}{3}$. Letting $\{z_n'\} = \{1 - \frac{1}{n}\} \to 1$, we have that

$$\lim_{n \to \infty} \operatorname{Re}\{f(z_n')\} = \frac{4}{3} = \sup_{|z|<1} \operatorname{Re}\{f(z)\}.$$

Similarly, $\displaystyle\lim_{n \to \infty} \operatorname{Re}\{f(z_n'')\} = -\frac{4}{3} = \inf_{|z|<1} \operatorname{Re}\{f(z)\}.$

Exercise 4.98. Use the `Sketch` option in *ComplexTool* to determine which of the following harmonic functions satisfy condition A.

(a) $f(z) = z + \frac{1}{2}\bar{z}^2$

(b) $\text{Re}\left[\frac{1}{2}\log\left(\frac{1+z}{1-z}\right)\right] + i\,\text{Im}\left[\frac{i}{2}\log\left(\frac{i+z}{i-z}\right)\right]$

(c) $\text{Re}\left(\frac{z}{1-z}\right) + i\,\text{Im}\left(\frac{z}{(1-z)^2}\right)$

(d) $\text{Re}\left[\frac{i}{2}\log\left(\frac{i+z}{i-z}\right)\right] + i\,\text{Im}\left[\frac{1}{2}\log\left(\frac{1+z}{1-z}\right)\right]$

(e) $\text{Re}\left(\frac{z + \frac{1}{3}z^3}{(1-z)^3}\right) + i\,\text{Im}\left(\frac{z}{(1-z)^2}\right)$

Try it out!

To prove a result about linear combinations of harmonic functions, we will need the following theorem by Hengartner and Schober [19] that employs condition A. However, we won't use Theorem 4.99 afterwards.

Theorem 4.99 (Hengartner and Schober). *Suppose f is analytic and non-constant in $\mathbb{D}$. Then*

$$\text{Re}\{(1 - z^2)f'(z)\} \geq 0, z \in \mathbb{D}$$

if and only if

1. *f is univalent in $\mathbb{D}$,*

2. *f is convex in the imaginary direction, and*

3. *condition A holds.*

We now study conditions under which f_3 is globally univalent.

Theorem 4.100. *Let $f_1 = h_1 + \overline{g_1}$, $f_2 = h_2 + \overline{g_2}$ be univalent harmonic mappings convex in the imaginary direction and $\omega_1 = \omega_2$. If f_1, f_2 satisfy condition A, then $f_3 = tf_1 + (1-t)f_2$ is convex in the imaginary direction (and univalent) $(0 \leq t \leq 1)$.*

Proof. To see that f_3 is locally univalent, use $g'_1 = \omega_1 h'_1$ and $g'_2 = \omega_2 h'_2 = \omega_1 h'_2$. Then

$$\omega_3 = \frac{tg'_1 + (1-t)g'_2}{th'_1 + (1-t)h'_2} = \frac{t\omega_1 h'_1 + (1-t)\omega_1 h'_2}{th'_1 + (1-t)h'_2} = \omega_1.$$

By Clunie and Sheil-Small's shearing theorem (see Theorem 4.52), we know that $h_j + g_j$ $(j = 1, 2)$ is univalent and convex in the imaginary direction. Also, $h_j + g_j$ satisfies condition A since $\text{Re}\{f_j\} = \text{Re}\{h_j + g_j\}$. Applying Theorem 4.99 we have

$$\text{Re}\{(1 - z^2)(h'_j(z) + g'_j(z))\} \geq 0, (j = 1, 2).$$

So

$$\begin{aligned}
\text{Re}\{(1-z^2)&(h'_3(z) + g'_3(z))\} \\
&= \text{Re}\{(1 - z^2)[t(h'_1(z) + g'_1(z)) + (1 - t)(h'_2(z) + g'_2(z))]\} \\
&= t\,\text{Re}\{(1 - z^2)(h'_1(z) + g'_1(z))\} + (1 - t)\,\text{Re}\{(1 - z^2)(h'_2(z) + g'_2(z))\} \geq 0.
\end{aligned}$$

By applying Theorem 4.99 in the other direction, we have that $h_3 + g_3$ is convex in the imaginary direction, and so by the shearing theorem, f_3 is convex in the imaginary direction.

Example 4.101. Let

$$f_1(z) = \text{Re}\left[\frac{i}{2}\log\left(\frac{1+z}{1-z}\right)\right] + i\,\text{Im}\left[-\frac{1}{2}\log\left(\frac{i+z}{i-z}\right)\right]$$

$$f_2(z) = \text{Re}\left[\frac{1}{2}\log\left(\frac{1+z}{1-z}\right)\right] + i\,\text{Im}\left[\frac{i}{2}\log\left(\frac{i+z}{i-z}\right)\right].$$

Notice f_1 maps $\mathbb{D}$ onto a square region (see Figure 4.33). The image is the same as for the harmonic square map in Example 4.72, but the function is different because f_1 has different arcs of the unit circle mapped to the vertices and the dilatation for f_1 is

FIGURE 4.33. Image of $\mathbb{D}$ under $f_1(z) = \text{Re}\left[\frac{i}{2}\log\left(\frac{1+z}{1-z}\right)\right] + i\,\text{Im}\left[-\frac{1}{2}\log\left(\frac{i+z}{i-z}\right)\right]$.

$\omega(z) = z^2$ which is different than the dilatation for the harmonic square map in Example 4.72. Condition A is satisfied for f_1 (see Example 4.104 for more details). f_2 maps $\mathbb{D}$ onto a region similar to a hypocycloid with four cusps except instead of cusps the domain has ends that extend to infinity (see Figure 4.34). The dilatation of f_2 is also $\omega(z) = z^2$ and

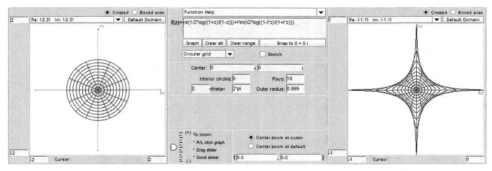

FIGURE 4.34. Image of $\mathbb{D}$ under $f_2(z) = \text{Re}\left[\frac{1}{2}\log\left(\frac{1+z}{1-z}\right)\right] + i\,\text{Im}\left[\frac{i}{2}\log\left(\frac{i+z}{i-z}\right)\right]$.

condition A is satisfied. By Theorem 4.100, $f_3 = tf_1 + (1-t)f_2$ is univalent. The image of $\mathbb{D}$ when $t = \frac{1}{2}$ is shown in Figure 4.35.

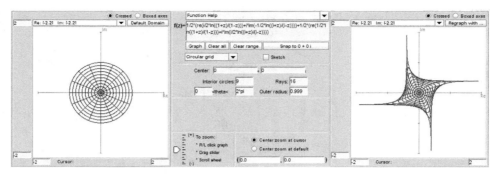

FIGURE 4.35. Image of $\mathbb{D}$ under $f_3(z) = \frac{1}{2} f_1(z) + \frac{1}{2} f_2(z)$.

Exploration 4.102. Let

$$f_1(z) = \mathrm{Re}\left[\frac{i}{\sqrt{3}} \log\left(\frac{1 + e^{-i\frac{\pi}{3}}z}{1 + e^{i\frac{\pi}{3}}z}\right)\right] + i\,\mathrm{Im}\left[\frac{1}{3}\log\left(\frac{1 + z + z^2}{1 - 2z + z^2}\right)\right],$$

$$f_2(z) = \mathrm{Re}\left(\frac{z}{1-z}\right) + i\,\mathrm{Im}\left(\frac{z}{(1-z)^2}\right).$$

Show that f_1 and f_2 satisfy the conditions of Theorem 4.100 and then use *ComplexTool* to plot images of $f_3 = t f_1 + (1 - t) f_2$ for various values of t. **Try it out!**

Large Project 4.103. Theorem 4.100 gives sufficient but not necessary conditions on f_1 and f_2 for the linear combination $f_3 = t f_1 + (1 - t) f_2$ to be univalent. That f_3 can be univalent when f_1 does not satisfy condition A is demonstrated by

$$f_1(z) = \mathrm{Re}\left(\frac{z + 1/3z^3}{(1-z)^3}\right) + i\,\mathrm{Im}\left(\frac{z}{(1-z)^2}\right),$$

$$f_2(z) = \mathrm{Re}\left(\frac{z}{1-z}\right) + i\,\mathrm{Im}\left(\frac{z}{(1-z)^2}\right).$$

Construct f_3 for various values of t and use *ComplexTool* to see the images of $\mathbb{D}$ under f_3.

The following functions suggest that several of the hypotheses of Theorem 4.100 can fail and f_3 can still be univalent:

$$f_1(z) = z - \frac{1}{m}\overline{z}^m \qquad \text{and} \qquad f_2(z) = z - \frac{1}{n}\overline{z}^n,$$

where $m, n \geq 2$. For various values of m, n, and t construct f_3 and use *ComplexTool* to see the images of $\mathbb{D}$ under f_3.

Investigate the examples and construct other examples in which f_3 is univalent but the hypotheses of Theorem 4.100 do not hold. Use them to make a conjecture about the hypotheses of a new theorem that guarantees f_3 will be univalent. Prove the new theorem. **Optional**

Let us look at an example that is surprising and is related to the nonconvex polygons described by Duren, McDougall, and Schaubroeck [15].

Example 4.104. Let $f_1 = h_1 + \overline{g_1}$ be the harmonic square map in Example 4.101, where

$$h_1(z) = \frac{i}{4} \log\left(\frac{1+z}{1-z}\right) - \frac{1}{4} \log\left(\frac{i+z}{i-z}\right)$$

$$g_1(z) = \frac{i}{4} \log\left(\frac{1+z}{1-z}\right) + \frac{1}{4} \log\left(\frac{i+z}{i-z}\right).$$

We can write this as

$$f_1(z) = \text{Re}\left[\frac{i}{2} \log\left(\frac{1+z}{1-z}\right)\right] + i \, \text{Im}\left[-\frac{1}{2} \log\left(\frac{i+z}{i-z}\right)\right].$$

Using the same approach as in Example 4.72, we see that $z = e^{i\theta} \in \partial\mathbb{D}$ is mapped to

$$u_1 + iv_1 = \begin{cases} z_1 = \frac{\pi}{2\sqrt{2}} \, e^{i\frac{\pi}{4}} & \text{if } \theta \in \left(\frac{-\pi}{2}, 0\right) \\ z_3 = \frac{\pi}{2\sqrt{2}} \, e^{i\frac{3\pi}{4}} & \text{if } \theta \in \left(0, \frac{\pi}{2}\right) \\ z_5 = \frac{\pi}{2\sqrt{2}} \, e^{i\frac{5\pi}{4}} & \text{if } \theta \in \left(\frac{\pi}{2}, \pi\right) \\ z_7 = \frac{\pi}{2\sqrt{2}} \, e^{i\frac{7\pi}{4}} & \text{if } \theta \in \left(\pi, \frac{3\pi}{2}\right). \end{cases}$$

So, f_1 maps $\mathbb{D}$ onto a square region with vertices at $z_1, z_3, z_5,$ and z_7 (see Figure 4.36). The dilatation for f_1 is $\omega = z^2$ and condition A is satisfied. For example, for a sequence of points, $\{z_n'\}$, in the fourth quadrant approaching 1,

$$\lim_{n\to\infty} \text{Re}\{f_1(z_n')\} = \frac{\pi}{2\sqrt{2}} = \sup_{|z|<1} \text{Re}\{f_1(z)\}$$

and for a sequence of points, $\{z_n''\}$, in the second quadrant approaching -1,

$$\lim_{n\to\infty} \text{Re}\{f_1(z_n'')\} = -\frac{\pi}{2\sqrt{2}} = \inf_{|z|<1} \text{Re}\{f_1(z)\}.$$

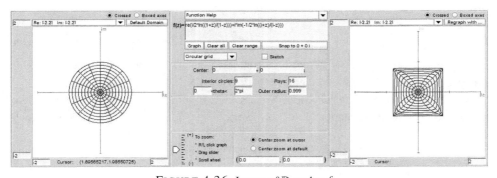

FIGURE 4.36. Image of $\mathbb{D}$ under f_1.

Let $f_2 = h_2 + \overline{g_2}$, where

$$h_2(z) = -\frac{1}{4}e^{-i\frac{3\pi}{4}} \log\left(\frac{e^{i\frac{\pi}{4}}+z}{e^{i\frac{\pi}{4}}-z}\right) - \frac{1}{4}e^{-i\frac{\pi}{4}} \log\left(\frac{e^{i\frac{3\pi}{4}}+z}{e^{i\frac{3\pi}{4}}-z}\right)$$

$$g_2(z) = \frac{1}{4}e^{i\frac{3\pi}{4}} \log\left(\frac{e^{i\frac{\pi}{4}}+z}{e^{i\frac{\pi}{4}}-z}\right) + \frac{1}{4}e^{i\frac{\pi}{4}} \log\left(\frac{e^{i\frac{3\pi}{4}}+z}{e^{i\frac{3\pi}{4}}-z}\right).$$

Then

$$f_2(z) = \mathrm{Re}\left\{\frac{i}{2\sqrt{2}}\left[\log\left(\frac{e^{i\frac{\pi}{4}}+z}{e^{i\frac{\pi}{4}}-z}\right)+\log\left(\frac{e^{i\frac{3\pi}{4}}+z}{e^{i\frac{3\pi}{4}}-z}\right)\right]\right\}$$

$$+ i\,\mathrm{Im}\left\{\frac{1}{2\sqrt{2}}\left[\log\left(\frac{e^{i\frac{\pi}{4}}+z}{e^{i\frac{\pi}{4}}-z}\right)-\log\left(\frac{e^{i\frac{3\pi}{4}}+z}{e^{i\frac{3\pi}{4}}-z}\right)\right]\right\},$$

and $z = e^{i\theta} \in \partial\mathbb{D}$ is mapped to

$$u_2 + iv_2 = \begin{cases} z_0 = \dfrac{\pi}{2\sqrt{2}} & \text{if } \theta \in (\frac{-\pi}{4}, \frac{\pi}{4}) \\[2mm] z_2 = \dfrac{i\pi}{2\sqrt{2}} & \text{if } \theta \in (\frac{\pi}{4}, \frac{3\pi}{4}) \\[2mm] z_4 = -\dfrac{\pi}{2\sqrt{2}} & \text{if } \theta \in (\frac{3\pi}{4}, \frac{5\pi}{4}) \\[2mm] z_6 = -\dfrac{i\pi}{2\sqrt{2}} & \text{if } \theta \in (\frac{5\pi}{4}, \frac{7\pi}{4}). \end{cases}$$

That is, f_2 maps $\mathbb{D}$ onto a rotated square region with vertices at z_0, z_2, z_4, and z_6 (see Figure 4.37) with $\omega = z^2$, and it also satisfies condition A.

FIGURE 4.37. Image of $\mathbb{D}$ under f_2.

By Theorem 4.100, $f_3 = tf_1 + (1-t)f_2$ is univalent. What is the image of $\mathbb{D}$ under f_3? Let's look at $t = \dfrac{1}{2}$. You might think that $f_3(\mathbb{D})$ would be an overlay of $f_1(\mathbb{D})$ and $f_2(\mathbb{D})$ (see Figure 4.38(a)), but it is not. It is the nonconvex star in Figure 4.38(b).

Why is the correct image the nonconvex star in Figure 4.38(b)? Let's look at where arcs of the unit circle are mapped under f_3. Notice $f_1(e^{i\theta})$ and $f_2(e^{i\theta})$ depend on which

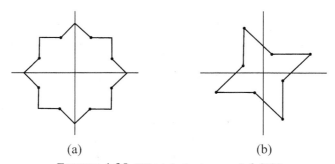

(a)   (b)

FIGURE 4.38. Which is the image of $f_3(\mathbb{D})$?

of eight arcs θ is in. If $\theta \in (-\frac{\pi}{4}, 0)$, then $f_1(e^{i\theta}) = z_1$ and $f_2(e^{i\theta}) = z_0$, and so in this interval $f_3(e^{i\theta}) = \frac{z_1 + z_0}{2}$ (that is, it is the midpoint between z_1 and z_0). Iif $\theta \in (0, \frac{\pi}{4})$, then $f_1(e^{i\theta}) = z_3$ and $f_2(e^{i\theta}) = z_0$, and $f_3(e^{i\theta}) = \frac{z_3 + z_0}{2}$. So

$$f_3(e^{i\theta}) = \begin{cases} w_1 = \frac{z_1 + z_0}{2} = \frac{\pi}{2\sqrt{2}} \cos \frac{\pi}{8} e^{i\frac{\pi}{8}} & \text{if } \theta \in (\frac{-\pi}{4}, 0) \\ w_2 = \frac{z_3 + z_0}{2} = \frac{\pi}{2\sqrt{2}} \cos \frac{3\pi}{8} e^{i\frac{3\pi}{8}} & \text{if } \theta \in (0, \frac{\pi}{4}) \\ w_3 = \frac{z_3 + z_2}{2} = \frac{\pi}{2\sqrt{2}} \cos \frac{\pi}{8} e^{i\frac{5\pi}{8}} & \text{if } \theta \in (\frac{\pi}{4}, \frac{\pi}{2}) \\ w_4 = \frac{z_5 + z_2}{2} = \frac{\pi}{2\sqrt{2}} \cos \frac{3\pi}{8} e^{i\frac{7\pi}{8}} & \text{if } \theta \in (\frac{\pi}{2}, \frac{3\pi}{4}) \\ w_5 = \frac{z_5 + z_4}{2} = \frac{\pi}{2\sqrt{2}} \cos \frac{\pi}{8} e^{i\frac{9\pi}{8}} & \text{if } \theta \in (\frac{3\pi}{4}, \pi) \\ w_6 = \frac{z_7 + z_4}{2} = \frac{\pi}{2\sqrt{2}} \cos \frac{3\pi}{8} e^{i\frac{11\pi}{8}} & \text{if } \theta \in (\pi, \frac{5\pi}{4}) \\ w_7 = \frac{z_7 + z_6}{2} = \frac{\pi}{2\sqrt{2}} \cos \frac{\pi}{8} e^{i\frac{13\pi}{8}} & \text{if } \theta \in (\frac{5\pi}{4}, \frac{3\pi}{2}) \\ w_8 = \frac{z_1 + z_6}{2} = \frac{\pi}{2\sqrt{2}} \cos \frac{3\pi}{8} e^{i\frac{15\pi}{8}} & \text{if } \theta \in (\frac{3\pi}{2}, \frac{7\pi}{4}). \end{cases}$$

The vertices w_1, w_3, w_5, and w_7 lie equally spaced on a circle of radius $r_{outer} = \frac{\pi}{2\sqrt{2}} \cos \frac{\pi}{8} \approx$ 1.026, and the vertices w_2, w_4, w_6, and w_8 lie equally spaced on a circle of radius $r_{inner} = \frac{\pi}{2\sqrt{2}} \cos \frac{3\pi}{8} \approx 0.425$. We can visualize the boundary of $f_3(\mathbb{D})$ by plotting the vertices $z_0, z_1, \ldots z_7$ and drawing the midpoints $w_1, \ldots, w_8$ (see Figure 4.39).

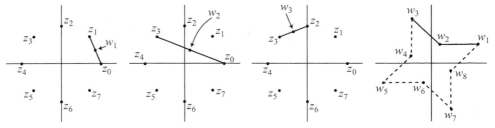

FIGURE 4.39. Visualizing the image of the boundary of $f_3(\mathbb{D})$.

We can also explore the linear combination of these two functions by using the applet, *LinComboTool* (see Figure 4.40). Open *LinComboTool*. Make sure that at the top of the page, the **Number of Polygonal Panels** is 2. In **Panel #1** enter the left end points of the intervals for the arcs of the unit circle used in f_1 (they need to be positive numbers). Then enter the real and imaginary values of the image of this arc under f_1. For example, if we take the interval $(0, \frac{\pi}{2})$ for Arc 1 (we start with this interval because we need to use nonnegative values), then for Arc 1 enter 0, for the real value of its image enter $\frac{\pi}{2\sqrt{2}} \cos \frac{3\pi}{4}$, and for the imaginary value of its image enter $\frac{\pi}{2\sqrt{2}} \sin \frac{3\pi}{4}$. Continue this for the other arcs noting that for Arc 4 we will enter $\frac{3\pi}{2}$ in the first box. If there are not enough boxes for the arcs, click on the **Add** button to add an arc. Similarly, click on **Remove** if there are too many boxes for the arcs. When you are done entering the points, click on **Graph** to produce the image $f_1(\mathbb{D})$. Then go to **Panel #2** and enter the points for f_2 and graph $f_2(\mathbb{D})$. After these are graphed, click on **Create LinCombogon** and the image will appear in the lower left-hand box (see Figure 4.41).

Exercise 4.105. Using *LinComboTool*, start with the same arc values and point values as in Example 4.104. You can change the value of t by sliding the red dot near the top of the

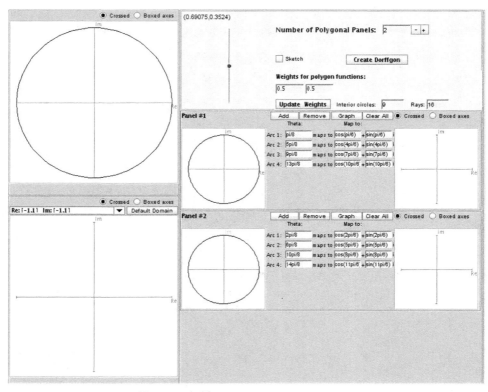

FIGURE 4.40. The applet *LinComboTool*.

page. Describe what happens as t varies from 0 to 1. In the example we showed that when $t = \frac{1}{2}$, $r_{outer} = \frac{\pi}{2\sqrt{2}} \cos \frac{\pi}{8} \approx 1.026$ and $r_{inner} = \frac{\pi}{2\sqrt{2}} \cos \frac{3\pi}{8} \approx 0.425$. Compute r_{outer} and r_{inner} for any t $(0 \leq t \leq 1)$. **Try it out!**

Remark 4.106. In Theorem 4.100, we do not need that $\omega_1 = \omega_2$. Looking over its proof we see that all we need is that f_3 is locally univalent. This occurs if

$$|\omega_3| = \left| \frac{tg_1' + (1-t)g_2'}{th_1' + (1-t)h_2'} \right| < 1. \tag{4.15}$$

Exercise 4.107. We can have one pair of functions f_1, f_2 mapping onto image domains G_1, G_2, and another pair of functions $\tilde{f}_1$, $\tilde{f}_2$ that map onto the same image domains G_1, G_2, but the linear combinations f_3 and $\tilde{f}_3$ map onto different image domains.

Repeat the steps in Example 4.104 using the same function for f_2 but replacing f_1 with the harmonic square map in Example 4.72, where

$$h_1(z) = \frac{1}{4} \log\left(\frac{1+z}{1-z}\right) + \frac{i}{4} \log\left(\frac{i+z}{i-z}\right)$$

$$g_1(z) = -\frac{1}{4} \log\left(\frac{1+z}{1-z}\right) + \frac{i}{4} \log\left(\frac{i+z}{i-z}\right).$$

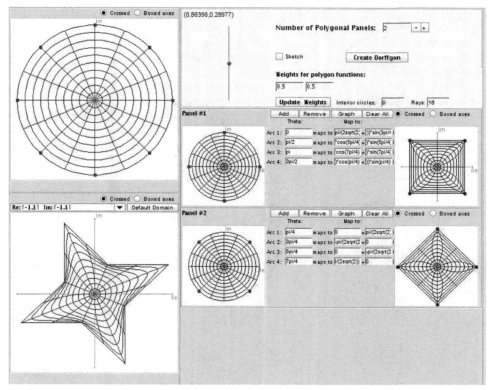

FIGURE 4.41. Image of $\mathbb{D}$ under f_3.

We have

$$h'_1(z) = \frac{1}{1 - z^4}, \qquad g'_1(z) = \frac{-z^2}{1 - z^4}$$

$$h'_2(z) = \frac{1}{1 + z^4}, \qquad g'_2(z) = \frac{z^2}{1 + z^4}.$$

(a) In this case, $\omega_1(z) = -z^2$ while $\omega_2(z) = z^2$. Using (4.15) in Remark 4.106, show that f_3 is locally univalent.

(b) Use *LinComboTool* find the image of $f_3(\mathbb{D})$ using this f_1 and f_2.

(c) Explain why this happens by using the approach in Example 4.104 to compute the new values of $w_1, \ldots, w_8$ and then use the visualization technique in the example to plot the vertices $z_0, \ldots, z_7$ and draw the midpoints $w_1, \ldots, w_8$.

Try it out!

Exercise 4.108. Repeat the steps in Exercise 4.107 using the same function for f_1 but replacing f_2 with the harmonic hexagon map that can be derived from (4.13) for h' and g'

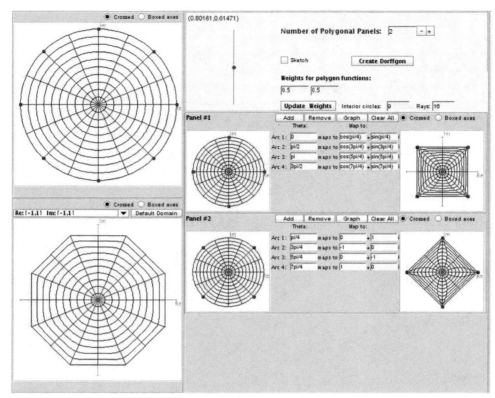

FIGURE 4.42. Image of $\mathbb{D}$ under f_3 in Exercise 4.107.

at the end of Example 4.72 , where

$$h_2'(z) = \frac{1}{1 - z^6} \Rightarrow$$

$$h_2(z) = \frac{1}{6} \log\left(\frac{1+z}{1-z}\right) + \frac{e^{\frac{-i\pi}{3}}}{6} \log\left(\frac{1 + e^{\frac{i\pi}{3}} z}{1 - e^{\frac{i\pi}{3}} z}\right) + \frac{e^{\frac{-i2\pi}{3}}}{6} \log\left(\frac{1 + e^{\frac{i2\pi}{3}} z}{1 - e^{\frac{i2\pi}{3}} z}\right)$$

$$g_2'(z) = \frac{-z^4}{1 - z^6} \Rightarrow$$

$$g_2(z) = -\frac{1}{6} \log\left(\frac{1+z}{1-z}\right) - \frac{e^{\frac{i\pi}{3}}}{6} \log\left(\frac{1 + e^{\frac{i\pi}{3}} z}{1 - e^{\frac{i\pi}{3}} z}\right) - \frac{e^{\frac{i2\pi}{3}}}{6} \log\left(\frac{1 + e^{\frac{i2\pi}{3}} z}{1 - e^{\frac{i2\pi}{3}} z}\right).$$

(a) In this case, $\omega_1(z) = -z^2$ while $\omega_2(z) = -z^4$. Using (4.15) in the remark, show that f_3 is locally univalent.

(b) Use *LinComboTool* find the image of $f_3(\mathbb{D})$ using this f_1 and f_2.

(c) Explain why this happens by using the approach in Example 4.104 to compute the new values of the vertices of $f_3(\mathbb{D})$.

Try it out!

We can generalize Theorem 4.100 to include the linear combination of n functions $f_1, \ldots, f_n$.

Theorem 4.109. *Let* $f_1 = h_1 + \overline{g_1}, \ldots, f_n = h_n + \overline{g_n}$ *be n univalent harmonic mappings convex in the imaginary direction and* $\omega_1 = \cdots = \omega_n$. *If* $f_1, \ldots, f_n$ *satisfy condition A, then* $F = t_1 f_1 + \cdots + t_n f_n$ *is convex in the imaginary direction, where* $0 \leq t_n \leq 1$ *and* $t_1 + \cdots + t_n = 1$.

Exercise 4.110. Prove Theorem 4.109. *Try it out!*

Exploration 4.111. Using Theorem 4.109, create maps in three different panels of *Lin-ComboTool*, where each map takes four arcs on the unit circle to four vertices of a square. Make sure that they satisfy the conditions of the theorem. Then click on the **Create LinCombogon** button to see the image domain. Explore this by using different maps in the panels. For an example, see Figure 4.43. *Try it out!*

Large Project 4.112. In Example 4.104 and Exercise 4.107 we took linear combinations of two harmonic square mappings and got fundamentally different images. Explore this with other n-gons. Use *LinComboTool* to determine what and how many fundamentally different (i.e., not rotations or scalings) images can be constructed when taking the linear

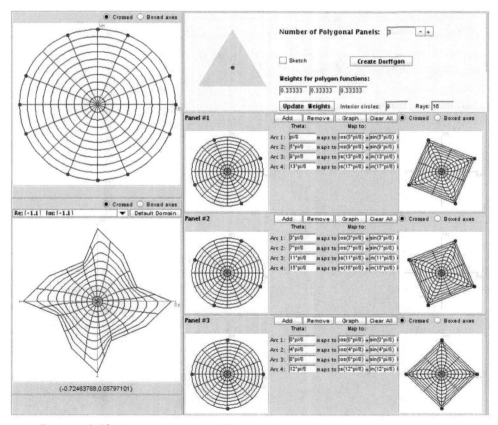

FIGURE 4.43. Example of image of $\mathbb{D}$ under the linear combination of three squares.

combination with $t = \frac{1}{2}$ of two harmonic 5-gon maps, 6-gon maps, and n-gon maps. Make sure that condition A holds and that $|\omega_3| < 1$. **Optional**

Large Project 4.113. In Exercise 4.108 we took linear combinations of a harmonic 4-gon mapping and a harmonic 6-gon mapping with dilatations $-z^2$ and $-z^4$, respectively. Use *LinComboTool* to determine what combinations are possible and what images can be constructed when taking linear combination with $t = \frac{1}{2}$ of a harmonic m-gon and n-gon, where $m < n$. Make sure that condition A holds and that $|\omega_3(z)| < 1$. **Optional**

For more on this topic see [10].

4.8   Convolutions

Another way of combining two univalent functions is with the Hadamard product or convolution. For analytic functions

$$f(z) = \sum_{n=0}^{\infty} a_n z^n \quad \text{and} \quad F(z) = \sum_{n=0}^{\infty} A_n z^n,$$

their convolution is defined as

$$f(z) * F(z) = \sum_{n=0}^{\infty} a_n A_n z^n.$$

Example 4.114. The convolution of the right half-plane function (see Example 4.22)

$$f(z) = \frac{z}{1-z} = \sum_{n=1}^{\infty} z^n$$

and the Koebe function (see Example 4.23)

$$F(z) = \frac{z}{(1-z)^2} = \sum_{n=1}^{\infty} n z^n$$

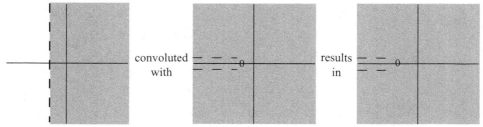

FIGURE 4.44. Right half-plane function convoluted with the Koebe function yields the Koebe function.

is

$$f(z) * F(z) = \frac{z}{1 - z} * \frac{z}{(1 - z)^2}$$

$$= \sum_{n=1}^{\infty} z^n * \sum_{n=1}^{\infty} n z^n$$

$$= (z + z^2 + z^3 + z^4 + \cdots) * (z + 2z^2 + 3z^3 + 4z^4 + \cdots)$$

$$= (z + 2z^2 + 3z^3 + 4z^4 + \cdots)$$

$$= \frac{z}{(1 - z)^2}.$$

Example 4.115. Take the Koebe function, $f(z) = \dfrac{z}{(1 - z)^2}$, and the horizontal strip map, $F(z) = \dfrac{1}{2} \log \left(\dfrac{1 + z}{1 - z} \right)$. What is their Hadamard product, $f(z) * F(z)$? We need to compute the Taylor series for F. To do so, we use

$$\log(1 - z) = \int \frac{-1}{1 - z} \, dz = -\int \sum_{n=0}^{\infty} z^n \, dz = \sum_{n=0}^{\infty} \frac{-1}{n + 1} z^{n+1}$$

and

$$\log(1 + z) = \sum_{n=0}^{\infty} (-1)^{n+1} \frac{1}{n + 1} z^{n+1}.$$

Hence,

$$\frac{1}{2} \log \left(\frac{1 + z}{1 - z} \right) = \sum_{n=0}^{\infty} (-1)^{n+1} \frac{1}{n + 1} z^{n+1} - \sum_{n=0}^{\infty} \frac{-1}{n + 1} z^{n+1}$$

$$= \sum_{n=0}^{\infty} \frac{1}{2n + 1} z^{2n+1}$$

$$= z + \frac{1}{3} z^3 + \frac{1}{5} z^5 + \cdots.$$

convoluted with   results in

FIGURE 4.45. The Koebe function convoluted with a horizontal strip map yields a double-slit map.

Thus,

$$f(z) * F(z) = \frac{z}{(1-z)^2} * \frac{1}{2} \log\left(\frac{1+z}{1-z}\right)$$

$$= \sum_{n=1}^{\infty} nz^n * \sum_{n=0}^{\infty} \frac{1}{2n+1} z^{2n+1}$$

$$= (z + 2z^2 + 3z^3 + 4z^4 + 5z^5 + \cdots) * (z + \frac{1}{3}z^3 + \frac{1}{5}z^5 + \cdots)$$

$$= z + z^3 + z^5 + \cdots .$$

Since $\dfrac{1}{1-z} = 1 + z + z^2 + z^3 + \cdots$, we have that $\dfrac{1}{1-z^2} = 1 + z^2 + z^4 + z^6 + \cdots$ and $\dfrac{z}{1-z^2} = z + z^3 + z^5 + \cdots$. That is,

$$f(z) * F(z) = \frac{z}{(1-z)^2} * \frac{1}{2} \log\left(\frac{1+z}{1-z}\right) = \frac{z}{1-z^2}.$$

Exercise 4.116. Let $f(z) = -\log(1-z)$ and $F(z) = \dfrac{z}{(1-z)^2}$. Determine $f(z) * F(z)$. ***Try it out!***

Proposition 4.117.

(a) *The right half-plane mapping,* $f(z) = \dfrac{z}{1-z}$, *acts as the convolution identity: if F is an analytic function, then* $\dfrac{z}{1-z} * F(z) = F(z)$.

(b) *The Koebe function,* $f(z) = \dfrac{z}{(1-z)^2}$, *acts as a differential operator: if $F(z)$ is an analytic function, then* $\dfrac{z}{(1-z)^2} * F(z) = zF'(z)$.

(c) *Convolution is commutative: if f_1 and f_2 are analytic functions, then $f_1 * f_2 = f_2 * f_1$.*

(d) *If f_1 and f_2 are analytic functions, then $(f_1(z) * f_2(z))' = zf_1'(z) * f_2(z)$.*

Exercise 4.118. Prove Proposition 4.117 (a)-(d). ***Try it out!***

If $f_1, f_2 \in S$, then $f_1 * f_2$ may not be in S. For example,

$$\frac{z}{(1-z)^2} * \frac{z}{(1-z)^2} = \sum_{n=1}^{\infty} nz^n * \sum_{n=1}^{\infty} nz^n$$

$$= \sum_{n=1}^{\infty} n^2 z^n \notin S.$$

Why do we know that $\displaystyle\sum_{n=1}^{\infty} n^2 z^n \notin S$?

However, we have a result by Ruscheweyh and Sheil-Small [25]. If the analytic function, $f \in S$, maps onto a domain that is convex, then we will denote that by writing $f \in K$.

Theorem 4.119 (Ruscheweyh and Sheil-Small). *If f_1, $f_2 \in K$, then $f_1 * f_2 \in K$.*

Now let's consider harmonic convolutions.

Definition 4.120. For harmonic univalent functions

$$f(z) = h(z) + \overline{g(z)} = z + \sum_{n=2}^{\infty} a_n z^n + \sum_{n=1}^{\infty} \overline{b_n}\, \overline{z}^n$$

$$F(z) = H(z) + \overline{G(z)} = z + \sum_{n=2}^{\infty} A_n z^n + \sum_{n=1}^{\infty} \overline{B_n}\, \overline{z}^n,$$

define the *harmonic convolution* by

$$f(z) * F(z) = h(z) * H(z) + \overline{g(z) * G(z)} = z + \sum_{n=2}^{\infty} a_n A_n z^n + \sum_{n=1}^{\infty} \overline{b_n B_n}\, \overline{z}^n.$$

If the harmonic function, $f \in S_H$, maps onto a domain that is convex, then we will write $f \in K_H$. As we mentioned it is known that if f_1, $f_2 \in K$, then $f_1 * f_2 \in K$. Is there a similar result for harmonic univalent convex mappings? There are a few known results about harmonic convolutions of functions on $\mathbb{D}$.

Theorem 4.121 (Clunie and Sheil-Small). *If $f \in K_H$ and $\varphi \in S$, then the functions*

$$f * (\alpha \overline{\varphi} + \varphi) \in S_H$$

map $\mathbb{D}$ onto a close-to-convex domain, where $|\alpha| \leq 1$.

Clunie and Sheil-Small posed the open problem (see [6]).

Open Problem 4.122. For $f \in K_H$, what are the harmonic functions F such $f * F \in K_H$?

As partial answers, there are the following results.

Theorem 4.123 (Ruscheweyh and Salinas). *Let g be analytic in $\mathbb{D}$. Then*

$$f \widetilde{*} g = \mathrm{Re}\{f\} * g + \overline{\mathrm{Im}\{f\} * g} \in K_H$$

for all $f \in K_H \iff$ for each $\gamma \in \mathbb{R}$, $g + i\gamma z g'$ is convex in the imaginary direction.

Theorem 4.124 (Goodloe). *Let f_m, $f_n \in K_H$ be the canonical harmonic functions that map $\mathbb{D}$ onto the regular m-gon and n-gon, respectively. Then $f_m * f_n \in K_H$ and the image of $\mathbb{D}$ is a p-gon, where $p = lcm(m, n)$.*

Exercise 4.125. Compute $f_k = f_4 * f_6$, where $f_4 = h_4 + \overline{g_4}$ is the canonical square map (see Example 4.101) given by

$$h_4(z) = \frac{1}{4} \log\left(\frac{1+z}{1-z}\right) + \frac{i}{4} \log\left(\frac{i+z}{i-z}\right) = \int \frac{1}{1-z^4}\, dz$$

$$g_4(z) = -\frac{1}{4} \log\left(\frac{1+z}{1-z}\right) + \frac{i}{4} \log\left(\frac{i+z}{i-z}\right) = \int \frac{-z^2}{1-z^4}\, dz.$$

and $f_6 = h_6 + \overline{g_6}$ is the canonical regular hexagon map (see Exercise 4.108) given by

$$h_6(z) = \frac{1}{6} \log\left(\frac{1+z}{1-z}\right) + \frac{e^{\frac{-i\pi}{3}}}{6} \log\left(\frac{1+e^{\frac{i\pi}{3}}z}{1-e^{\frac{i\pi}{3}}z}\right) + \frac{e^{\frac{-i2\pi}{3}}}{6} \log\left(\frac{1+e^{\frac{i2\pi}{3}}z}{1-e^{\frac{i2\pi}{3}}z}\right)$$

$$= \int \frac{1}{1-z^6}\, dz$$

$$g_6(z) = -\frac{1}{6} \log\left(\frac{1+z}{1-z}\right) - \frac{e^{\frac{i\pi}{3}}}{6} \log\left(\frac{1+e^{\frac{i\pi}{3}}z}{1-e^{\frac{i\pi}{3}}z}\right) - \frac{e^{\frac{i2\pi}{3}}}{6} \log\left(\frac{1+e^{\frac{i2\pi}{3}}z}{1-e^{\frac{i2\pi}{3}}z}\right)$$

$$= \int \frac{-z^4}{1-z^6}\, dz.$$

Sketch $f_k(\mathbb{D})$ using *ComplexTool*. **Try it out!**

In considering Open Problem 4.122, let's look at a simple problem: if $f_1, f_2 \in K_H$, then when is $f_1 * f_2 \in S_H$?

We need Lewy's theorem that $f = h + \overline{g}$ with $h'(z) \neq 0$ in $\mathbb{D}$ is locally univalent and sense-preserving if and only if $\omega(z) = g'(z)/h'(z)| < 1, \forall z \in \mathbb{D}$.

Theorem 4.126. *Let $f_1 = h_1 + \overline{g}_1$, $f_2 = h_2 + \overline{g}_2 \in K_H$ with $h_k(z) + g_k(z) = \frac{z}{1-z}$ for $k = 1, 2$. If $f_1 * f_2$ is locally univalent and sense-preserving, then $f_1 * f_2 \in S_H$ and is convex in the direction of the real axis.*

Proof. Since $h(z) + g(z) = \dfrac{z}{1-z}$ and $F(z) * \dfrac{z}{1-z} = F(z)$ for any analytic function F, we have that

$$h_2 - g_2 = (h_1 + g_1) * (h_2 - g_2)$$
$$= h_1 * h_2 - h_1 * g_2 + h_2 * g_1 - g_1 * g_2$$
$$h_1 - g_1 = (h_1 - g_1) * (h_2 + g_2)$$
$$= h_1 * h_2 + h_1 * g_2 - h_2 * g_1 - g_1 * g_2.$$

Thus,

$$h_1 * h_2 - g_1 * g_2 = \tfrac{1}{2}[(h_1 - g_1) + (h_2 - g_2)]. \tag{4.16}$$

We will now show that $(h_1 - g_1) + (h_2 - g_2)$ is convex in the direction of the real axis. We have

$$h'(z) - g'(z) = \left(h'(z) + g'(z)\right)\left(\frac{h'(z) - g'(z)}{h'(z) + g'(z)}\right)$$

$$= \left(h'(z) + g'(z)\right)\left(\frac{1 - \omega(z)}{1 + \omega(z)}\right) = \frac{p(z)}{(1-z)^2},$$

where $\mathrm{Re}\{p(z)\} > 0, \forall z \in \mathbb{D}$. So, letting $\varphi(z) = z/(1-z)^2$, we have

$$\mathrm{Re}\left\{ \frac{z[(h_1'(z) - g_1'(z)) + (h_2'(z) - g_2'(z))]}{\varphi(z)} \right\} = \mathrm{Re}\left\{ \frac{\frac{z}{(1-z)^2}[p_1(z) + p_2(z)]}{\frac{z}{(1-z)^2}} \right\}$$

$$= \mathrm{Re}\left\{ p_1(z) + p_2(z) \right\} > 0.$$

Therefore, by Theorem 4.87 in Section 4.6 and (4.16), $h_1 * h_2 - g_1 * g_2$ is convex in the direction of the real axis.

Since we assumed that $f_1 * f_2$ is locally univalent, we apply Clunie and Sheil-Small's shearing theorem (see Theorem 4.52) to get that $f_1 * f_2 = h_1 * h_2 - g_1 * g_2$ is convex in the direction of the real axis. $\square$

It is known (see [9]), that for any right half-plane mapping $f = h + \overline{g} \in K_H$,

$$h(z) + g(z) = \frac{z}{1-z}.$$

Hence, Theorem 4.126 applies to harmonic right half-plane mappings.

Example 4.127. Let $f_0 = h_0 + \overline{g}_0$ be the canonical right half-plane mapping given in Example 4.40 with $h_0(z) + g_0(z) = \frac{z}{1-z}$ with $\omega(z) = -z$. Then

$$h_0(z) = \frac{z - \frac{1}{2}z^2}{(1-z)^2}$$

$$g_0(z) = -\frac{\frac{1}{2}z^2}{(1-z)^2}.$$

Let $f_1 = h_1 + \overline{g}_1$, where $h_1(z) + g_1(z) = \frac{z}{1-z}$ with $\omega(z) = z$. Then

$$h_1(z) = \frac{1}{4} \log\left(\frac{1+z}{1-z}\right) + \frac{1}{2}\frac{z}{1-z}$$

$$g_1(z) = -\frac{1}{4} \log\left(\frac{1+z}{1-z}\right) + \frac{1}{2}\frac{z}{1-z}.$$

Note that f_1 is a right half-strip mapping (see Figure 4.46).

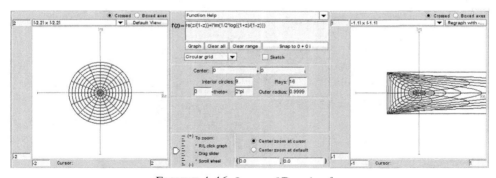

FIGURE 4.46. Image of $\mathbb{D}$ under f_1.

Let $F_1 = f_0 * f_1 = H_1 + \overline{G}_1$. We have

$$H_1(z) = h_0(z) * h_1(z) = \frac{1}{2}\left[h_1(z) + zh_1'(z)\right] = \frac{1}{8} \log\left(\frac{1+z}{1-z}\right) + \frac{\frac{3}{4}z - \frac{1}{4}z^3}{(1-z)^2(1+z)}$$

$$G_1(z) = g_0(z) * g_1(z) = \frac{1}{2}\left[g_1(z) - zg_1'(z)\right] = -\frac{1}{8} \log\left(\frac{1+z}{1-z}\right) + \frac{\frac{1}{4}z - \frac{1}{2}z^2 - \frac{1}{4}z^3}{(1-z)^2(1+z)},$$

with

$$\widetilde{\omega}(z) = -z\left(\frac{2z^2 + z + 1}{z^2 + z + 2}\right).$$

The image of $\mathbb{D}$ under $F_1 = f_0 * f_1$ is shown in Figure 4.47.

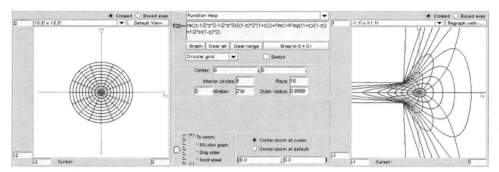

FIGURE 4.47. Image of $\mathbb{D}$ under $F_1 = f_0 * f_1$.

Exercise 4.128. Compute $F = H + \overline{G}$, where $F = f_0 * f_0$. Sketch $F(\mathbb{D})$ using *ComplexTool*. **Try it out!**

In the rest of this section we will consider for which $\omega = g'/h'$ is $f = h + \overline{g}$ locally univalent. Let

$$f_0(z) = h_0(z) + \overline{g_0(z)} = \frac{z - \frac{1}{2}z^2}{(1-z)^2} - \overline{\frac{\frac{1}{2}z^2}{(1-z)^2}}$$

be the canonical right half-plane mapping used in Example 4.127.

Also, as mentioned after the proof of Theorem 4.126, the collection of functions $f = h + \overline{g} \in S_H^O$ that map $\mathbb{D}$ onto the right half-plane, $R = \{w : \mathrm{Re}(w) > -1/2\}$, have the form

$$h(z) + g(z) = \frac{z}{1-z}.$$

We will use the following method to prove that local univalency holds:

Method 1. (Cohn's Rule, see [23, p 375]) Given a polynomial

$$f(z) = a_0 + a_1 z + \cdots + a_n z^n$$

of degree n, let

$$f^*(z) = z^n \, \overline{f(1/\overline{z})} = \overline{a_n} + \overline{a_{n-1}}z + \cdots + \overline{a_0}z^n.$$

Denote by p and s the number of zeros of f inside the unit circle and on it, respectively. If $|a_0| < |a_n|$, then

$$f_1(z) = \frac{\overline{a_n}\,f(z) - a_0\,f^*(z)}{z}$$

is of degree $n - 1$ with $p_1 = p - 1$ and $s_1 = s$ the number of zeros of f_1 inside the unit circle and on it, respectively.

Theorem 4.129. *Let* $f = h + \overline{g} \in K_H^O$ *with* $h(z) + g(z) = \frac{z}{1-z}$ *and* $\omega(z) = e^{i\theta}z^n$ *($n \in \mathbb{N}$ and $\theta \in \mathbb{R}$). If $n = 1, 2$, then $f_0 * f \in S_H^O$ is convex in the direction of the real axis.*

Proof. Let the dilatation of $f_0 * f$ be given by $\widetilde{\omega} = (g_0 * g)'/(h_0 * h)'$. By Theorem 4.126 and by Lewy's theorem, we need only show that $|\widetilde{\omega}(z)| < 1, \forall z \in \mathbb{D}$.

If F is analytic in $\mathbb{D}$ and $F(0) = 0$, then

$$h_0(z) * F(z) = \frac{1}{2}\big[F(z) + zF'(z)\big]$$

$$g_0(z) * F(z) = \frac{1}{2}\big[F(z) - zF'(z)\big].$$

Since $g'(z) = \omega(z)h'(z)$, we know $g''(z) = \omega(z)h''(z) + \omega'(z)h'(z)$. Hence,

$$\widetilde{\omega}(z) = -\frac{zg''(z)}{2h'(z) + zh''(z)} = \frac{-z\omega'(z)h'(z) - z\omega(z)h''(z)}{2h'(z) + zh''(z)}.$$

Using $h(z) + g(z) = \frac{z}{1-z}$ and $g'(z) = \omega(z)h'(z)$, we can solve for $h'(z)$ and $h''(z)$ in terms of z and $\omega(z)$:

$$h'(z) = \frac{1}{(1 + \omega(z))(1 - z)^2}$$

$$h''(z) = \frac{2(1 + \omega(z)) - \omega'(z)(1 - z)}{(1 + \omega(z))^2(1 - z)^3}.$$

Substituting h' and h'' into the equation for $\widetilde{\omega}$, we get

$$\begin{aligned}
\widetilde{\omega}(z) &= \frac{-z\omega'(z)h'(z) - z\omega(z)h''(z)}{2h'(z) + zh''(z)} \\
&= -z\frac{\omega^2(z) + [\omega(z) - \frac{1}{2}\omega'(z)z] + \frac{1}{2}\omega'(z)}{1 + [\omega(z) - \frac{1}{2}\omega'(z)z] + \frac{1}{2}\omega'(z)z^2}.
\end{aligned} \tag{4.17}$$

If $\omega(z) = e^{i\theta}z$, then (4.17) yields

$$\widetilde{\omega}(z) = -ze^{2i\theta}\frac{\left(z^2 + \frac{1}{2}e^{-i\theta}z + \frac{1}{2}e^{-i\theta}\right)}{\left(1 + \frac{1}{2}e^{i\theta}z + \frac{1}{2}e^{i\theta}z^2\right)} = -ze^{2i\theta}\frac{p(z)}{q(z)}.$$

Note that $q(z) = z^2 \overline{p(1/\overline{z})}$. Hence, if z_0 is a zero of p, then $\frac{1}{\overline{z_0}}$ is a zero of q. So,

$$\widetilde{\omega}(z) = -ze^{2i\theta}\frac{(z + A)(z + B)}{(1 + \overline{A}z)(1 + \overline{B}z)}.$$

Using Method 1, we have

$$p_1(z) = \frac{\overline{a_2}p(z) - a_0p^*(z)}{z} = \frac{3}{4}z + \left(\frac{1}{2}e^{-i\theta} - \frac{1}{4}\right).$$

So, p_1 has one zero at $z_0 = \frac{1}{3} - \frac{2}{3}e^{-i\theta} \in \mathbb{D}$. By Cohn's Rule, p has two zeros, namely A and B, with $|A|, |B| < 1$.

Next, consider the case in which $\omega(z) = e^{i\theta}z^2$. Then

$$|\widetilde{\omega}(z)| = |z^2| \left| \frac{z^3 + e^{-i\theta}}{1 + e^{i\theta}z^3} \right| = |z|^2 < 1.$$

$\square$

Example 4.130. Let $f_2 = h_2 + \overline{g}_2$ be the harmonic mapping in $\mathbb{D}$ such that $h_1(z) + g_2(z) = \frac{z}{1-z}$ and $\omega_2(z) = -z^2$. Then we can compute

$$h_2(z) = \frac{1}{8} \ln \left(\frac{1+z}{1-z} \right) + \frac{1}{2} \frac{z}{1-z} + \frac{1}{4} \frac{z}{(1-z)^2},$$

$$g_2(z) = -\frac{1}{8} \ln \left(\frac{1+z}{1-z} \right) + \frac{1}{2} \frac{z}{1-z} - \frac{1}{4} \frac{z}{(1-z)^2},$$

and the image of $\mathbb{D}$ under f_2 is the right half-plane, $R = \{w \in \mathbb{C} \,|\, \text{Re}\{w\} \geq -\frac{1}{2}\}$. Note that $f_2(e^{it}) = \frac{1}{2} + i\frac{\pi}{16}$, if $0 < t < \pi$, and $f_2(e^{it}) = \frac{1}{2} - i\frac{\pi}{16}$, if $\pi < t < 2\pi$ (see Figure 4.48).

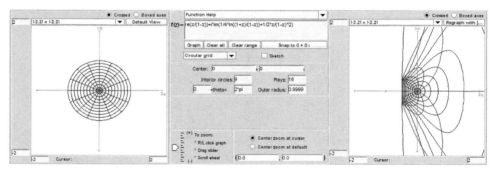

FIGURE 4.48. Image of $\mathbb{D}$ under f_2.

Let

$$F_2 = h_0 * h_2 + \overline{g_0 * g_2} = H_2 + \overline{G_2}.$$

Then we can compute that

$$H_2(z) = \frac{1}{16} \ln \left(\frac{1+z}{1-z} \right) + \frac{1}{4} \frac{z}{1-z} + \frac{1}{8} \frac{z}{(1-z)^2} + \frac{1}{2} \frac{z}{(1-z)^3(1+z)}$$

$$G_2(z) = -\frac{1}{16} \ln \left(\frac{1+z}{1-z} \right) + \frac{1}{4} \frac{z}{1-z} - \frac{1}{8} \frac{z}{(1-z)^2} + \frac{1}{2} \frac{z^3}{(1-z)^3(1+z)}.$$

(4.18)

It can be shown analytically that $F_2(\mathbb{D})$ is the entire complex plane except for two half-lines $x \pm \frac{\pi}{16}i, x \leq -\frac{1}{4}$. This is not clear if we use *ComplexTool* with the standard settings to view this image (see Figure 4.49). However, using both this image and the image of the unit circle (see Figure 4.50), it seems reasonable. To graph the image of $\partial\mathbb{D}$ in *ComplexTool*, change the settings in the middle box of *ComplexTool* to **Interior circles: 0** and **Rays: 0**.

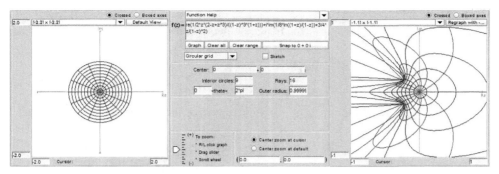

FIGURE 4.49. Image of $\mathbb{D}$ under $F_2 = f_0 * f_2$.

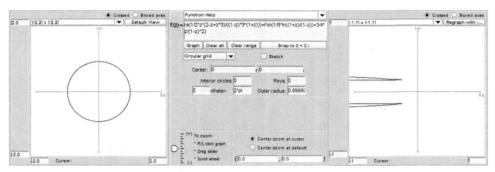

FIGURE 4.50. Image of $\partial\mathbb{D}$ under $F_2 = f_0 * f_2$.

Remark 4.131. If we assume the hypotheses of Theorem 4.129 with the exception of making $n \geq 3$, then for each n we can find $\omega(z) = e^{i\theta}z^n$ such that $f_0 * f \notin S_H^O$. For example, if n is odd, let $\omega(z) = -z^n$ and then (4.17) yields

$$\widetilde{\omega}(z) = -z^n \frac{z^{n+1} + \left(\frac{n}{2} - 1\right)z - \frac{n}{2}}{1 + \left(\frac{n}{2} - 1\right)z^n - \frac{n}{2}z^{n+1}}.$$

It suffices to show that for some point $z_0 \in \mathbb{D}$, $|\widetilde{\omega}(z_0)| > 1$. Let $z_0 = -\frac{n}{n+1} \in \mathbb{D}$. Then

$$\widetilde{\omega}(z_0) = \left(\frac{n}{n+1}\right)^n \frac{\left(\frac{n}{n+1}\right)^{n+1} - \left(\frac{n}{2} - 1\right)\left(\frac{n}{n+1}\right) - \frac{n}{2}}{1 - \left(\frac{n}{2} - 1\right)\left(\frac{n}{n+1}\right)^n - \left(\frac{n}{2}\right)\left(\frac{n}{n+1}\right)^{n+1}}$$

$$\tag{4.19}$$

$$= 1 + \frac{\left[\left(\frac{n+1}{n}\right)^n - \left(\frac{n}{n+1}\right)^{n+1}\right] + \left[1 - \frac{n}{n+1}\right]}{\left(\frac{n}{2} - 1\right) + \left(\frac{n}{2}\right)\left(\frac{n}{n+1}\right) + \left(\frac{n+1}{n}\right)^n}.$$

Note that $\left[\left(\frac{n+1}{n}\right)^n - \left(\frac{n}{n+1}\right)^{n+1}\right] + \left[1 - \frac{n}{n+1}\right] > 0$. Also, $\left(\frac{n}{2} - 1\right) + \left(\frac{n}{2}\right)\left(\frac{n}{n+1}\right) + \left(\frac{n+1}{n}\right)^n > 0$ since $\left(\frac{n}{2} - 1\right) + \left(\frac{n}{2}\right)\left(\frac{n}{n+1}\right) > n - \frac{3}{2} > e$ and $\left(\frac{n+1}{n}\right)^n$ converges to e as $n \to \infty$. Thus, if $n \geq 5$ is odd, $|\widetilde{\omega}(z_0)| > 1$. If $n = 3$, it is easy to compute that $|\widetilde{\omega}(z_0)| = \left(\frac{3}{4}\right)^3 \left|\frac{3^4 - \frac{1}{2} \cdot 3 \cdot 4^3 - \frac{3}{2} \cdot 4^4}{4^4 - \frac{1}{2} \cdot 3^3 \cdot 4 - \frac{3}{2} \cdot 3^4}\right| > 2$. If n is even, let $\omega(z) = z^n$ and $z_0 = -\frac{n}{n+1}$. This simplifies to the same $\widetilde{\omega}(z_0)$ given (4.19) and the argument also holds for $n \geq 6$. If $n = 4$, $|\widetilde{\omega}(z_0)| = \left(\frac{4}{5}\right)^4 \left|\frac{4^5 - 4 \cdot 5^4 - 2 \cdot 5^5}{5^5 - 4^4 \cdot 5 - 2 \cdot 4^5}\right| > 15$.

Exploration 4.132. Using *ComplexTool*, graph $\widetilde{\omega}(\mathbb{D})$ given in (4.17) for $\omega(z) = -z^n$, where $n = 1, 2, 3, 4$. Explain how the images support Theorem 4.129 and Remark 4.131. *Try it out!*

Theorem 4.133. *Let $f = h + \overline{g} \in K_H^O$ with $h(z) + g(z) = \frac{z}{1-z}$ and $\omega(z) = \frac{z+a}{1+az}$ with $a \in (-1, 1)$. Then $f_0 * f \in S_H^O$ is convex in the direction of the real axis.*

Proof. Using $\omega(z) = \frac{z+a}{1+az}$, where $-1 < a < 1$, we have

$$
\widetilde{\omega}(z) = -z \frac{\left(z^2 + \frac{1+3a}{2}z + \frac{1+a}{2}\right)}{\left(1 + \frac{1+3a}{2}z + \frac{1+a}{2}z^2\right)}
$$

$$
= -z \frac{f(z)}{f^*(z)}
$$

$$
= -z \frac{(z + A)(z + B)}{(1 + \overline{A}z)(1 + \overline{B}z)}
$$

$$
= -z \frac{p(z)}{q(z)}.
$$

Using Method 1,

$$
p_1(z) = \frac{\overline{a_2}p(z) - a_0 p^*(z)}{z} = \frac{(a+3)(1-a)}{4}z + \frac{(1+3a)(1-a)}{4}.
$$

So p_1 has one zero at $z_0 = -\frac{1+3a}{a+3}$ that is in the unit circle since $-1 < a < 1$. Thus, $|A|$, $|B| < 1$.   □

Large Project 4.134. In Theorem 4.126, we require that the convolution function satisfy the dilatation condition

$$
|\omega(z)| = \left| \frac{g'(z)}{h'(z)} \right| < 1, \forall z \in \mathbb{D}.
$$

Determine ω functions for which the dilatation condition holds and ones for which it does not hold. See Theorem 4.129, Theorem 4.133, and Remark 4.131 for examples. **Optional**

Large Project 4.135. Similar to the right half-plane map, $f = h + \overline{g}$ is an asymmetric vertical strip map if $h(z) + g(z) = \frac{1}{2i \sin \alpha} \log \left(\frac{1+ze^{i\alpha}}{1+ze^{-i\alpha}} \right)$, where $0 < \alpha < \pi$. Theorem 4.126 can be stated in terms of asymmetric vertical strip mappings instead of right half-plane mappings.

   Theorem. *Let $f = h + \overline{g} \in K_H^O$ with $\omega = g'/h'$ be such that $h + g = \frac{1}{2i \sin \alpha} \log \left(\frac{1+ze^{i\alpha}}{1+ze^{-i\alpha}} \right)$, where $0 < \alpha < \pi$. Then $f_0 * f \in S_H^O$ and is convex in the direction of the real axis.*

   Determine ω functions for which the dilatation condition holds for this theorem and ones for which it does not hold. **Optional**

4.9   Conclusion

We have presented an introduction to harmonic univalent mappings and described a few topics to entice a beginner. Our emphasis has been on the geometric aspects of harmonic univalent mappings that students can explore using the exercises, the exploratory problems, and the projects along with the applets. There are more interesting and deeper topics in harmonic univalent mappings. Here is a short list along with some resources: coefficient estimates and conjectures ([6], [12], [27]); a generalized Riemann mapping theorem ([12], [18]); properties of special classes of functions such as convex, close-to-convex, starlike, and typically real ([6], [12], [27]); harmonic polynomials ([5], [28], [31]); extremal problems ([12]); harmonic meromorphic functions ([20], [29]); inner mapping radius ([3], [11]); and multiply connected domains ([12], [14]). Another topic is the connection between harmonic mappings and minimal surfaces. It is discussed in Chapter 2. In addition, there are several nice general resources that can be used to learn more about harmonic univalent functions. These include Peter Duren's book [12], Clunie and Sheil-Small's original article [6], Bshouty and Hengartner's article [4], and Schober's article [26]. Bshouty and Hengartner complied a list of open problems and conjectures [3].

4.10 Additional Exercises

Anamorphosis and Möbius Maps

Exercise 4.136. Determine the explicit image of horizontal lines $\operatorname{Im} z = k$ under the inversion map $f(z) = \frac{1}{z}$ and analytically prove your result.

Exercise 4.137. Analytically determine the image of the left half-plane $\{z \mid \operatorname{Re}\{z\} < 0\}$ under the map

$$M(z) = \frac{z - i}{z}.$$

Use *ComplexTool* to check your answer.

Exercise 4.138. The strip $D = \{z \mid 0 < \operatorname{Im}\{z\} < 1\}$ can be thought of as the intersection of the domains $D_1 = \{z \mid 0 < \operatorname{Im}\{z\}\}$ and $D_2 = \{z \mid \operatorname{Im}\{z\} < 1\}$. Analytically determine the image of the strip D under the Möbius transformation $M(z) = \frac{z-i}{z+i}$.

Exercise 4.139. Determine a Möbius transformation that maps the left half-plane $\{z \mid \operatorname{Re}\{z\} < 0\}$ onto the disk $\{z \mid |z - 1| < 1\}$. Use *ComplexTool* to check your answer.

Exercise 4.140. Prove that if z_1, z_2, z_3 are distinct points and w_1, w_2, w_3 are distinct points, then the Möbius transformation T satisfying $T(z_1) = w_1$, $T(z_2) = w_2$, $T(z_3) = w_3$ is unique.

Exercise 4.141. Let $M(z) = \frac{az+b}{cz+d}$ with $ad \neq bc$ be a Möbius transformation. Prove that M is a rotation if and only if $M(0) = 0$ and M preserves distances. Preserving distances means $|z_1 - z_2| = |M(z_1) - M(z_2)|$.

The Family S of Analytic, Normalized, Univalent Functions

Exploration 4.142.

(a) Using *ComplexTool* guess the smallest $k > 0$ such that $(z + k)^2$ is univalent in $\mathbb{D}$.

(b) Prove your guess.

(c) Using *ComplexTool* guess the smallest $k > 0$ such that $(z + k)^3$ is univalent in $\mathbb{D}$.

(d) Prove your guess.

Exercise 4.143. Show that $f(z) = z + a_3 z^3$ is univalent in $\mathbb{D} \iff |a_3| \le \frac{1}{3}$. Determine $f(\mathbb{D})$ analytically when $a_3 = -\frac{1}{3}$.

Exercise 4.144. Work out the details to show that $\frac{z}{(1-z)^2} = \sum_{n=1}^{\infty} n z^n = z + 2z^2 + 3z^3 + \cdots$.

Exercise 4.145. Determine $f(\mathbb{D})$ analytically $f(z) = \dfrac{z - cz^2}{(1 - z)^2}$, where $0 < c < 1$.

Exercise 4.146. Prove that $f(z) = \dfrac{1}{2} \log\left(\dfrac{1 + z}{1 - z}\right)$ is univalent. Determine $f(\mathbb{D})$ analytically.

Exploration 4.147. Let

$$f_c(z) = \frac{1}{2c}\left[\left(\frac{1 + z}{1 - z}\right)^c - 1\right].$$

(a) Show that if $c = 2$, then $f_c(z)$ is the Koebe function.

(b) Show that if $c = 1$, then $f_c(z)$ is the right half-plane mapping.

(c) Use *ComplexTool* to view the image of $\mathbb{D}$ under f_c for various values c, $0 < c < 2$. For what values of c does f_c appear to be univalent?

Exercise 4.148. Find the image of $\mathbb{D}$ analytically under the univalent function $f(z) = \dfrac{z}{1 - z^2}$.

The Family S_H of Normalized, Harmonic, Univalent Functions

Exercise 4.149. Determine if $f(x, y) = u(x, y) + iv(x, y) = (x^3 + xy^2) + i(x^2 y + y^3)$ is complex-valued harmonic.

Exercise 4.150. Prove that $f(x, y) = u(x, y) + iv(x, y)$ is harmonic $\iff \dfrac{\partial^2 f}{\partial z\, \partial \overline{z}} = 0$.

Exercise 4.151. Rewrite $f(x, y) = u(x, y) + iv(x, y) = (x - \frac{1}{2}x^2 + \frac{1}{2}y^2) + i(y - xy)$ in terms of z and $\overline{z}$ and determine if f is analytic.

Exercise 4.152. Prove that for all functions $f \in S_H^O$, the sharp inequality $|b_2| \le \frac{1}{2}$ holds.

Exercise 4.153. Verify that the image of $\mathbb{D}$ under the harmonic function $f(z) = z + \frac{1}{2}z^2$ is a hypocycloid with three cusps.

Exercise 4.154. If a domain is convex in the direction $e^{i\varphi}$ for every value of $\varphi \in [0, \pi)$, then it is called a *convex* domain. For example, a disk is a convex domain. For which values of $n = 1, 2, 3, \ldots$ do the following functions map $\mathbb{D}$ onto a convex domain?

(a) $f(z) = z^n$,

(b) $f(z) = z - \frac{1}{n}z^n$ (see Example 4.19 and Definition 4.20),

(c) $f(z) = \dfrac{z}{(1-z)^n}$ (see Examples 4.22 and 4.23 to get you started).

The Shearing Technique

Exploration 4.155. Let $f = h + \overline{g}$ with $h(z) - g(z) = z - \frac{1}{n}z^n$ and $\omega(z) = z^{n-1}$. Use *ShearTool* to sketch the graph of $f(\mathbb{D})$ for different values of n and then compute h and g explicitly so that $f \in S_H^O$.

Exercise 4.156. Let $f = h + \overline{g}$ with $h(z) - g(z) = \frac{z}{(1-z)^2}$ and $\omega(z) = -z$. Compute h and g explicitly so that $f \in S_H^O$ and determine $f(\mathbb{D})$.

Exercise 4.157. Let $f = h + \overline{g}$ with $h(z) - g(z) = \frac{z}{(1 - z)^2}$ and $\omega(z) = z\frac{z + \frac{1}{2}z}{1 + \frac{1}{2}z}$.

(a) Show that $|\omega(z)| < 1, \forall z \in \mathbb{D}$.

(b) Compute h and g explicitly so that $f \in S_H^O$.

(c) Show that $f(\mathbb{D})$ is a slit domain like the Koebe domain. Determine where the tip of the slit is located.

(d) What is the significance of this example in relationship to the Riemann mapping theorem?

Exercise 4.158. Let $f = h + \overline{g}$ with $h(z) + g(z) = \frac{z}{1 - z}$ and $\omega(z) = e^{i\theta}z$, where $\theta \in [0, 2\pi)$. Use *ShearTool* to sketch the graph of $f(\mathbb{D})$ for different values of n and compute h and g explicitly so that $f \in S_H^O$.

Exploration 4.159. We can find harmonic functions $f_n = h_n + \overline{g}_n$ that map onto regular n-gons by generalizing the ideas from Example 4.72. Use *ShearTool* to explore the images of $\mathbb{D}$ under $f = h + \overline{g}$, where f comes from shearing

$$h_n(z) - g_n(z) = \sum_{k=0}^{n-1} \frac{-2\cos\left(\frac{2\pi k}{n}\right)}{n} \log\left(1 - ze^{i\frac{2\pi k}{n}}\right)$$

with $\omega(z) = -z^{n-2}$.

Exploration 4.160. Let $f = h + \overline{g}$ with $h(z) - g(z) = \frac{1}{2}\log\left(\frac{1 + z}{1 - z}\right)$ and $\omega(z) = m^2z^2$,

where $m = e^{i\theta}(0 \le \theta \le \frac{\pi}{2})$. Use *ShearTool* to sketch the graph of $f(\mathbb{D})$ for different values of n and compute h and g explicitly so that $f \in S_H^O$ Note: This Exploration fits nicely with minimal surfaces, because when $m = i$, f lifts to a canonical minimal surface, Scherk's doubly-periodic, and when $m = 1$, f lifts to a different canonical minimal surface, the helicoid.

Properties of the Dilatation

Exploration 4.161. Shear $h(z) - g(z) = \frac{z}{1-z}$ using $\omega(z) = az$, where $-1 \le a \le 1$ and sketch $f(\mathbb{D})$ using *ShearTool*. Describe what happens to $f(\mathbb{D})$ as a varies.

Exploration 4.162. Shear $h(z) - g(z) = \frac{z}{1-z}$ using $\omega(z) = z^n$, where $n = 1, 2, 3, 4, 5$ and sketch $f(\mathbb{D})$ using *ShearTool*. Describe what happens to $f(\mathbb{D})$ as a varies.

Exploration 4.163. Shear $h(z) - g(z) = \log\left(\frac{1-z}{1+z}\right)$ using $\omega(z) = e^{i\pi n/6}z$, where $n = 0, \ldots, 6$ and sketch $f(\mathbb{D})$ using *ShearTool*. Describe what happens to $f(\mathbb{D})$ as n varies.

Exploration 4.164. Shear $h(z) - g(z) = \frac{z}{1-z} + ae^{\frac{z+1}{z-1}}$ using $\omega(z) = e^{\frac{z+1}{z-1}}$, where $-0.5 \leq a \leq 0.5$ and sketch $f(\mathbb{D})$ using *ShearTool*. Describe what happens to $f(\mathbb{D})$ as a varies.

Exercise 4.165. Let $h_\alpha(z) = \frac{z}{1+ze^{-i\alpha}}$, where $0 < \alpha < \pi$, and $\omega_\alpha(z) = e^{-i\left(\frac{z+e^{-i\alpha}}{1+ze^{-i\alpha}}\right)}$. Compute $f_\alpha = h_\alpha + \overline{g}_\alpha$ and show that $f_\alpha \in S_H^O$. Use *ComplexTool* to sketch $f_\alpha(\mathbb{D})$ for various values of α. As α approaches 0, you should get the image shown in Figure 4.28.

Exercise 4.166. Let $h_\gamma(z) = \frac{1}{2i \sin \gamma} \log\left(\frac{1+ze^{i\gamma}}{1+ze^{-i\gamma}}\right)$, where $\frac{\pi}{2} \leq \gamma < \pi$, and $\omega_\gamma(z) = e^{-\left(\frac{2\sin(\pi-\gamma)}{(\pi-\gamma)}h(z)-1\right)}$. Compute $f_\gamma = h_\gamma + \overline{g}_\gamma$ and show that $f_\gamma \in S_H^O$. Use *ComplexTool* to sketch $f_\gamma(\mathbb{D})$ for various values of γ. As γ approaches π, you should get the image shown in Figure 4.28, but each $f_\gamma(\mathbb{D})$ is different than any $f_\alpha(\mathbb{D})$ in Exercise 4.165.

Harmonic Linear Combinations

Exercise 4.167. Let

$$f_1(z) = \text{Re}\left\{ -z - 2\log(1-z) \right\} + i\,\text{Im}\left\{z\right\},$$

$$f_2(z) = \text{Re}\left\{\frac{z + 1/3z^3}{(1-z)^3}\right\} + i\,\text{Im}\left\{\frac{z}{(1-z)^2}\right\}.$$

(a) Show that f_1 can be derived by shearing $h(z) - g(z) = z$ with $g'(z) = zh'(z)$.

(b) Use *ComplexTool* to plot the image of $\mathbb{D}$ under f_1. Recall that f_2 is the harmonic Koebe function. What is the image of $\mathbb{D}$ under f_2?

(c) Use *ComplexTool* to see that $f_3 = tf_1 + (1-t)f_2$ is not univalent for at least one value of t, $0 \leq t \leq 1$. Why does this not contradict Theorem 4.100?

Exploration 4.168. Let

$$f_1(z) = \text{Re}\left\{\frac{z}{(1-z)^2}\right\} + i\,\text{Im}\left\{\frac{1}{2}\log\left(\frac{1+z}{1-z}\right)\right\},$$

$$f_2(z) = \text{Re}\left\{\frac{i}{2}\log\left(\frac{1-iz}{1+iz}\right)\right\} + i\,\text{Im}\left\{\frac{1}{2}\log\left(\frac{1+z}{1-z}\right)\right\}.$$

Show that f_1 and f_2 satisfy the conditions of Theorem 4.100 and then use *ComplexTool* to plot images of $f_3 = tf_1 + (1-t)f_2$ for various values of t.

Exploration 4.169. Let

$$f_1(z) = \text{Re}\left\{\frac{z}{1-z} - \frac{1}{2}e^{\frac{z+1}{z-1}}\right\} + i\,\text{Im}\left\{\frac{z}{1-z} + \frac{1}{2}e^{\frac{z+1}{z-1}}\right\},$$

$$f_2(z) = \text{Re}\left\{z - \frac{1}{4}z^2 - \frac{1}{4}(z-1)^2 e^{\frac{z+1}{z-1}}\right\} + i\,\text{Im}\left\{z - \frac{1}{4}z^2 + \frac{1}{4}(z-1)^2 e^{\frac{z+1}{z-1}}\right\}.$$

Show that f_1 and f_2 satisfy the conditions of Theorem 4.100 and then use *ComplexTool* to plot images of $f_3 = tf_1 + (1-t)f_2$ for various values of t.

Exercise 4.170. Repeat the steps in Example 4.104 using the same function for f_1 but replacing f_2 with the harmonic square map in Example 4.72, where

$$h_2(z) = \frac{1}{4} \log \left(\frac{1+z}{1-z} \right) + \frac{i}{4} \log \left(\frac{i+z}{i-z} \right)$$

$$g_2(z) = -\frac{1}{4} \log \left(\frac{1+z}{1-z} \right) + \frac{i}{4} \log \left(\frac{i+z}{i-z} \right).$$

(a) In this case, $\omega_1(z) = z^2$ while $\omega_2(z) = -z^2$. Using (4.15) in Remark 4.106, show that f_3 is locally univalent.

(b) Use *LinComboTool* find the image of $f_3(\mathbb{D})$ using this f_1 and f_2.

(c) Explain why this happens by using the approach in Example 4.104 to compute the new values of $w_1, \ldots, w_8$ and then use the visualization technique in the example to plot the vertices $z_0, \ldots, z_7$ and draw the midpoints $w_1, \ldots, w_8$.

Exploration 4.171. Using *LinComboTool*, start with the same arc values and corresponding point values as in Example 4.104. In **Panel #1** increase the arc values by increments of $\frac{\pi}{16}$ while not changing the point values, and decrease the arc values in **Panel #2** by the same amount. You can do this either by changing the value in the **Arc** n box or by using the cursor to move the four blue dots the same amount in the same direction in **Panel #1** and the same amount in the opposite direction in **Panel #2** on the unit circle of the domain in each panel. Describe how the image domain changes as the arc values in **Panel #1** increase by a total of $\frac{\pi}{4}$ and decrease in **Panel #2** by the same amount.

Exploration 4.172. Using *LinComboTool*, create a map in **Panel #1** that maps three arcs on the unit circle to three vertices of an equilateral triangle. Then create a second map in **Panel #2** that maps three different arcs on the unit circle to three vertices of a rotated equilateral triangle. Make sure that the maps satisfy the conditions of Theorem 4.100. Click on the **Create LinCombogon** button to see the image domain. Explore this idea by using different maps in the panels to get at least three different image domains.

Exploration 4.173. Using *LinComboTool*, create a map in **Panel #1** that maps six arcs on the unit circle to six vertices of a regular hexagon. Then create a second map in **Panel #2** that maps six different arcs on the unit circle to six vertices of a rotated regular hexagon. Make sure that the maps satisfy the conditions of Theorem 4.100. Click on the **Create LinCombogon** button to see the image domain. Explore this idea by using different maps in the panels to get at least three different image domains.

Convolutions

Exercise 4.174. Let

$$f(z) = \int \frac{1}{1-z^2} \, dz = \frac{1}{2} \log \left(\frac{1+z}{1-z} \right)$$

and

$$F(z) = \int \frac{1}{1-z^3} \, dz = \frac{1}{3} e^{\frac{i5\pi}{3}} \log \left(1 + e^{\frac{i\pi}{3}} z \right) + \frac{1}{3} e^{\frac{i5\pi}{3}} \log \left(1 + e^{\frac{i5\pi}{3}} z \right) - \frac{1}{3} \log \left(1 - z \right).$$

Using (4.13) at the end of Example 4.72, determine $f * F$ and the image of $\mathbb{D}$ under this convolution. In general, what is $f * F$ when $f'(z) = \frac{1}{1-z^m}$ and $F(z) = \frac{1}{1-z^n}$?

Exercise 4.175. In Theorem 4.121, let f be the canonical right half-plane mapping $f_0 \in K_H$ and let $\varphi(z) = \frac{z}{(1-z)^2} \in S$. Compute $F = f_0 * (\overline{\varphi} + \varphi)$ and use *ComplexTool* to sketch $F(\mathbb{D})$.

Exercise 4.176. Derive the expressions for H_2 and G_2 given in (4.18).

Small Project 4.177. Compute $f_a = h_a + \overline{g}_a$, where $h_a(z) + g_a(z) = \frac{z}{1-z}$ and $\omega(z) = \frac{z+a}{1+az}$. From Theorem 4.133, we know that $F_a = f_0 * f_a \in S_H$ for $-1 < a < 1$. Compute F_a and use *ComplexTool* to sketch $F_a(\mathbb{D})$ for various values of a, $-1 < a < 1$. Describe what happens as a varies between 1 and -1.

4.11   Bibliography

[1] Lars Ahlfors, *Conformal Invariants: Topics in Geometric Function Theory*, McGraw-Hill, Inc., New York, 1973.

[2] Zach Boyd, Michael Dorff, Rachel Messick, Matthew Romney, and Ryan Viertel, Harmonic univalent mappings with singular inner function dilatation, preprint.

[3] Daoud Bshouty and Walter Hengartner (editors), Problems and conjectures for harmonic mappings, from a workshop held at the Technion, Haifa, 1995.

[4] ———, Univalent harmonic mappings in the plane, *Handbook of Complex Analysis: Geometric Function Theory*, Vol. 2, 479–506, Elsevier, Amsterdam, 2005.

[5] Daoud Bshouty and Abdallah Lyzzaik, On Crofoot-Sarason's conjecture for harmonic polynomials, *Comput. Methods Funct. Theory* **4** (2004), no. 1, 35–41.

[6] James Clunie and Terry Sheil-Small, Harmonic univalent functions, *Ann. Acad. Sci. Fenn. Ser. A.I Math.* **9** (1984), 3–25.

[7] Louis de Branges, A proof of the Bieberbach conjecture, *Acta Math.* **154** (1985), no. 1–2, 137-152.

[8] Michael Dorff, Convolutions of planar harmonic convex mappings, *Complex Variables Theory Appl.*, **45** (2001), no. 3, 263–271.

[9] ———, Harmonic mappings onto asymmetric vertical strips, in *Computational Methods and Function Theory 1997*, (N. Papamichael, St. Ruscheweyh, and E. B. Saff, eds.), 171–175, World Sci. Publishing, River Edge, NJ, 1999.

[10] Michael Dorff, Missy Lucas, Rachel Messick, and Chad Witbeck, Convex combinations of harmonic mappings and minimal graphs, preprint.

[11] Michael Dorff, and Ted Suffridge, The inner mapping radius of harmonic mappings of the unit disk, *Complex Variables Theory Appl.* **33** (1997), no. 1–4, 97–103.

[12] Peter Duren, *Harmonic Mappings in the Plane*, Cambridge Tracts in Mathematics, 156, Cambridge University Press, Cambridge, 2004.

[13] ———, *Univalent Functions*, Springer-Verlag, New York, 1983.

[14] Peter Duren, and Walter Hengartner, Harmonic mappings of multiply connected domains, *Pacific J. Math.* **180** (1997), no. 2, 201–220.

[15] Peter Duren, Jane McDougall, and Lisabeth Schaubroeck, Harmonic mappings onto stars, *J. Math. Anal. Appl.* **307** (2005), no. 1, 312–320.

[16] Mary Goodloe, Hadamard products of convex harmonic mappings, *Complex Var. Theory Appl.* **47** (2002), no. 2, 81–92.

[17] Paul Greiner, Geometric properties of harmonic shears, *Comput. Methods Funct. Theory*, **4** (2004), no. 1, 77–96.

[18] Walter Hengartner and Glenn Schober, Harmonic mappings with given dilatation, *J. London Math. Soc.* (2) **33** (1986), no. 3, 473–483.

[19] ——, On schlicht mappings to domains convex in one direction, *Comment. Math. Helv.* **45** (1970), 303–314.

[20] ——, Univalent harmonic functions, *Trans. Amer. Math. Soc.* **299** (1987), no. 1, 1–31.

[21] Richard Laugesen, Planar harmonic maps with inner and Blaschke dilatations, *J. London Math. Soc. (2)* **56** (1997), 37–48.

[22] Christian Pommerenke, On starlike and close-to-convex functions, *Proc. London Math. Soc. (3)* **13** (1963), 290–304.

[23] Qazi Ibadur Rahman and Gerhard Schmeisser, *Analytic Theory of Polynomials*, London Mathematical Society Monographs New Series, 26, Oxford University Press, Oxford, 2002.

[24] Stephan Ruscheweyh and Luis Salinas, On the preservation of direction-convexity and the Goodman-Saff conjecture, *Ann. Acad. Sci. Fenn., Ser. A. I. Math.* **14** (1989), 63–73.

[25] Stephan Ruscheweyh and Terry Sheil-Small, Hadamard products of schlicht functions and the Polya-Schoenberg conjecture, *Comment. Math. Helv.*, **48** (1973), 119–135.

[26] Glenn Schober, Planar harmonic mappings, *Computational Methods and Function Theory* (Valparaso, 1989), 171–176, Lecture Notes in Math., **1435**, Springer, Berlin, 1990.

[27] Terry Sheil-Small, Constants for planar harmonic mappings, *J. London Math. Soc.* (2) **42** (1990), no. 2, 237–248.

[28] Ted Suffridge, Harmonic univalent polynomials, *Complex Var. Theory Appl.*, **35** (1998), no. 2, 93–107.

[29] John Thompson, A family of meromorphic harmonic mappings, *Complex Var. Theory Appl.* **48** (2003), no. 8, 627–648.

[30] Allen Weitsman, Harmonic mappings whose dilatations are singular inner functions, unpublished manuscript.

[31] Alan Wilmshurst, The valence of harmonic polynomials, *Proc. Amer. Math. Soc.* **126** (1998), no. 7, 2077–2081.

5

Mappings to Polygonal Domains

Jane M. McDougall and Lisbeth E. Schaubroeck (text),
James S. Rolf (software)

5.1 Introduction

A rich source of problems in analysis is determining when, and how, we can create a one-to-one function of a particular type from one region onto another. In this chapter, we consider the problem of mapping the unit disk $\mathbb{D}$ onto a polygonal domain by two different classes of functions. For analytic functions we give an overview and examples of the well known Schwarz-Christoffel transformation. We then diverge from analytic function theory and consider the Poisson integral formula to find harmonic functions that will serve as mapping functions onto polygonal domains. Proving that harmonic functions are univalent requires us to explore some less widely known theory of harmonic functions and to use relatively new techniques.

Because of the Riemann mapping theorem, we can simplify our mapping problem for either class of function to asking when we can map the unit disk $\mathbb{D} = \{z : |z| < 1\}$ univalently onto a target region. This is because if we want to map one domain (other than the entire set of complex numbers) onto another, we can first map it to $\mathbb{D}$ by an analytic function, and then apply an analytic or harmonic mapping from $\mathbb{D}$ to the other domain (recall that the composition of a harmonic function with an analytic function is harmonic).

We begin in Section 5.2 by using the Schwarz-Christoffel formula to find univalent analytic maps onto polygonal domains, and so set the stage for the corresponding problem for harmonic functions with the Poisson integral formula in Section 5.3. Perhaps because of their importance in applications, many first books on complex analysis introduce Schwarz-Christoffel mappings through examples, without emphasis on subtleties of the deeper theory. Our approach will be the same, and the examples we include are chosen to bring together ideas found elsewhere in this book, such as the shearing technique from Chapter 4 and the construction of minimal surfaces from Chapter 2.

We also include an example of a Schwarz-Christoffel map onto a regular star, a polygonal domain that is highly symmetric but not convex. The problem of using the Poisson integral formula to construct a univalent harmonic function onto a non-convex domain is not at all well understood. In Section 5.3, after developing the theory for convex domains, we carefully construct univalent harmonic maps onto regular star domains, and lead the student to further investigation.

We use the term *univalent* for one-to-one, and take *domain* to mean "open connected set in the complex plane." The applets used in this chapter are

1. *ComplexTool*, to plot the image of domains in $\mathbb{C}$ under complex-valued functions.

2. *PolyTool*, to visualize the harmonic function that is the extension of a particular kind of boundary correspondence. The user can dynamically change the boundary correspondence and watch the harmonic function change.

3. *StarTool*, to examine the functions that map the unit disk $\mathbb{D}$ onto an n-pointed star. The user can modify the shape of the star (by changing n and r) and the boundary correspondence (by changing p).

5.2   Schwarz-Christoffel Maps

In this section we consider conformal maps from the unit disk and the upper half-plane onto various simply connected polygonal domains. By the Riemann mapping theorem, we can map the unit disk conformally onto any simply connected domain that is a proper subset of the complex numbers with a mapping function that is essentially unique.

While the Riemann mapping theorem tells us that there exists a univalent analytic function to map $\mathbb{D}$ onto our domain, finding an actual mapping function is no easy task. Even for a simple domain such as a square, the mapping function from the disk cannot be expressed in terms of elementary functions. One situation in which this problem is relatively simple is in mapping a region bounded by a circle or line in the complex plane to another such region using fractional linear transformations. For this problem, we need only pick three points on the bounding line or circle in the domain, and map them (in order) to three arbitrarily chosen points on the boundary of the target region (see for example [10], [14], [18], or [19]). The selection of three pairs of points determines the fractional linear transformation completely, and works, for example, when finding a conformal map from $\mathbb{D}$ onto any planar region bounded by a line or circle.

Exercise 5.1. Show that the linear fractional transformation (or Möbius transformation) $z' = \phi(z) = i\frac{1+z}{1-z}$ maps the unit disk to the upper half-plane by finding the images of three boundary points. Then show that its inverse function $\phi^{-1}(z') = z = \frac{z'-i}{z'+i}$ maps the upper half-plane to the unit disk. ***Try it out!***

How can a mapping function be found when the target region is more complicated? This question is relevant to solving the heat equation or the study of fluid flows, as discussed in Chapter 3.

The Schwarz-Christoffel transformation frequently enables us to find a function mapping onto a polygonal domain. In most texts the formula is presented as a mapping from the upper half-plane $\mathbb{H}$ onto the target polygon. We now develop this formula—for a more thorough treatment see [12], [3], and [15]. Suppose our target polygon has interior angles $\alpha_k \pi$ and exterior angles $\beta_k \pi$, where $\alpha_k + \beta_k = 1$ and $\alpha_k > 0$. The *exterior angle* measures the angle through which a bug, traversing the polygon in the counterclockwise direction, would turn at each vertex. The exterior angle is signed, following the usual convention that a counterclockwise rotation is positive while a clockwise rotation is negative. We obtain the exterior angle by extending the side on which the bug approaches the vertex, and then seeing through what angle the bug must rotate to get to the next side of the polygon. For example, in Figure 5.1 the angle marked by $\beta_2\pi$ is $-\frac{\pi}{2}$, so $\beta_2 = -1/2$, which coincides with the description of $\beta_2\pi$ as a clockwise turn on the boundary. We can see from $\alpha_2 + \beta_2 = 1$ that $\alpha_2 = 3/2$. For a simple closed polygon (that is, one with no self-intersections), it is always possible to describe the interior and exterior angles using coefficients α_k and β_k as described. We refer to a vertex such as the one with $\beta_1 > 0$ in Figure 5.1 as a *convex corner* and a vertex such as the one with $\beta_2 < 0$ as a *non-convex corner*.

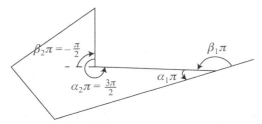

FIGURE 5.1. A sample polygon with both a convex and a non-convex corner.

The Schwarz-Christoffel formula for mapping the upper half-plane $\mathbb{H}$ to the polygon with exterior angles described by coefficients β_k as above is

$$f(z) = A_1 \int_0^z \frac{1}{(w - x_1)^{\beta_1} (w - x_2)^{\beta_2} \cdots (w - x_n)^{\beta_n}} dw + A_2, \quad z \in \mathbb{H}. \qquad (5.1)$$

The real values x_i are preimages of the n vertices of the polygon, which we will refer to from now on as *prevertices*. Different choices of the constants A_1 and A_2 rotate, scale, or translate the target n-gon.

In (5.1), we use w as the variable of integration, and the limits of integration are chosen to make the definite integral into a function of z. The (arbitrary) choice of 0 as a fixed point might have to be altered if it corresponds to a point of discontinuity of the integrand.

Exercise 5.2. You may be familiar with the sine and arcsine functions on the complex plane. Verify that the Schwarz-Christoffel mapping of $\mathbb{H}$ onto the infinite half strip described by $|\operatorname{Re}(z)| < \frac{\pi}{2}$ and $\operatorname{Im}(z) > 0$ is given by the arcsine function. Use the prevertices $x_1 = -1, x_2 = 1$ in formula 5.1. ***Try it out!***

We can observe that the angles at the vertices are represented in the formula, but nowhere do we see an accomodation for the side lengths of the target polygon. They are influenced by the choice of prevertices x_i, but in a nonlinear and non-obvious way. Many texts include a map onto a triangle as a first example, but from elementary geometry we know that the ratios of the side lengths of a triangle are determined by its interior angles. To illustrate how side length depends on the choice of prevertices, we first consider the Schwarz-Christoffel map of the upper half-plane onto a rectangle. We will make a somewhat arbitrary choice of prevertices, and then evaluate the resulting integral to determine the target rectangle. We will find that just as the prevertices are arbitrarily chosen, so are the side lengths of our target rectangle. In computing the integral, we will encounter a first example of a special function known as an *elliptic integral*.

Example 5.3. We map the upper half-plane $\mathbb{H}$ onto a rectangle. We choose prevertices $x_1 = -3, x_2 = -1, x_3 = 1$, and $x_4 = 3$. Since our target image is a rectangle, all exterior angles are $\pi/2$, and $\beta_i = 1/2$ for each i. Using (5.1), we obtain the mapping function

$$f(z) = A_1 \int_0^z \frac{1}{(w-1)^{1/2}(w-3)^{1/2}(w+1)^{1/2}(w+3)^{1/2}} dw + A_2, \quad z \in \mathbb{H}.$$

The constant A_1 allows us to scale and rotate the image of $\mathbb{H}$ and A_2 allows for a translation. By choosing $A_1 = 1$ and $A_2 = 0$ we simplify to

$$f(z) = \int_0^z \frac{1}{\sqrt{(w^2-1)(w^2-9)}} dw. \tag{5.2}$$

The choice of constants does not affect the *aspect ratio* (ratio of adjacent sides) of the rectangle. However the integral cannot be evaluated using techniques in standard calculus texts. Instead it is a special function known as an *elliptic integral* (of the first kind, with parameter $k = \frac{1}{3}$).

Definition 5.4. An *elliptic integral* of the first kind is an integral of the form

$$F(\phi, k) = \int_0^{\sin\phi} \frac{1}{\sqrt{(1-w^2)(1-k^2 w^2)}} dw.$$

An alternate form is $F(\phi, k) = \int_0^\phi \frac{1}{\sqrt{1-k^2 \sin^2\theta}} d\theta$.

The change of variables $w = \sin\theta$, $dw = \cos\theta d\theta = \sqrt{1-w^2} d\theta$ shows that the integrals in Definition 5.4 are identical. (The computer algebra system Mathematica uses an alternate form, representing the integral by **EllipticF[ϕ,m]**, where $m = k^2$.)

Exercise 5.5. Carry out the change of variables $w = \sin\theta$, $dw = \cos\theta d\theta = \sqrt{1-w^2} d\theta$ to show that

$$F(\phi, k) = \int_0^{\sin\phi} \frac{1}{\sqrt{(1-w^2)(1-k^2 w^2)}} dw = \int_0^\phi \frac{1}{\sqrt{1-k^2 \sin^2\theta}} d\theta.$$

Try it out!

Returning to our integral from (5.2), we manipulate the integrand to recognize it as an elliptic integral of the first kind.

$$\begin{aligned}
f(z) &= \int_0^z \frac{1}{\sqrt{(w^2-1)(w^2-9)}} dw \\
&= \int_0^z \frac{1}{\sqrt{\frac{1}{1/9}}} \frac{1}{\sqrt{(w^2-1)(\frac{1}{9}w^2-1)}} dw \\
&= \frac{1}{3} \int_0^z \frac{1}{\sqrt{(1-w^2)(1-\frac{1}{9}w^2)}} dw \\
&= \frac{1}{3} F(\arcsin z, \frac{1}{3}).
\end{aligned}$$

As we will see, our choice for the prevertices ($-3, -1, 1$, and 3) directly affects the aspect ratio of the rectangle.

Exercise 5.6. Follow these steps to determine the aspect ratio of the rectangle that is the image of $\mathbb{H}$ under the function

$$f(z) = \frac{1}{3}F(\arcsin z, \frac{1}{3}).$$

(a) Explain why the integral

$$K_1 = \int_0^1 \frac{1}{\sqrt{(1 - w^2)\left(1 - \frac{1}{9}w^2\right)}} dw$$

is a real number. Hint: use the geometry of the integrand. Conclude that

$$f(1) = \frac{1}{3}F\left(\arcsin 1, \frac{1}{3}\right) = \frac{1}{3}F\left(\pi/2, \frac{1}{3}\right) = K_1/3.$$

By symmetry, show that $f(-1) = -K_1/3$. Thus the length of one side of the rectangle is $2|K_1|/3$.

(b) Determine that since $f(3)$ is the next vertex of the target rectangle (moving counterclockwise), then $f(3) = f(1) + iK_2$ where K_2 is a real constant. Combine this with

$$f(3) = \frac{1}{3}F\left(\arcsin 3, \frac{1}{3}\right)$$

to show that $iK_2 = \frac{1}{3}\left(F(\arcsin 3, \frac{1}{3}) - F(\pi/2, \frac{1}{3})\right)$. The choice of sign for K_2 could be positive or negative, depending on our choice of $\sqrt{-1}$. For consistency with our choice of angles, K_2 should be positive (Mathematica uses the "wrong" branch of the square root for the function on the real axis).

(c) Combine the findings above to determine that the aspect ratio of the rectangle is

$$\frac{2F\left(\pi/2, \frac{1}{3}\right)}{F\left(\arcsin 3, \frac{1}{3}\right) - F\left(\pi/2, \frac{1}{3}\right)} \approx 1.279...$$

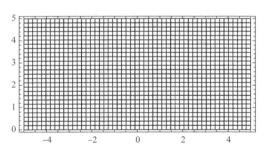 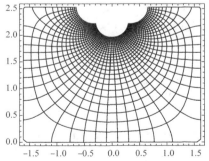

FIGURE 5.2. A portion of upper half-plane (left) and a portion of the target rectangle (right) to which it maps under the function of Exercise 5.3.

The mapping for Example 5.3 is illustrated in Figure 5.2. Only a portion of the upper half-plane and its image are shown. The figure caption explains why the target rectangle is incompletely filled in the upper central area. However, we can see that the aspect ratio is approximately what we calculated.

Small Project 5.7. Rework Example 5.3 for a more general situation. Use the prevertices $x_1 = -\lambda$, $x_2 = -1$, $x_3 = 1$, and $x_4 = \lambda$, where $\lambda > 1$. You can find an equation involving λ as a variable that, chosen correctly, would force the target rectangle to be a square. The equation cannot be explicitly solved, but its solution can be approximated numerically. This is a standard problem in some introductory complex analysis books (see for instance example 22 of section 14 in [18]).

One observation we can make based on this example is that while it is straightforward to write down a mapping function that has the correct angles, there is no simple way to prescribe the side lengths. Also, there are only a few cases when our integral can be expressed in terms of elementary functions, and in general the integral is not easy to evaluate. To find and evaluate specific Schwarz-Christoffel mappings, it is usually helpful to use symmetry of the target polygon (and of the prevertices) to simplify the computations.

The theory of Schwarz-Christoffel maps is intricate, and there is no guarantee that the Schwarz-Christoffel formula will result in a univalent function (see [9]). All we can say for sure is that a map from the upper half plane to a simply connected polygonal domain that is conformal must take the form of (5.1) for some choice of constants and prevertices (for details, see [3]).

A simplification to the Schwarz-Christoffel formula that is frequently employed is to set one of the prevertices to be ∞, which removes one factor from the denominator in the integrand. Due to the symmetry of our target region, a rectangle, we have not used this simplification in our examples here. Rather, we made use of symmetry in Exercise 5.6 by choosing the prevertices to be ± 1 and ± 3. However we were unable to find explicit values of the prevertices to map onto a square. In addition to the problem of prescribing the lengths of the sides of the target polygon, a further problem arises for target polygons more complicated than a rectangle. Typically we will produce an integral that cannot be evaluated, even if we utilize special functions.

In mapping onto a regular polygon, these two problems can be resolved if we obtain a Schwarz-Christoffel formula that maps from the unit disk instead of the upper half plane, using symmetrically placed points ζ_i on the unit circle in place of the x_i in our existing formula. This modified formula can be found by precomposing our mapping function (5.1) with the Möbius transformation from the unit disk to the upper half-plane discussed in Exercise 5.1. The prevertices can be chosen to be the symmetrically placed nth roots of unity. If appropriate symmetry is exhibited by the angles of the target polygon, then the side lengths of the target polygon will be equal. The following exercise asks you to show that the transformed formula from the upper half plane turns out to have the same form as our original Schwarz-Christoffel formula.

Exercise 5.8. Carry out the change of variables $w = \phi(z) = i\frac{1+z}{1-z}$, which maps the disk in the z-plane to the upper half w-plane (see Exercise 5.1). Show that the Schwarz-Christoffel formula retains the same form as (5.1). *Try it out!*

The Schwarz-Christoffel map that we will use on the unit disk is then

$$f(z) = C_1 \int_0^z \frac{1}{(w - \zeta_1)^{\beta_1}(w - \zeta_2)^{\beta_2} \cdots (w - \zeta_n)^{\beta_n}} dw + C_2, \ z \in \mathbb{D}, \qquad (5.3)$$

where $\beta_i \pi$ is the exterior angle of the ith vertex of the target polygon, and the pre-images

ζ_i of the vertices are on the unit circle. We use ζ_i instead of x_i to emphasize that the prevertices are not on the real axis. As with (5.1), the complex constants C_1 and C_2 with $C_1 \neq 0$ rotate, resize, and translate the polygon.

Example 5.9. We obtain the Schwarz-Christoffel map onto a regular n-gon. The exterior angles of a regular n-gon are $2\pi/n$, so $\beta_i = 2/n$, and the mapping function is

$$f(z) = \int_0^z \frac{1}{[(w - \zeta_1)(w - \zeta_2) \cdots (w - \zeta_n)]^{2/n}} dw.$$

Suppose that the ζ_i are the nth roots of unity. Because

$$\prod_{i=1}^{n} (w - \zeta_i) = w^n - 1$$

the integral becomes $\int_0^z \frac{1}{(w^n-1)^{2/n}} dw$. By factoring out $(-1)^{2/n}$ we can adjust the multiplicative constant and choose our mapping function to be

$$f(z) = (-1)^{-2/n} \int_0^z \frac{1}{(w^n - 1)^{2/n}} dw = \int_0^z \frac{1}{(1 - w^n)^{2/n}} dw. \tag{5.4}$$

Here f has been defined from the Schwarz-Christoffel formula with choices of constant $C_1 = (-1)^{-2/n}$ (which rotates the figure by $4\pi/n$ radians) and $C_2 = 0$. The last formula cannot be evaluated using the usual methods from calculus, but can once again be evaluated using special functions known as hypergeometric functions.

5.2.1 Basic Facts about Hypergeometric Functions

The integral in the last example cannot be expressed in terms of elementary functions, but can be evaluated and plotted using a computer algebra system by using some facts about power series known as *hypergeometric functions*. Hypergeometric functions have many applications, and can be used to evaluate the integrals obtained above. We introduce the most widely utilized hypergeometric functions—the so-called "two F ones," where each coefficient of the power series is a rational function with numerator and denominator of the second order in n.

Definition 5.10. The *hypergeometric function* $_2F_1(a, b; c; z)$ is the power series

$$_2F_1(a, b; c; z) = \sum_{n=0}^{\infty} \frac{(a)_n (b)_n}{(c)_n} \frac{z^n}{n!},$$

where a, b, and $c \in \mathbb{C}$ and

$$(x)_n = x(x + 1) \cdots (x + n - 1)$$

is the *shifted factorial*, or *Pochhammer symbol*.

Exercise 5.11. Use algebra to check that $(x)_{n+1} / (x)_n = x + n$. **Try it out!**

If we compute the ratio of successive terms in the geometric series $\sum_{n=0}^{\infty} r^n z^n$ we obtain the ratio rz. In the next exercise we carry out the computation for a hypergeometric series.

Exercise 5.12. Show that the ratio of successive terms in the series $_2F_1(a, b; c; z)$ is

$$\frac{(a+n)(b+n)z}{(c+n)(n+1)}.$$

Try it out!

The formula obtained motivates the use of the term "hypergeometric." In a geometric series the ratio of successive terms is a constant times z, whereas in a hypergeometric function the ratio is a rational function of n, multiplied by z.

Exercise 5.13. Apply the ratio test to show the hypergeometric function $_2F_1(a, b; c; z)$ converges on compact subsets of the unit disk. *Try it out!*

Functions which are useful or widely applicable typically earn the status of "special function." A number of well known special functions can be written as hypergeometric series. For example,

$$\log \frac{1+z}{1-z} = 2z\,_2F_1\left(1/2, 1; 3/2; z^2\right)$$
$$(1-z)^{-a} = \,_2F_1(a, b; b; z)$$
$$\arcsin z = z\,_2F_1\left(1/2, 1/2; 1; z^2\right).$$

The functions $\sin(z)$ and $\cos(z)$ themselves can be obtained as limiting cases of $_2F_1$ series. Now we consider an example that involves a $_2F_1$ hypergeometric series, $z\,_2F_1(\frac{1}{2}, \frac{1}{4}; \frac{5}{4}; z^4)$. We will see that it is a Schwarz-Christoffel transformation that maps the unit disk onto a square.

Exercise 5.14. Use Definition 5.10 to find the first few terms in the series of $z\,_2F_1(\frac{1}{2}, \frac{1}{4}; \frac{5}{4}; z^4)$. The following table gives the first values of the Pochhammer symbols that are needed. If you graph your result using *ComplexTool*, you should get a picture similar to Figure 5.3.

n	$(1/2)_n$	$(1/4)_n$	$(5/4)_n$	Coefficient of z^{4n+1}
0	1	1	1	1
1	1/2	1/4	5/4	1/10
2	3/4	5/16	45/16	1/24
3	15/8	45/64	585/64	5/208
4	105/16	585/256	9945/256	35/2176
5	945/32	9945/1024	208845/1024	3/256

Try it out!

The rate of convergence in this example is fast—even with just a few non-zero terms of the series, we obtain a map whose image is approximately a square. We now see how to derive the formula for a Schwarz-Christoffel map onto the square.

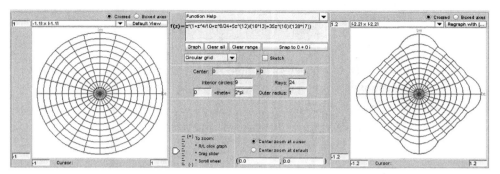

FIGURE 5.3. *ComplexTool* image of an approximation of the conformal map (using the first five terms).

Definition 5.15 (Euler representation). The hypergeometric function $_2F_1\,(a,b;c;z)$ can be written in integral form as

$$_2F_1\,(a,b;c;z) = \frac{\Gamma(c)}{\Gamma(b)\Gamma(c-b)} \int_0^1 \frac{t^{b-1}(1-t)^{c-b-1}}{(1-tz)^a}\,dt,$$

which is known as the *Euler representation* of the $_2F_1$ function.

The symbol Γ stands for the *gamma function*, defined for z in the right half plane by

$$\Gamma(z) = \int_0^\infty t^{z-1}e^{-t}dt.$$

It can be extended analytically to the whole plane except for the negative integers $-1, -2, -3, \dots$.

Exercise 5.16. For integer values of n, the gamma function is related to the factorial by $\Gamma(n) = (n-1)!$. Prove this by directly evaluating $\Gamma(1) = 1$ and then showing that $\Gamma(z+1) = z\Gamma(z)$. Hint: use integration by parts. ***Try it out!***

Exercise 5.17. In this exercise, you will show that Definitions 5.10 and 5.15 are equivalent.

(a) Expand $(1-tz)^{-a}$ using the binomial theorem to obtain a power series

$$(1-tz)^{-a} = \sum_{n=0}^\infty \frac{(a)_n}{n!}t^n z^n.$$

(b) Use the formula (see [13], Theorem 7, p. 19 for a proof)

$$\Gamma\,(p)\,\Gamma\,(q) = \Gamma\,(p+q)\cdot\int_0^1 t^{p-1}\,(1-t)^{q-1}\,dt$$

(where p and q have positive real parts) to show

$$\int_0^1 t^{n+b-1}(1-t)^{c-b-1}dt = \frac{\Gamma\,(b+n)\,\Gamma\,(c-b)}{\Gamma\,(c+n)}.$$

The integral on the right is also an important special function known as the *beta function* in the variables p and q.

(c) Show

$$\frac{(b)_n}{(c)_n} = \frac{\Gamma\,(c)}{\Gamma\,(b)\,\Gamma\,(c-b)}\frac{\Gamma\,(b+n)\,\Gamma\,(c-b)}{\Gamma\,(c+n)}.$$

(d) Put the previous facts together to obtain the required formula. Substitute the power series for the denominator, and interchange summation and integral to show that

$$\int_0^1 \frac{t^{b-1}(1-t)^{c-b-1}}{(1-tz)^a}dt = \frac{\Gamma\,(b)\,\Gamma\,(c-b)}{\Gamma\,(c)}\,_2F_1\,(a,b;c;z)\,.$$

Try it out!

It takes a little work to establish the relationship of the beta function with the gamma function used in part (b). For a detailed exposition and an introduction to special functions in general, see [13].

Example 5.18. For a square (a regular 4-gon), the Schwarz-Christoffel map from the unit disk is given by $z\,_2F_1(\frac{1}{2},\frac{1}{4};\frac{5}{4};z^4)$. The numbers may seem arbitrary, but when we apply the Euler integral representation we will see that they are the numbers we need (after a transformation) to evaluate the integral. With $n=4$ in the integral representation (5.4) we obtain

$$\int_0^z \frac{1}{\sqrt{1-w^4}}dw.$$

Let $a = 1/4$, $b = 1/2$, and $c = 5/4$. We use the Euler integral representation for $_2F_1\,(a,b;c;z)$ to get

$$_2F_1\left(1/2,1/4;5/4;z^4\right) = \frac{\Gamma(5/4)}{\Gamma(1/4)\Gamma(1)}\int_0^1 \frac{t^{-3/4}(1-t)^0}{(1-tz^4)^{1/2}}dt$$

$$= 1/4\int_0^1 \frac{1}{t^{3/4}}\frac{1}{\sqrt{1-tz^4}}dt.$$

Change variables by letting $w^4 = tz^4$ (so $t = (w^4/z^4)$) and $4w^3dw = z^4dt$. Then

$$_2F_1\left(1/2,1/4;5/4;z^4\right) = 1/4\int_0^z \frac{z^3}{w^3}\frac{1}{\sqrt{1-w^4}}\frac{4w^3dw}{z^4}$$

$$= \frac{1}{z}\int_0^z \frac{1}{\sqrt{1-w^4}}dw.$$

Thus

$$f\,(z) = \int_0^z \frac{1}{\sqrt{1-w^4}}dw = z\,_2F_1\left(1/2,1/4;5/4;z^4\right).$$

Contrast the map in Figure 5.4, where the domain is the (bounded) unit disk, with the map onto a rectangle in Figure 5.2. An advantage of the disk map is that we see the entire mapping domain and that the entire target square is filled. Rotational and reflectional symmetry is obtained by using the symmetrically placed *nth* roots of unity on the unit circle as prevertices.

Exercise 5.19. Show that the conformal map from the disk onto the regular n-gon is, up to rotations, translations and scalings, given by $z\,_2F_1\,(2/n,1/n;(n+1)/n;z^n)$. **Try it out!**

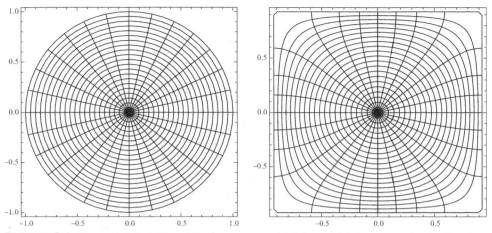

FIGURE 5.4. The unit disk (left) and the target square (right) to which it maps under the function of Exercise 5.18.

For readers who have worked through Chapters 2 or 4, we describe another situation where the technique of integration illustrated in Example 5.18 can be useful. In Chapter 4, Section 4.5 discusses the shear construction, and in Chapter 2, Section 2.6 discusses the shear construction and its relationship with minimal surfaces.

We now examine the conformal map onto a non-convex polygon.

Small Project 5.20. Construct a non-convex hexagon P with $\beta_1 = \beta_4 = -1/3$ and $\beta_2 = \beta_3 = \beta_5 = \beta_6 = 2/3$. (Draw it!) Find the representation for the Schwarz-Christoffel transformation 5.3 that maps the unit disk $\mathbb{D}$ onto such a hexagon P, with the prevertices ζ_i being the sixth roots of unity. Specifically, let $\zeta_1 = 1$ and $\zeta_4 = -1$, with the other sixth roots of unity going in order counterclockwise around the circle. Verify that the analytic function $f(z) : \mathbb{D} \to P$ is given by

$$f(z) = z \, {}_2F_1\left(\frac{2}{3}, \frac{1}{6}; \frac{7}{6}; z^6\right) - \frac{z^3}{3} \, {}_2F_1\left(\frac{2}{3}, \frac{1}{2}; \frac{3}{2}; z^6\right).$$

In the language of Chapter 4, let $h(z) - g(z)$ be $f(z)$ and let the dilatation be $\omega(z) = z^2$. Find the harmonic function $h(z) + \overline{g(z)}$. Verify that by using

$$h(z) = z \, {}_2F_1\left(\frac{2}{3}, \frac{1}{6}; \frac{7}{6}; z^6\right)$$

and

$$g(z) = \frac{z^3}{3} \, {}_2F_1\left(\frac{2}{3}, \frac{1}{2}; \frac{3}{2}; z^6\right),$$

we obtain $\omega(z) = \frac{h'}{g'} = z^2$. Use a computer algebra system to create a picture of the image of $\mathbb{D}$ under $h(z) + \overline{g(z)}$.

If you have studied Chapter 2, you can find the minimal surface that lifts from this harmonic function. You should find that it is defined by

$$x_1 = \operatorname{Re}\left(z\,_2F_1\left(\frac{2}{3},\frac{1}{6};\frac{7}{6};z^6\right) + \frac{z^3}{3}\,_2F_1\left(\frac{2}{3},\frac{1}{2};\frac{3}{2};z^6\right)\right)$$

$$x_2 = \operatorname{Im}\left(z\,_2F_1\left(\frac{2}{3},\frac{1}{6};\frac{7}{6};z^6\right) - \frac{z^3}{3}\,_2F_1\left(\frac{2}{3},\frac{1}{2};\frac{3}{2};z^6\right)\right)$$

$$x_3 = \operatorname{Im}\left(z^2\,_2F_1\left(\frac{2}{3},\frac{1}{3};\frac{3}{2};z^6\right)\right).$$

We now examine the conformal map onto a symmetric non-convex polygon in the shape of a star. In the next section we intend to find harmonic maps onto the same figure, and will find that while harmonic mappings onto polygons with convex corners are relatively easy to construct, there is little or no theory when non-convex corners are involved.

Example 5.21. Suppose we want to map onto a (non-convex) m-pointed star, so there are $2m$ vertices. The interior angles alternate between $\pi\alpha_1$ and $\pi\alpha_2$ where $\alpha_1 < 1 < \alpha_2$ (see Figure 5.5). Assuming we have a non-convex star, exterior angles alternate between a positive value β_1 and a negative value β_2.

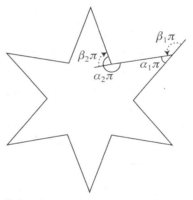

FIGURE 5.5. Interior and exterior angles of a symmetric star.

Since the exterior angles of a simple closed polygon must add to 2π, $\beta_1 + \beta_2$ must satisfy $m(\beta_1 + \beta_2) = 2$ so $\beta_1 + \beta_2 = 2/m$. For odd indices i we have $\beta_i = \beta_1 > 0$ and for even i, $\beta_i = \beta_2 < 0$. Using the Schwarz-Christoffel formula for the unit disk we obtain the mapping function

$$f(z) = \int_0^z \frac{\prod\limits_{i\text{ even}}(w - \zeta_i)^{-\beta_2}}{\prod\limits_{i\text{ odd}}(w - \zeta_i)^{\beta_1}}\,dw.$$

Letting ζ_i be nth roots of unity,

$$\prod_{\substack{i\text{ even}}}^{n}(z - \zeta_i) = z^m - 1 \text{ and } \prod_{\substack{i\text{ odd}}}^{n}(z - \zeta_i) = z^m + 1,$$

so we can rewrite the mapping function as

$$f(z) = \int_0^z \frac{(w^m - 1)^{-\beta_2}}{(w^m + 1)^{\beta_1}} dw.$$

Disregarding constants chosen to expand or rotate the figure as necessary, we can set

$$f(z) = \int_0^z \frac{(1 - w^m)^{-\beta_2}}{(1 + w^m)^{\beta_1}} dw,$$

a conformal mapping from the disk onto the star shape in Figure 5.5.

Example 5.22. This appears as an exercise in [12], Chapter V. Prove that the integral that maps the unit disk onto a 5-pointed star with interior angles alternating between $\pi/5$ and $7\pi/5$ is given by

$$f(z) = \int_0^z \frac{(1 - w^5)^{2/5}}{(1 + w^5)^{4/5}} dw.$$

The corresponding exterior angles are $4\pi/5$ and $-2\pi/5$, so $\beta_1 = 4/5$, $\beta_2 = -2/5$, and $m = 5$.

To compute this integral we use the Appell F_1 function of two variables defined below.

Definition 5.23. The *Appell F_1 function* is defined by

$$F_1(a; b_1, b_2; c; x, y) = \sum_{n=0}^{\infty} \sum_{m=0}^{\infty} \frac{(a)_{n+m} (b_1)_m (b_2)_n}{m! n! (c)_{n+m}} x^m y^n,$$

In Mathematica we can evaluate the Appell F_1 function using the command **AppelF1(a,b₁,b₂,c,x,y)**. Furthermore, there is an integral representation formula, analogous formula to that of the hypergeometric function, given by

$$F_1(a; b_1, b_2; c; x, y)$$
$$= \frac{\Gamma(c)}{\Gamma(a)\Gamma(c-a)} \int_0^1 u^{a-1} (1-u)^{c-a-1} (1-ux)^{-b_1} (1-uy)^{-b_2} du.$$

We omit the proof, but a derivation of the integral representation can be found in [2, Chapter 9]. Substituting we find that

$$F_1\left(1/5; 4/5, -2/5; 6/5; z^5, -z^5\right)$$
$$= \frac{\Gamma(6/5)}{\Gamma(1/5)\Gamma(1)} \int_0^1 u^{-4/5} (1-u)^0 \left(1 - uz^5\right)^{-4/5} \left(1 + uz^5\right)^{2/5} du$$
$$= 1/5 \int_0^1 \frac{\left(1 - uz^5\right)^{2/5}}{u^{4/5} \left(1 + uz^5\right)^{4/5}} du.$$

To evaluate the integral in the Schwarz-Christoffel formula we change variables, letting $w^5 = uz^5$ (so $5w^4 dw = z^5 du$). Then

$$F_1\left(1/5; 4/5, -2/5; 6/5; z^5, -z^5\right) = 1/5 \int_0^z \frac{z^4}{w^4} \frac{\left(1 - w^5\right)^{2/5}}{\left(1 + w^5\right)^{4/5}} \frac{5w^4 dw}{z^5}$$
$$= \frac{1}{z} \int_0^z \frac{\left(1 - w^5\right)^{2/5}}{\left(1 + w^5\right)^{4/5}} dw.$$

Thus

$$f(z) = \int_0^z \frac{(1 - w^5)^{2/5}}{(1 + w^5)^{4/5}} dw = z\, F_1\left(1/5; 4/5, -2/5; 6/5; z^5, -z^5\right);$$

the mapping is illustrated in Figure 5.6.

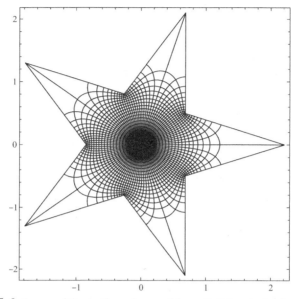

FIGURE 5.6. Image of the conformal map of the unit disk onto the 5-pointed star.

Exercise 5.24. Show that the conformal map from the disk onto the m-pointed star with exterior angle $\beta_1 > 0$ and $\beta_2 = 2/m - \beta_1$ (up to rotations, translations and scalings) is given by $z\, F_1\left(1/n; \beta_1, \beta_2, (n + 1)/n; z^n, -z^n\right)$ where F_1 is the Appell F_1 function. **Try it out!**

5.3   The Poisson Integral Formula

While the Schwarz-Christoffel formula gives analytic and thus angle-preserving (conformal) functions from $\mathbb{D}$ to a polygon, it often starts with an integral that requires advanced mathematics to evaluate. If our goal is not an analytic function, we could work with the Poisson integral formula, which gives a harmonic function from the unit disk to the target domain. The definition of a harmonic function is

Definition 5.25. A real-valued function $u(x, y)$ is *harmonic* provided that

$$u_{xx} + u_{yy} = 0.$$

A complex-valued function $f(z) = f(x + iy) = u(x, y) + iv(x, y)$ is harmonic if u and v are harmonic.

The definition of a complex-valued harmonic function does not require that u and v be harmonic conjugates, so while all analytic functions are harmonic, a complex-valued harmonic function is not necessarily analytic. The functions we work with for the rest of the chapter will not be analytic, and thus not conformal.

You may be familiar with the Poisson integral formula as a way of constructing a real-valued harmonic function that satisfies certain boundary conditions. For example, if the boundary conditions give the temperature of the boundary of a perfectly insulated plate, then the harmonic function gives the steady-state temperature of its interior. Another application is finding electrostatic potential given boundary conditions. We give a brief summary. For more detailed discussion, see [14] or [19].

Theorem 5.26 (Poisson Integral Formula). *Let the complex-valued function $\hat{f}(e^{i\theta})$ be piecewise continuous and bounded for θ in $[0, 2\pi]$. Then the function $f(z)$ defined by*

$$f(z) = \frac{1}{2\pi} \int_0^{2\pi} \frac{1 - |z|^2}{|e^{it} - z|^2} \hat{f}(e^{it}) dt \tag{5.5}$$

is the unique harmonic function in the unit disk that satisfies the boundary condition

$$\lim_{r \to 1} f(re^{i\theta}) = \hat{f}(e^{i\theta})$$

for all θ where $\hat{f}$ is continuous.

We present the proof when the boundary function $\hat{f}(e^{i\theta})$ is the real part of a function that is analytic on a disk with radius larger than 1. This proof can be found in any standard complex analysis textbook, for example, [10] or [14]. The full result appears in [1, Chapter 6] and [10, Chapter 8].

Proof. Cauchy's integral formula tells us that if we have a function $f(z)$ that is analytic inside and on the circle $|z| = R$, then, for $|z| < R$,

$$f(z) = \frac{1}{2\pi i} \int_{|\zeta| = R} \frac{f(\zeta)}{\zeta - z} d\zeta. \tag{5.6}$$

We use the Greek letter zeta (ζ) as the variable of integration in the integral. In the discussion that follows, we will be thinking about evaluating the function $f(z)$ at some fixed value of z, so the variable under consideration is now ζ. We observe, for reasons that will soon be obvious, that for fixed z with $|z| < 1$, the function

$$\frac{f(\zeta) \, \overline{z}}{1 - \zeta \, \overline{z}}$$

is analytic in the variable ζ on and inside $|\zeta| < 1$, since the denominator is non-zero. (Exercise: Why is the denominator non-zero?) Thus, by the Cauchy integral theorem, we know that

$$\frac{1}{2\pi i} \int_{|\zeta| = 1} \frac{f(\zeta) \, \overline{z}}{1 - \zeta \, \overline{z}} = 0. \tag{5.7}$$

Combining (5.6) and (5.7), we see that

$$f(z) = \frac{1}{2\pi i} \int_{|\zeta|=1} \frac{f(\zeta)}{\zeta - z} d\zeta + 0$$

$$= \frac{1}{2\pi i} \int_{|\zeta|=1} \left(\frac{f(\zeta)}{\zeta - z} + \frac{f(\zeta)\,\bar{z}}{1 - \zeta\,\bar{z}} \right) d\zeta$$

$$= \frac{1}{2\pi i} \int_{|\zeta|=1} \frac{1 - \zeta\,\bar{z} + \bar{z}(\zeta - z)}{(\zeta - z)(1 - \zeta\,\bar{z})} f(\zeta) d\zeta$$

$$= \frac{1}{2\pi i} \int_{|\zeta|=1} \frac{1 - |z|^2}{(\zeta - z)(1 - \zeta\,\bar{z})} f(\zeta) d\zeta.$$

Parameterize the circle $|\zeta| = 1$ by $\zeta(t) = e^{it}$, giving $d\zeta = i e^{it} dt$ so

$$f(z) = \frac{1}{2\pi i} \int_0^{2\pi} \frac{1 - |z|^2}{(e^{it} - z)(1 - e^{it}\,\bar{z})} f(e^{it}) i e^{it} dt$$

$$= \frac{1 - |z|^2}{2\pi} \int_0^{2\pi} \frac{1}{(e^{it} - z)e^{it}(e^{-it} - \bar{z})} f(e^{it}) e^{it} dt$$

$$= \frac{1 - |z|^2}{2\pi} \int_0^{2\pi} \frac{f(e^{it})}{(e^{it} - z)(e^{-it} - \bar{z})} dt.$$

Taking the real part of both sides gives us a harmonic function since the real part of an analytic function is harmonic, and also yields (5.5).

$\square$

Exercise 5.27. Verify that the Poisson kernel, $\dfrac{1 - |z|^2}{|e^{it} - z|^2}$, can be written as

$$\mathrm{Re}\left(\frac{e^{it} + z}{e^{it} - z} \right) = \mathrm{Re}\left(\frac{1 + ze^{-it}}{1 - ze^{-it}} \right).$$

Try it out!

The integral in (5.5) is, in general, difficult to integrate. However, if there is an arc on which $\hat{f}(e^{it})$ is constant, the integration is easy.

Exercise 5.28. Verify that

$$\frac{1}{2\pi} \int_a^b K\,\mathrm{Re}\left(\frac{e^{it} + z}{e^{it} - z} \right) dt = K \frac{b - a}{2\pi} + \frac{K}{\pi} \arg\left(\frac{1 - ze^{-ib}}{1 - ze^{-ia}} \right)$$

$$= \frac{K}{\pi} \left[\arg\left(\frac{e^{ib} - z}{e^{ia} - z} \right) - \frac{b - a}{2} \right]. \tag{5.8}$$

Hint: Interchange integration and taking the real part of a function. The identity $\frac{1+w}{1-w} = 1 + \frac{2w}{1-w}$ will be helpful. *Try it out!*

The last formulation of (5.8) can be visualized geometrically. Consider a situation with e^{ia} and e^{ib} on the unit circle and z inside the unit disk. Then the vector from z to e^{ia} is $e^{ia} - z$ and the vector from z to e^{ib} is $e^{ib} - z$, so that the angle between them is $\arg\left(\dfrac{e^{ib} - z}{e^{ia} - z} \right)$, as is shown if Figure 5.7.

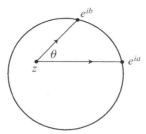

FIGURE 5.7. Geometric interpretation of $\theta = \arg\left(\dfrac{e^{ib} - z}{e^{ia} - z}\right)$.

Example 5.29. Assume that the unit disk is a thin insulated plate with a temperature along the boundary of 50 degrees for the top semicircle and 20 degrees along the bottom semicircle. From physics, we know that the function that describes the temperature within the unit disk is harmonic. Find it.

Solution: Apply (5.8) for $a_1 = 0, b_1 = \pi, K_1 = 50$ and then add it to the result where $a_2 = \pi, b_2 = 2\pi, K_2 = 20$, to get

$$f(z) = \frac{1}{2\pi}\left(70\pi + 60\arg\left(\frac{1+z}{1-z}\right)\right).$$

(70 is $50 + 20$ and $60 = 2(50 - 20)$.) As z ranges across the unit disk, $\frac{1+z}{1-z}$ covers the right half-plane (you can check this by graphing $\frac{1+z}{1-z}$ using *ComplexTool*), so its argument is between $-\pi/2$ and $\pi/2$. This gives values for $f(z)$ between 20 and 50, which makes sense. The solution gives the average temperature plus or minus half the difference between the maximum and minimum temperatures.

Exercise 5.30. Referring to

$$f(z) = \frac{1}{2\pi}\left(70\pi + 60\arg\left(\frac{1+z}{1-z}\right)\right)$$

in Example 5.29, find $f(0)$, $f(i/2)$, and $f(-i/2)$. Do your answers make sense?

Using the result of Exercise 5.28, we can see that computing the Poisson integral formula for a piecewise constant boundary is simple. Many of its applications come from having the boundary correspondence remain constant on arcs of the unit circle.

Most introductory analysis books give the Poisson integral formula for real-valued $\hat{f}(e^{i\theta})$. It can also be applied to create a harmonic function for complex-valued $\hat{f}(e^{i\theta})$, but the univalence of the harmonic function is not at all apparent. Let's see what happens if we use the Poisson integral formula with complex boundary values.

Example 5.31. The simplest example is obtained by letting the first third of the unit circle (that is, the arc from 0 to $e^{i2\pi/3}$) map to 1, the next third to $e^{i2\pi/3}$, and the last third to

$e^{i4\pi/3}$. Let's work through the details of this integration, working from (5.8). We compute

$$f(z) = \frac{1}{2\pi}\left((\frac{2\pi}{3} - 0) + 2\arg\left(\frac{1 - ze^{-i2\pi/3}}{1 - ze^0}\right)\right.$$

$$+ e^{i2\pi/3}(\frac{4\pi}{3} - \frac{2\pi}{3}) + 2e^{i2\pi/3}\arg\left(\frac{1 - ze^{-i4\pi/3}}{1 - ze^{-i2\pi/3}}\right)$$

$$+ e^{i4\pi/3}(2\pi - \frac{4\pi}{3}) + 2e^{i4\pi/3}\arg\left(\frac{1 - ze^{-2\pi i}}{1 - ze^{-i4\pi/3}}\right)\bigg)$$

$$= \frac{2\pi}{3(2\pi)}\left(1 + e^{i2\pi/3} + e^{i4\pi/3}\right)$$

$$+ \frac{1}{\pi}\left(\arg\left(\frac{1 - ze^{-i2\pi/3}}{1 - ze^0}\right) + e^{i2\pi/3}\arg\left(\frac{1 - ze^{-i4\pi/3}}{1 - ze^{-i2\pi/3}}\right)\right.$$

$$+ e^{i4\pi/3}\arg\left(\frac{1 - ze^{-2\pi i}}{1 - ze^{-i4\pi/3}}\right)\bigg)$$

$$= 0 + \frac{1}{\pi}\left(\arg\left(\frac{1 - ze^{-i2\pi/3}}{1 - z}\right) + e^{i2\pi/3}\arg\left(\frac{1 - ze^{-i4\pi/3}}{1 - ze^{-i2\pi/3}}\right)\right.$$

$$+ e^{i4\pi/3}\arg\left(\frac{1 - z}{1 - ze^{-i4\pi/3}}\right)\bigg).$$

Figure 5.8 shows the image of the unit disk as graphed in *ComplexTool*. It appears to be one-to-one on the interior of the unit disk. It is not one-to-one on the boundary! (Entering this formula into *ComplexTool* is unwieldy, so the function is one of the **Pre-defined functions**, the one called **Harmonic Triangle**. We will soon use the *PolyTool* applet to graph similar functions.)

FIGURE 5.8. *ComplexTool* image of the harmonic function mapping to the triangle.

Exercise 5.32. Find a formula that maps the unit disk harmonically to the interior of a convex regular n-gon. *Try it out!*

Small Project 5.33. Refer to Chapter 4 and its discussion of the shear construction. Find the pre-shears of the polygonal mappings in Exercise 5.32. What analytic function do you shear to get the polygonal function? A first step is to determine the dilatation of the function. See [4] for more details.

Exercise 5.34. For a non-convex example, show that the function that maps quarters of the unit circle to the four vertices $\{1, i, -1, \frac{i}{2}\}$ is

$$\frac{3i}{8} + \frac{1}{\pi} \left(\arg \left(\frac{1 + iz}{1 - z} \right) + i \arg \left(\frac{1 + z}{1 + iz} \right) - \arg \left(\frac{1 - iz}{1 + z} \right) + \frac{i}{2} \arg \left(\frac{1 - z}{1 - iz} \right) \right).$$

When we graph this in *ComplexTool*, it appears not to be one-to-one. Furthermore, $f(0) = \frac{3i}{8}$, so that the image of $\mathbb{D}$ is not the interior of the polygon. This is another of the **Predefined functions** in *ComplexTool*. *Try it out!*

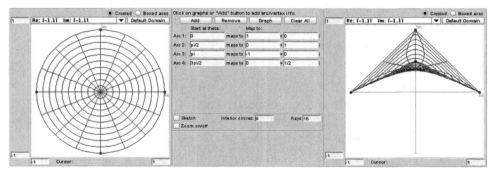

FIGURE 5.9. The *PolyTool* Applet.

At this point, you should start using the *PolyTool* applet. In it, you can specify which arcs of the unit circle will map to which points in the range, and the applet will compute and graph the harmonic function defined by extending the boundary correspondence to a function on $\mathbb{D}$. When you open this applet, you see a circle on the left and a blank screen on the right. You can create a harmonic function that maps portions of the boundary of the circle to vertices of a polygon in one of two ways. First, you can click on the unit circle in the left panel to denote an arc endpoint, continue choosing arc endpoints there, and then choose the target vertices by clicking in the right graph. (As you click, text boxes between the panels fill with information about where you clicked.) Once you have the boundary correspondence you want, click the **Graph** button. Second, you can click the **Add** button to get text boxes for input. For example, to create the function in Exercise 5.34, click **Add**, then fill in the first row of boxes for **Arc 1:** with **0 maps to 1+0i**. When you want another set of arcs, click **Add** again. The **Arc** boxes denote the starting point of the arc (i.e., for the arc from 0 to $\pi/3$, use 0 in the **Arc** box). Continue filling, and when you are ready to compute the Poisson integral to get the harmonic function, click the **Graph** button. Once you have a function graphed, you can drag around either the arc endpoints (in the domain on the left) or the target points (in the range on the right) and watch the function change.

Exploration 5.35. Are there ways of rearranging the boundary conditions to make the function in Exercise 5.34 univalent? What if the bottom half of the unit circle gets mapped to $i/2$ and the top half of the unit circle is divided into thirds for the other three vertices?

This new function isn't univalent, but is in some sense is closer to univalent than the mapping in Exercise 5.34. Can a modification be made so that it is univalent? *Try it out!*

Exercise 5.36. This is an extension of Exploration 5.35. Sheil-Small [16] proved (by techniques other than those discussed so far) that the harmonic extension of the boundary correspondence below maps the unit disk univalently onto the desired shape:

arc from	to	maps to
$-i$	i	i
i	$-3/5 + 4i/5$	-1
$-3/5 + 4i/5$	$-3/5 - 4i/5$	$i/2$
$-3/5 - 4i/5$	$-i$	1

For this function, first convince yourself that it appears to be univalent, and then find $f(z)$. *Try it out!*

We will be working with harmonic functions that are extensions of a piecewise constant boundary correspondence as in Example 5.36. For future discussions, we make the definition:

Definition 5.37. Let $\{e^{it_k}\}$ be a partition of $\partial\mathbb{D}$, where $t_0 < t_1 < \ldots < t_n = t_0 + 2\pi$. Let $\hat{f}(e^{it}) = v_k$ for $t_{k-1} < t < t_k$. We call *the harmonic extension of this step function* (as defined by the Poisson integral formula) $f(z)$.

Example 5.38. To understand the definition better, we demonstrate how the notation in Definition 5.37 is used for the function in Exercise 5.36. Since the arc from $-i$ to i can be thought of as the arc along the unit circle from $e^{-i\pi/2}$ to $e^{i\pi/2}$, we say that $t_0 = -\pi/2$ and $t_1 = \pi/2$. These points map to the vertex at i, so $v_1 = i$. Finding the next arc set is more difficult, because we need to find the angle t_2 that goes with the point $z = \frac{-3}{5} + \frac{4}{5}i = e^{it_2}$. We can get only a numerical estimate, found by $\pi + \arctan\left(\frac{4/5}{-3/5}\right) = \pi + \arctan(-4/3) \approx 2.2143$. (We add π because values of the arctangent function are in the first or fourth quadrant, while the angle is in the second quadrant.) Thus we have $t_2 \approx 2.2143$ and $v_2 = -1$. Continuing, we have $t_3 \approx 4.0689$ and $v_3 = i/2$. We finish with $t_4 = 3\pi/2$ and $v_4 = 1$. As in the definition, $t_4 = t_0 + 2\pi$. Then from Definition 5.37 the function $f(z)$ is the harmonic function that appears to be univalent when graphed in *PolyTool*.

Exercise 5.39. Combine the result of Exercise 5.28 (on p. 286) with Definition 5.37 to show that the function $f(z)$ in Definition 5.37 can be written

$$f(z) = v_1\left(\frac{t_1 - t_0}{2\pi}\right) + \frac{v_1}{\pi}\arg\left(\frac{1 - ze^{-it_1}}{1 - ze^{-it_0}}\right)$$

$$+ v_2\left(\frac{t_2 - t_1}{2\pi}\right) + \frac{v_1}{\pi}\arg\left(\frac{1 - ze^{-it_2}}{1 - ze^{-it_1}}\right)$$

$$+ \cdots + v_n\left(\frac{t_n - t_{n-1}}{2\pi}\right) + \frac{v_n}{\pi}\arg\left(\frac{1 - ze^{-it_n}}{1 - ze^{-it_{n-1}}}\right).$$

Try it out!

Since the Poisson integral formula gives rise to a harmonic function, we must learn some basics of the theory of harmonic functions before proceeding.

5.4 Harmonic Function Theory

Chapter 4 gives a detailed explanation of harmonic functions, as does [5]. Much of that material will be helpful for our investigations, so we repeat it here.

5.4.1 The Basics

A harmonic function f can be written as $f = h + \overline{g}$, where h and g are analytic functions. The analytic dilatation $\omega(z) = \frac{g'(z)}{h'(z)}$ is a measure of how much the harmonic function does not preserve angles. A dilatation of $\omega(z) \equiv 0$ means that the function is analytic, and so must be conformal. A dilatation with modulus near 1 indicates that the function distorts angles greatly. (For more intuition about the dilatation, see Section 4.6 of Chapter 4.) A result of Lewy states that a harmonic function has nonzero Jacobian (denoted $J_f(z) = |h'|^2 - |g'|^2$) if it is locally univalent. This result is in line with our understanding of the relationship between local univalence and a nonvanishing derivative for analytic functions.

Theorem 5.40 (Lewy's Theorem). *For a harmonic function f defined on a domain Ω, if f locally univalent in Ω, then $J_f(z) \neq 0$ for all $z \in \Omega$.*

This is equivalent to Lewy's theorem in Chapter 4.

A consequence of Lewy's theorem is that if a function is locally univalent in Ω, then its analytic dilatation satisfies either $|\omega(z)| < 1$ for all $z \in \Omega$, or $|\omega(z)| > 1$ for all $z \in \Omega$. In our work, we will study only functions that are locally univalent in a domain Ω and satisfy $|\omega(z)| < 1$ for all $z \in \Omega$. They are called *sense-preserving* because they preserve the orientation of curves in Ω.

Exercise 5.41. Verify that the condition that $J_f(z) \neq 0$ is equivalent to $|\omega(z)| \neq 1$ as long as $h'(z) \neq 0$. Conclude that a function that is locally univalent and sense-preserving has $J_f(z) > 0$ and $|\omega(z)| < 1$. *Try it out!*

Of particular interest is determining how to split the argument function, which is harmonic and sense-preserving, into h and $\overline{g}$.

Exercise 5.42. Show that $f(z) = K \arg(z)$ has canonical decomposition
$$h(z) = \frac{1}{2i} K \log(z) \quad \text{and} \quad g(z) = \frac{1}{2i} \overline{K} \log(z).$$
Try it out!

Another consequence of the canonical decomposition of a harmonic function is that we can write the analytic functions h and g defined in a domain Ω in terms of their power series expansions, centered at $z_0 \in \Omega$, as

$$f(z) = a_0 + \sum_{k=n}^{\infty} a_k(z - z_0)^k + \overline{b_0 + \sum_{k=m}^{\infty} b_k(z - z_0)^k}. \tag{5.9}$$

If f is sense-preserving, then we have that either $m > n$ or $m = n$ with $|b_n| < |a_n|$. In either case, when f is represented by (5.9), we say that f has a zero of order n at z_0.

Exercise 5.43. In this exercise, we prove that the zeros of a sense-preserving harmonic function are isolated.

(a) Assume that $f(z)$ is a sense-preserving locally univalent function with series expansion as given in (5.9). Show that if $f(z_0) = 0$, there exists a positive δ and a function ψ such that for $0 < |z - z_0| < \delta$ we can write

$$f(z) = h(z) + \overline{g(z)} = a_n(z - z_0)^n(1 + \psi(z)) \qquad (5.10)$$

where

$$\psi(z) = \frac{a_{n+1}}{a_n}(z - z_0) + \frac{a_{n+2}}{a_n}(z - z_0)^2 + \cdots + \overline{b_m(z - z_0)^m}\frac{1}{a_n(z - z_0)^n} + \cdots,$$

where the ellipses denote a continuation in m.

(b) Show that (a) implies that $|\psi(z)| < 1$ for z sufficiently close to z_0, since $m \geq n$ and $|b_n/a_n| < 1$ if $m = n$.

(c) Show that (b) implies that the zeros of a sense-preserving harmonic function are isolated, since $f(z) \neq 0$ near z_0 (except, of course, at z_0).

Try it out!

5.4.2   The Argument Principle

The argument principle for analytic functions gives a nice way to describe the number of zeros and poles inside a contour. It is in many introductory complex analysis courses. However, we explore this topic to emphasize the geometric nature of the result before studying the related result for harmonic functions. We first provide a formal definition of the winding number of the image of a contour about the origin.

Definition 5.44. The *winding number* of the image under $f(z)$ of a simple closed contour Γ about the origin is the net change in argument of $f(z)$ as z traverses Γ in the positive (counterclockwise) direction, divided by 2π. It can be denoted by $\frac{1}{2\pi}\Delta_\Gamma \arg f(z)$.

To explore the relationship between the winding number of the image of a contour about the origin and the number of zeros and poles contained within the contour, do the following exploration.

Exploration 5.45. Open *ComplexTool*. Change the **Interior circles** to 1 and **Rays** to 0. This will graph only the boundary of the circle. As you graph the following functions, examine how the image of the circle winds around, or encloses, the origin. Count how many times the image of the circle winds around the origin, counting the counterclockwise direction as positive and the clockwise direction as negative. If the image of the circle winds around the origin once, you know that there must be a zero of $f(z)$ inside that circle. (Think about this last sentence and make sure you understand it.)

(a) Graph $f(z) = z^2$, using a circle of radius 1. Use the **Sketch** button and trace around the circle in the domain to get a feeling for how many times its image winds around the origin. You should already know the answer. (Any other radius will work too. Why is that?)

(b) Graph $f(z) = z(z - 0.3)$. Use circles of radius $0.2, 0.3$, and 0.5. (You may have to zoom in on the image of the circle of radius 0.2 to understand what it is doing.) You may check the **Vary radius** checkbox and use the slider to change the radius of the circle.

(c) Graph $f(z) = z^4 - 6z + 3$. Use circles of radius 0.9, 1.5, 1.7, and 2.

(d) Graph $f(z) = \frac{z^4 - 6z + 3}{z - 1}$, using circles of radius 0.9 and 1.5.

(e) Graph $f(z) = \frac{z^4 - 6z + 3}{(z-1)^2}$, using circles of radius 0.9, 1.5, and 2.

(f) Go back through the previous three items, now changing the function while keeping the radius fixed.

(g) For the previous functions, find the locations of the zeros and poles, paying attention to how far they are from the origin.

(h) Make up your own function and do some more experiments.

Based on the exploration, what is your connection between the winding number of the image of a circle about the origin and the number of zeros and poles inside the circle? *Try it out!*

Theorem 5.46 (Argument Principle for Analytic Functions). *Let C be a simple closed contour lying entirely within a domain D. Suppose f is analytic in D except at a finite number of poles inside C and that $f(z) \neq 0$ on C. Then*

$$\frac{1}{2\pi i} \oint_C \frac{f'(z)}{f(z)} dz = N_0 - N_p,$$

where N_0 is the number of zeros of f inside C and N_p is the number of poles of f inside C, counted according to their order or multiplicity.

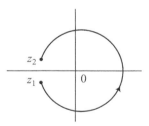

FIGURE 5.10. A path around a branch cut.

Before proving Theorem 5.46, we explore the connection between the winding number and the integral $\frac{1}{2\pi i} \oint_C \frac{f'(z)}{f(z)} dz$. We start with a related integral, $\int_{z_1}^{z_2} \frac{f'(z)}{f(z)} dz$, where z_1 and z_2 are points close to each other, but lying on opposite sides of a branch cut of $\log f(z)$, and we take a counterclockwise path along C from z_1 to z_2. (See Figure 5.10.) We have:

$$\int_{z_1}^{z_2} \frac{f'(z)}{f(z)} dz = \log f(z)|_{z_1}^{z_2}$$

$$= \ln|f(z_2)| - \ln|f(z_1)| + i(\arg(f(z_2)) - \arg(f(z_1))).$$

When we take the limit as $z_1 \to z_2$, we get that $\ln|f(z_2)| - \ln|f(z_1)| \to 0$ and $i(\arg(f(z_2)) - \arg(f(z_1))) \to 2\pi i \cdot$ (the winding number). Think about this last statement and make sure you understand it.

Proof. The proof of the argument principle relies on the Cauchy integral formula and deformation of contours. We begin by deforming the contour C to a series of smaller contours around the isolated zeros and poles of f. If there are no zeros or poles, then $\frac{f'(z)}{f(z)}$ is analytic, so the integral is zero, as desired. We analyze the zeros and poles individually, and add the results to get the conclusion. More formally, when f has zeros or poles inside C they must be isolated, and because f is analytic on C, there are only a finite number of distinct zeros or poles inside C. Denote the zeros and poles by z_j, for $j = 1, 2, \ldots, n$. Let γ_j be a circle of radius $\delta > 0$ centered at z_j, where δ is chosen small enough that the circles γ_j lie in D and do not meet each other. Join each circle γ_j to C by a Jordan arc λ_j in D. Consider the closed path Γ formed by moving around C in the positive direction while making a detour along each λ_j to γ_j, running once around this circle in the negative direction, then returning along λ_j to C. See Figure 5.11.

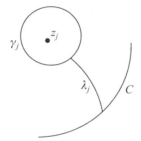

FIGURE 5.11. The contour Γ.

This curve Γ contains no zeros or poles of f, so

$$\Delta_\Gamma \arg f(z) = \frac{1}{2\pi i} \oint_\Gamma \frac{f'(z)}{f(z)} dz = 0$$

by the argument above. When considering the total change in argument of $f(z)$ along Γ, the contributions of the arcs λ_j along Γ cancel out, so that

$$\Delta_C \arg f(z) = \sum_{j=1}^{n} \Delta_{\gamma_j} \arg f(z),$$

where the circles γ_j are now traversed in the positive direction. Thus now we may consider each γ_j and sum the results.

Suppose that f has a zero of order n at $z = z_j$. Then $f(z) = (z - z_j)^n f_n(z)$, where $f_n(z)$ is an analytic function satisfying $f_n(z_j) \neq 0$, as shown in any introductory complex analysis book. Then

$$f'(z) = n(z - z_j)^{n-1} f_n(z) + (z - z_j)^n f_n'(z)$$

and

$$\begin{aligned}
\frac{f'(z)}{f(z)} &= \frac{n(z - z_j)^{n-1} f_n(z) + (z - z_j)^n f_n'(z)}{(z - z_j)^n f_n(z)} \\
&= \frac{n}{z - z_j} + \frac{f_n'(z)}{f_n(z)}.
\end{aligned}$$

When we integrate this along γ_j, we get $n(2\pi i) + 0$, because $\frac{f_n'(z)}{f_n(z)}$ is analytic inside the contour.

Suppose that f has a pole of order m at $z = z_k$. This means that f can be written as $f(z) = (z - z_k)^{-m} f_m(z)$, where f_m is analytic an nonzero at $z = z_k$. Then, as previously, we have

$$
\begin{aligned}
\frac{f'(z)}{f(z)} &= \frac{-m(z - z_k)^{-m-1} f_m(z) + (z - z_k)^{-m} f_m'(z)}{(z - z_k)^{-m} f_m(z)} \\
&= \frac{-m}{z - z_k} + \frac{f_m'(z)}{f_m(z)}.
\end{aligned}
$$

When we integrate this along γ_k, we get $-m(2\pi i) + 0$.

Summing over $j = 1, 2, \ldots n$ gives us the integral over Γ and the result. $\qquad\square$

There are many versions of the argument principle for harmonic functions. We need only the simple proof presented in this section, developed by Duren, Hengartner, and Laugesen [6].

Theorem 5.47 (Argument Principle for Harmonic Functions). *Let D be a Jordan domain with boundary C. Suppose f is sense-preserving harmonic function on D, continuous in $\overline{D}$ and $f(z) \neq 0$ on C. Then $\Delta_C \arg f(z) = 2\pi N$, where N is the number of zeros of $f(z)$ in D, counted according to multiplicity.*

Proof. Suppose that f has no zeros in D. This means that $N = 0$ and the origin is not an element of $f(D \cup C)$. A fact from topology says that in this case, $\Delta_C \arg f(z) = 0$, and the theorem is proved. We prove this fact. Let ϕ be a homeomorphism of the closed unit square S onto $D \cup C$ with the restriction of ϕ to the boundary, $\hat{\phi} : \partial S \to C$, also a homeomorphism. See Figure 5.12.

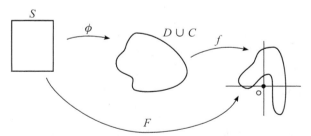

FIGURE 5.12. The composition of f and ϕ.

The composition $F = f \circ \phi$ is a continuous mapping of S onto the plane with no zeros, and we want to prove that $\Delta_{\partial S} \arg F(z) = 0$. Subdivide S into finitely many small squares S_j so that on each S_j, the argument of F varies by at most $\pi/2$. Then, since $F(S_j)$ cannot enclose the origin, $\Delta_{\partial S_j} \arg F(z) = 0$. When we consider $\Delta_{\partial S} \arg F(z)$, it is the sum $\sum_j \Delta_{\partial S_j} \arg F(z)$ because the contributions to the sum from the boundaries of each S_j cancel out, except where the boundary of S_j agrees with the boundary of S. Thus $\Delta_{\partial S} \arg F(z) = 0$, as desired.

Now suppose f has zeros in D. Because they are isolated (as proven in Exercise 5.43), and because f is not zero on C, there is only a finite number of distinct zeros in D. We proceed as in the proof of the analytic argument principle, and denote the zeros by z_j for $j = 1, 2, \ldots, n$. Let γ_j be a circle of radius $\delta > 0$ centered at z_j, where δ is chosen so small that the circles γ_j lie in D and do not meet each other. Join each circle γ_j to C by a Jordan arc λ_j in D. Consider the closed path Γ formed by moving around C in the positive direction while making a detour along each λ_j to γ_j, running once around this circle in the negative direction, then returning along λ_j to C. (See Figure 5.11.) This curve Γ contains no zeros of f, so $\Delta_\Gamma \arg F(z) = 0$ by the first case in this proof. When considering the total change in argument along Γ of $f(z)$, the contributions of the arcs λ_j along Γ cancel out, so that

$$\Delta_C \arg f(z) = \sum_{j=1}^{n} \Delta_{\gamma_j} \arg f(z),$$

where each of the circles γ_j is now traversed in the positive direction. Thus now we may consider each γ_j and sum the results.

Suppose that f has a zero of order n at a point z_0. Then, as in Exercise 5.43 on page 291, f can be locally written as

$$f(z) = a_n(z - z_0)^n(1 + \psi(z))$$

where $a_n \neq 0$ and $|\psi(z)| < 1$ on a sufficiently small circle γ defined by $|z - z_0| = \delta$. This shows that

$$\Delta_\gamma \arg f(z) = n\Delta_\gamma \arg(z - z_0) + \Delta_\gamma \arg(1 + \psi(z)) = 2\pi n. \qquad (5.11)$$

Therefore, if f has zeros of order n_j at the points z_j, the conclusion is that

$$\Delta_C \arg f(z) = \sum_{j=1}^{n} \Delta_{\gamma_j} \arg f(z) = 2\pi \sum_{j=1}^{n} n_j = 2\pi N,$$

and the theorem is proved.

$\square$

Exercise 5.48. Justify why $\Delta_\gamma \arg(1 + \psi(z)) = 0$ in (5.11). ***Try it out!***

In the next section, we look at the polygonal maps defined in Section 5.3. We will use the argument principle for harmonic functions in the proof of the Rado-Kneser-Choquet theorem.

5.5   Rado-Kneser-Choquet Theorem

As you examine the image of the unit disk using the examples in Section 5.3, you may notice that some of the functions seem to be one-to-one on the interior of the domain, while others do not seem to be univalent. Look again at the examples, and compare functions that map to convex domains with functions that map to non-convex domains.

Exploration 5.49. Make a conjecture about when functions are one-to-one, using the exercises from Section 5.3. Do this before reading the Rado-Kneser-Choquet theorem! ***Try it out!***

In general, we completely understand the behavior of harmonic extensions (as defined in Definition 5.37) that map to convex regions:

Theorem 5.50 (Rado-Kneser-Choquet Theorem). *Let Ω be a subset of $\mathbb{C}$ that is a bounded convex domain whose boundary is a Jordan curve Γ. Let $\hat{f}$ map $\partial\mathbb{D}$ continuously onto Γ and suppose that $\hat{f}(e^{it})$ runs once around Γ monotonically as e^{it} runs around $\partial\mathbb{D}$. Then the harmonic extension given in the Poisson integral formula is univalent in $\mathbb{D}$ and defines a harmonic mapping of $\mathbb{D}$ onto Ω.*

For the proof of this important theorem, we use the following lemma.

Lemma 5.51. *Let ψ be a real-valued function harmonic in $\mathbb{D}$ and continuous in $\overline{\mathbb{D}}$. Suppose ψ has the property that, after a rotation of coordinates, $\psi(e^{it}) - \psi(e^{-it}) \geq 0$ on the interval $[0, \pi]$, with strict inequality $\psi(e^{it}) - \psi(e^{-it}) > 0$ on a subinterval $[a, b]$ with $0 \leq a < b \leq \pi$. Then ψ has no critical points in $\mathbb{D}$.*

The condition on ψ may seem mysterious at first, so we will discuss it. One kind of function for which the property holds is that ψ is at most bivalent on $\partial\mathbb{D}$. What does "at most bivalent" mean? Univalent means that a function is one-to-one. *Bivalent* means that a function is two-to-one, or that there may be $z_1 \neq z_2$ such that $f(z_1) = f(z_2)$, but that if $f(z_1) = f(z_2) = f(z_3)$, then at least two of z_1, z_2, z_3 are equal. Another kind of function ψ described in Lemma 5.51 is one that is continuous on $\partial\mathbb{D}$ where $\psi(e^{it})$ rises from a minimum at $e^{-i\alpha}$ to a maximum at $e^{i\alpha}$, then decreases to its minimum at $e^{-i\alpha}$ as e^{it} runs around the unit circle, without having any other local extrema but allowing arcs of constancy. See Figure 5.13.

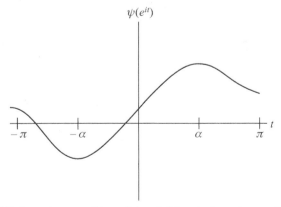

FIGURE 5.13. Boundary condition of ψ satisfying the hypotheses of Lemma 5.51.

Proof of Lemma 5.51. To show that ψ has no critical points in $\mathbb{D}$, we show that $\frac{\partial\psi}{\partial z} \neq 0$ in $\mathbb{D}$. This is equivalent to saying that

$$\frac{1}{2}\left(\frac{\partial\psi}{\partial x} - i\frac{\partial\psi}{\partial y}\right) \neq 0.$$

We will simplify the proof by proving that $\psi_z(0) \neq 0$, and claim that will be sufficient. If z_0 is a point in $\mathbb{D}$, consider the function $\varphi(z) = \frac{z_0 - z}{1 - \overline{z_0} z}$ that is a conformal self-map of $\mathbb{D}$ with $\varphi(0) = z_0$, and let $F(\zeta) = \psi(\varphi(\zeta))$. Observe that F is harmonic in $\mathbb{D}$, continuous in $\overline{\mathbb{D}}$, and satisfies the same condition about $F(e^{it}) - F(e^{-it})$ as ψ does. Applying the chain rule to $F(\zeta)$ gives $F_\zeta(\zeta) = \psi_z(\varphi(\zeta))\varphi'(\zeta)$, since φ is analytic and thus has $\varphi_{\overline{\zeta}} = 0$. (In general, the chain rule is more complicated for harmonic functions. Here, since φ is analytic, it has takes its familiar form.) Substituting 0 for ζ gives $F_\zeta(0) = \psi_z(z_0)\varphi'(0)$, implying that if $F_\zeta(0) = 0$ then $\psi_\zeta(z_0) = 0$. Thus when we have proven that $\psi_z(0) \neq 0$, we will be able to generalize to $\psi_z(z_0) \neq 0$ for all z_0 in $\mathbb{D}$.

We use the Poisson integral formula to prove that $\psi_z(0) \neq 0$. Substituting in ψ (or $\hat{\psi}(e^{it}) = \lim_{r \to 1} \psi(re^{it})$ on $\partial \mathbb{D}$) gives

$$\psi(z) = \frac{1}{2\pi} \int_0^{2\pi} \frac{1 - |z|^2}{|e^{it} - z|^2} \hat{\psi}(e^{it}) dt = \frac{1}{2\pi} \int_0^{2\pi} \frac{1 - z\overline{z}}{(e^{it} - z)(e^{-it} - \overline{z})} \hat{\psi}(e^{it}) dt.$$

From

$$\frac{\partial}{\partial z}\left(\hat{\psi}(e^{it}) \frac{1 - z\overline{z}}{(e^{it} - z)(e^{-it} - \overline{z})}\right) = \frac{\hat{\psi}(e^{it})}{e^{-it} - \overline{z}} \frac{\partial}{\partial z}\left(\frac{1 - z\overline{z}}{e^{it} - z}\right)$$

$$= \left(\frac{\hat{\psi}(e^{it})}{e^{-it} - \overline{z}}\right) \cdot \left(\frac{e^{it}(e^{-it} - \overline{z})}{(e^{it} - z)^2}\right)$$

$$= \hat{\psi}(e^{it})\left(\frac{e^{it}}{(e^{it} - z)^2}\right)$$

we get

$$\psi_z(0) = \frac{1}{2\pi} \int_0^{2\pi} \hat{\psi}(e^{it}) e^{-it} dt.$$

From the hypotheses of the lemma, we know that there is a $t \in (0, \pi)$ such that $\psi(e^{it}) - \psi(e^{-it}) > 0$. Thus

$$\text{Im } \psi_z(0) = \text{Im}\left(\frac{1}{2\pi} \int_0^{2\pi} \hat{\psi}(e^{it}) e^{-it} dt\right)$$

$$= -\frac{1}{2\pi} \int_0^{2\pi} \hat{\psi}(e^{it}) \sin(t) dt.$$

Since $\hat{\psi}$ is periodic,

$$\text{Im } \psi_z(0) = -\frac{1}{2\pi}\left(\int_0^{\pi} \hat{\psi}(e^{it}) \sin(t) dt + \int_{-\pi}^0 \hat{\psi}(e^{it}) \sin(t) dt\right)$$

$$= -\frac{1}{2\pi}\left(\int_0^{\pi} \hat{\psi}(e^{it}) \sin(t) dt - \int_0^{\pi} \hat{\psi}(e^{-it}) \sin(t) dt\right)$$

$$= -\frac{1}{2\pi} \int_0^{\pi} (\hat{\psi}(e^{it}) - \hat{\psi}(e^{-it})) \sin(t) dt < 0.$$

The last inequality relies on the fact that $\sin(t)$ is non-negative on $[0, \pi]$. We have shown that $\text{Im } \psi_z(0) \neq 0$, proving the lemma. $\square$

Proof of Theorem 5.50. Without loss of generality, assume that $\hat{f}(e^{it})$ runs around Γ counterclockwise as t increases. (Otherwise, take conjugates.) We will show that if f is not locally univalent in $\mathbb{D}$, then Lemma 5.51 gives a contradiction.

Suppose that $f = u + iv$ is not locally univalent, so that the Jacobian of f vanishes at z_0 in $\mathbb{D}$. This means that $\begin{pmatrix} u_x & v_x \\ u_y & v_y \end{pmatrix}$ has a determinant of 0 at z_0, so the system of equations

$$au_x + bv_x = 0$$
$$au_y + bv_y = 0$$

has a nonzero solution (a, b). Thus the real-valued harmonic function $\psi = au + bv$ has a critical point at z_0 (since $(a, b) \neq (0, 0)$). However, the hypothesis of Theorem 5.50 implies that ψ satisfies the hypothesis of Lemma 5.51. Thus we have a contradiction, so f must be locally univalent.

Now that we see that f is locally univalent, we apply the argument principle to show that f is univalent in $\mathbb{D}$. Since f is sense-preserving on $\partial\mathbb{D}$ and locally univalent, f is sense-preserving throughout $\mathbb{D}$. If f is not univalent, there are points z_1 and z_2 in $\mathbb{D}$ such that $f(z_1) = f(z_2)$. That would imply that $f(z) - f(z_1)$ has two zeros in $\mathbb{D}$, so that the winding number of $f(z) - f(z_1)$ about the origin is 2, which contradicts the hypotheses about the boundary correspondence. This completes the proof.

$\square$

Exercise 5.52. Give a detailed proof that the hypothesis of Theorem 5.50 implies that ψ satisfies the hypothesis of Lemma 5.51. *Try it out!*

The description of $\hat{f}$ in Theorem 5.50 does not require that it be one-to-one on $\partial\mathbb{D}$, but permits arcs of constancy. Furthermore, the Rado-Kneser-Choquet theorem is true when $\hat{f}$ has jump discontinuities as long as the image of $\partial\mathbb{D}$ is not contained in a straight line. This requires some additional justification, so we state it separately as a corollary.

Corollary 5.53. *Let $f(z)$ be defined as in Definition 5.37 on p. 290. Suppose the vertices $v_1, v_2, \ldots, v_n$, when traversed in order, define a convex polygon whose interior is denoted by Ω. Then $f(z)$ is univalent in $\mathbb{D}$ and defines a harmonic mapping from $\mathbb{D}$ onto Ω.*

Here is some motivation for the proof of Corollary 5.53. Take a sequence of functions $\hat{f}_m(e^{it})$ that are continuous and converge to the boundary correspondence $\hat{f}(e^{it})$ of Definition 5.37. (One can be described by having $\hat{f}_m(e^{it}) = v_k$ for t-values in the interval $(t_{k-1} + \frac{t_k - t_{k-1}}{2m}, t_k - \frac{t_k - t_{k-1}}{2m})$ while for t-values in the interval $(t_k - \frac{t_k - t_{k-1}}{2m}, t_k + \frac{t_{k+1} - t_k}{2m})$, $\hat{f}_m(e^{it})$ maps the interval linearly to the segment between v_k and v_{k+1}.) Each $\hat{f}_m(e^{it})$ satisfies the conditions of the Rado-Kneser-Choquet theorem, and so extends to a univalent harmonic function, $f_m(z)$, in the unit disk. But the sequence f_m converges uniformly on compact subsets of $\mathbb{D}$, so the entire sequence converges uniformly to f in $\mathbb{D}$. Therefore, $f(z)$ inherits the univalence from the sequence. The fact that the limit function is univalent is not immediately apparent. Full details may be found in [8].

The theorem does not guarantee anything about univalence if the domain Ω is not convex. In fact, we may expect that univalence will not be achieved. For example, look at Exercise 5.34 on page 289.

Exploration 5.54. Extend Exploration 5.35 on page 289. Instead of modifying the boundary correspondence, start with the correspondence in Exercise 5.36. Then move the vertex at $i/2$ closer to i. A nice picture comes from having the vertex set be $\{1, i, -1, \frac{9i}{10}\}$. There we see the lack of univalence clearly. ***Try it out!***

5.5.1 Boundary behavior

In this section, we explore what seems to be true in some of the above examples, that there appears to be interesting boundary behavior of our harmonic extensions of step functions. Examine this behavior in the next exploration.

Exploration 5.55. Using *ComplexTool* or *PolyTool*, graph the function from Example 5.31 that maps $\mathbb{D}$ to a triangle. Investigate the behavior of the boundary using the sketching tool of the applet. Approach the break point between arcs, such as $z = 1$, along different paths. Approach radially, then approach along a line that is not a radius of the circle. Observe how the paths that approach 1 cause the image of the path to approach different points along the line segment that makes up a portion of the boundary of the range. (As you get close to an arc endpoint, the image of the sketch may jump to a vertex: here, examine where the image is immediately before that jump.) In *PolyTool* and *ComplexTool*, you can hit the **Graph** button to clear previous sketching but keep the polygonal map. Repeat the exercise with some other examples of polygonal functions. Try to answer:

(a) Given a point ζ on the boundary of the polygon, is it possible to find a path γ approaching $\partial \mathbb{D}$ such that $\gamma(z)$ approaches ζ?

(b) As you approach an arc endpoint in $\partial \mathbb{D}$ radially, what point on the boundary of the polygon do you approach?

Try it out!

As you performed the exploration, you probably discovered some of the known properties of the boundary behavior of harmonic extensions of step functions, originally proven by Hengartner and Schober [8], who proved a more general form of the next theorem. We state the theorem as it applies to the step functions of Definition 5.37. The *cluster set* of f at a point e^{it_k} is the set of limits of sequences $\{z_n\}$, where z_n is inside Γ, and $\lim_{n \to \infty} (z_n) = e^{it_k}$.

Theorem 5.56. *Let f be the harmonic extension of a step function $\hat{f}(e^{it_k})$ in Definition 5.37. Denote by Γ the polygon defined by the vertices v_k. By definition, the radial limits $\lim_{r \to 1} f(re^{it})$ lie on Γ for all t except those in $\{t_k\}$. The unrestricted limit*

$$\hat{f}(e^{it}) = \lim_{z \to e^{it}} f(z)$$

exists at every point $e^{it} \in \partial \mathbb{D} \setminus \{e^{it_1}, e^{it_2}, ..., e^{it_n}\}$ and lies on Γ. Furthermore,
(a) the one-sided limits as $t \to t_k$ are

$$\lim_{t \to t_k^-} \hat{f}(e^{it}) = v_k \qquad and \qquad \lim_{t \to t_k^+} \hat{f}(e^{it}) = v_{k+1}$$

(b) the cluster set of f at $e^{it_k} \in \{e^{it_1}, e^{it_2}, ..., e^{it_n}\}$ is the linear segment joining v_k to v_{k+1}.

Proof. Part (a) of the theorem follows directly from the definition of f and the Poisson integral formula. Since the boundary correspondence is defined in Definition 5.37, the limits follow. Now we need to show part (b). Let us consider e^{it_k}. If z approaches e^{it_k} along the circular arc

$$\arg\left(\frac{e^{it_k} + z}{e^{it_k} - z}\right) = \frac{\lambda\pi}{2}, \qquad -1 < \lambda < 1, \tag{5.12}$$

then $f(z)$ converges to

$$\frac{1}{2}(1 - \lambda)v_k + \frac{1}{2}(1 + \lambda)v_{k+1}.$$

Therefore the cluster set of f at t_k is the line segment joining v_k and v_{k+1}, and part (b) is proven.

$\square$

Exercise 5.57. Use analytic geometry to see that (5.12) is a circular arc. Hint: Consider the cases where z is either on the circle or is the center of the circle for some intuition. *Try it out!*

Exercise 5.58. Theorem 5.56 holds for non-univalent mappings. Go back to previous examples and identify the line segment that connnects the vertices. Regraph the example from Exercise 5.34. Using the **Sketch** utility in either *ComplexTool* or *PolyTool*, check that the limit as you approach one of the t_k does appear to be that line segment. *Try it out!*

5.6   Star Mappings

From the Rado-Kneser-Choquet theorem, we see that harmonic functions mapping the unit disk $\mathbb{D}$ to convex polygons are well understood. That is, if we define a harmonic function mapping from the unit disk to a convex polygon as in Definition 5.37, the function is univalent in $\mathbb{D}$. Theorem 5.56 describes the boundary behavior fully, showing that the limit of the function as we approach one of the break points between vertex pre-images, t_k, gives the line segment joining the vertices.

Non-convex polygons are not nearly as well understood. We first examine non-convex polygons in their simplest mathematical form: the regular stars.

Definition 5.59. By an n-pointed *star*, or r-star, we mean an equilateral $2n$-gon with the vertex set

$$\left\{r\alpha^{2k}, \alpha^{2k+1} : k = 1, 2, \ldots, n \text{ and } \alpha = e^{i\pi/n}\right\},$$

where r is some real constant.

When $r = 1$, the n-pointed star is a regular $2n$-gon and when $r < \cos(\pi/n)$ or $r > \sec(\pi/n)$ it is a strictly non-convex $2n$-gon. Our preimages of the vertices of the $2n$-gon will be arcs centered at the $2n$th roots of unity (this is different from our previous examples).

Example 5.60. We will find a harmonic mapping of the unit disk into the 0.3-star. That is, we will find the harmonic extension of the following boundary correspondence:

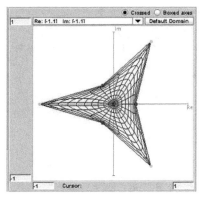

FIGURE 5.14. The 0.3-star for $n = 3$.

for t from	to	$\hat{f}(e^{it})$
$-\pi/6$	$\pi/6$	0.3
$\pi/6$	$\pi/2$	$e^{i\pi/3}$
$\pi/2$	$5\pi/6$	$0.3e^{i2\pi/3}$
$5\pi/6$	$7\pi/6$	-1
$7\pi/6$	$3\pi/2$	$0.3e^{i4\pi/3}$
$3\pi/2$	$11\pi/6$	$e^{i5\pi/3}$

After going through details as in previous examples, we discover that the harmonic extension is

$$f(z) = \frac{1}{\pi}\left[0.3\arg\left(\frac{1 - ze^{-i\pi/6}}{1 - ze^{i\pi/6}}\right)\right.$$

$$+ e^{i\pi/3}\arg\left(\frac{1 + iz}{1 - ze^{-i\pi/6}}\right) + 0.3e^{i2\pi/3}\arg\left(\frac{1 - ze^{-i5\pi/6}}{1 + iz}\right)$$

$$- \arg\left(\frac{1 - ze^{-i7\pi/6}}{1 - ze^{-i5\pi/6}}\right) + 0.3e^{i4\pi/3}\arg\left(\frac{1 - iz}{1 - ze^{-7i\pi/6}}\right)$$

$$\left.+ e^{i5\pi/3}\arg\left(\frac{1 - ze^{-i11\pi/6}}{1 - ze^{-i3\pi/2}}\right)\right].$$

Graph this using *ComplexTool* (it is one of the **Pre-defined functions**). It appears to be univalent, but we have not yet proved that.

Exercise 5.61. Prove that if $f(z)$ is the harmonic extension to the r-star as defined in Definition 5.59, then $f(0) = 0$. Interpret this result geometrically. ***Try it out!***

Exercise 5.62. Modify the function in Exercise 5.60 to have $r = 0.15$ and see whether it appears univalent. To graph it in *ComplexTool*, choose the previous star as one of the **Pre-defined functions** and then modify the equation in the function box. ***Try it out!***

To work with these stars, we may sometimes want to vary the boundary correspondence. That is, we may want not to split $\partial\mathbb{D}$ evenly among the $2n$ vertices. It will become useful to us to have an unequal correspondence in the boundary arcs but maintain some symmetry. To do this, we still consider arcs centered at the $2n$-th roots of unity, but alternate between larger and smaller arcs. If we examine the geometry, we realize that an even split would make each arc have length $\frac{2\pi}{2n} = \pi/n$. Two consecutive arcs would together have length $2\pi/n$. To maintain some symmetry but let the arcs alternate in size, we want two consecutive arcs to add to $2\pi/n$, but not split evenly. We introduce a parameter p, with $0 < p < 1$, to split consecutive arcs into $p2\pi/n$ and $(1-p)2\pi/n$, whose sum is still $2\pi/n$. We make the definition:

Definition 5.63. Let $n \geq 2$ be an integer, r a positive real number, and $\alpha = e^{i\pi/n}$. Define a *boundary correspondence from $\partial\mathbb{D}$ the vertices of the r-star* for all but a finite number of points on $\partial\mathbb{D}$ by mapping arcs with endpoints $\{\alpha e^{-ip\pi/n}, \alpha e^{ip\pi/n}, 0 \leq k \leq n-1\}$ as:

$$\hat{f}(e^{it}) = \begin{cases} r\alpha^{2k}, & e^{it} \in (\alpha^{2k}e^{-ip\pi/n}, \alpha^{2k}e^{ip\pi/n}) \\ \alpha^{2k+1}, & e^{it} \in (\alpha^{2k+1}e^{-i(1-p)\pi/n}, \alpha^{2k+1}e^{i(1-p)\pi/n}) \end{cases} . \tag{5.13}$$

Let f be the Poisson extension of $\hat{f}$.

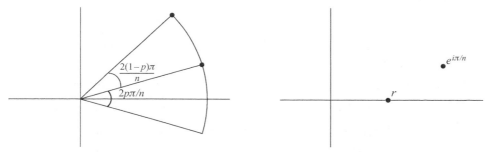

FIGURE 5.15. The first two arcs and their images from Definition 5.63. The dots on the left-hand side indicate points of discontinuity of the boundary correspondence.

The arc $(\alpha^{2k}e^{-ip\pi/n}, \alpha^{2k}e^{ip\pi/n})$ centered at α^{2k} is mapped to the vertex $r\alpha^{2k}$ and the arc $(\alpha^{2k+1}e^{-i(1-p)\pi/n}, \alpha^{2k+1}e^{i(1-p)\pi/n})$ centered at α^{2k+1} is mapped to the vertex α^{2k+1}.

Exercise 5.64. Show that the interval in (5.13),

$$(\alpha^{2k+1}e^{-i(1-p)\pi/n}, \alpha^{2k+1}e^{i(1-p)\pi/n}),$$

can be written as $(\alpha^{2k}e^{ip\pi/n}, \alpha^{2k+2}e^{-ip\pi/n})$. *Try it out!*

Now use the *StarTool* applet. The default for this applet is the 3-pointed star from Example 5.60. The arcs and their target vertices are colored (with a light blue arc mapping to a light blue vertex, for example). The default p-value is 0.5, which corresponds to evenly spaced arcs. You can use the slider bars (the plus/minus buttons for n) or type in the text boxes to change the values for n, p, and r. The maximum n-value allowed 18, which is sufficient for the explorations below. As with *ComplexTool*, there is the option to **Sketch** on the graph to see the mapping properties of the stars. There is also an option to **Show**

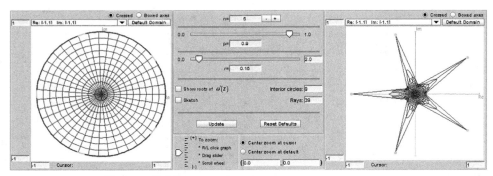

FIGURE 5.16. A star with $n = 5$, $r = 0.15$, and $p = 0.9$.

roots of $\omega(z)$. The roots of $\omega(z)$ can be helpful but are not essential for the first explorations. In general, try to see what happens for small values of n, such as $4, 5, 6$, or 7.

Exploration 5.65. Explore different values of n and r. See if you can find a pattern for the univalence of the star function.

(a) What is the relationship between the r that you choose and the p-value necessary for univalence? Is there a range of p that works?

(b) What happens as p goes to 0 or 1?

(c) For a given p-value, can you determine the minimal r? That is, how small can you make r and maintain univalence?

(d) Is there a minimum r-value, one so that no p-value that will give univalence?

(e) As you change n, what happens to the p that you need to achieve univalence?

(f) What is the relationship between r, n, and p? (It is unlikely that you will be able to answer this question now, but make some conjectures.)

Try it out!

5.7   Dilatations of Polygonal Maps are Blaschke Products

We now use the tools of harmonic functions to study the polygonal maps. To understand when the star maps are univalent, we first examine their dilatations. As we discover, the dilatation of a polygonal map is a Blaschke product.

Exercise 5.66. For the function generated in Exercise 5.34 on page 289:

(a) Find the formulas for $h(z)$ and $g(z)$. (Use the result of Exercise 5.42.)

(b) Find the derivatives $h'(z)$ and $g'(z)$, verifying that they simplify to

$$h'(z) = \frac{1}{(2\pi i)(1 - z^4)} \left((\frac{3}{2} - \frac{3}{2}i)z^2 - 3iz + (\frac{5}{2} + \frac{5}{2}i) \right)$$

and

$$g'(z) = \frac{1}{(2\pi i)(1 - z^4)} \left((\frac{5}{2} - \frac{5}{2}i)z^2 + 3iz + (\frac{3}{2} + \frac{3}{2}i) \right).$$

(c) Show that the zeros of $g'(z)$ are $z_1 \approx 0.9245 - 0.9245i$ and $z_2 \approx -0.3245 + 0.3245i$ and that the zeros of $h'(z)$ are $1/\overline{z_1}$ and $1/\overline{z_2}$. Thus we are able to write the dilatation as

$$\omega(z) = C\left(\frac{z_1 - z}{1 - \overline{z_1}z}\right)\left(\frac{z_2 - z}{1 - \overline{z_2}z}\right),$$

where C is a constant.

(d) What are $|z_1|$, $|z_2|$, and $|C|$? Remember these when you read Theorem 5.81 in Section 5.8.

Try it out!

Motivated by the results of Exercise 5.66, we examine functions of a particular form.

Exploration 5.67. We examine the properties of functions of the form
$$B_{z_0}(z) = \frac{z_0 - z}{1 - \overline{z_0}z}.$$

(a) Using *ComplexTool*, graph the image of the unit disk under $B_{0.5}(z) = \frac{0.5-z}{1-0.5z}$, $B_{0.5\exp(i\pi/4)}(z) = \frac{0.5e^{i\pi/4}-z}{1-0.5e^{-i\pi/4}z}$, and $B_{0.5i}(z) = \frac{0.5i-z}{1+0.5iz}$. What is $B(0)$? What is the image of the unit disk under $B(z)$? Does $B(z)$ appear to be univalent?

(b) Graph the image of the unit disk under $B_{2\exp(i\pi/4)}(z) = \frac{2e^{i\pi/4}-z}{1-2e^{-i\pi/4}z}$. What is $B(0)$? What is the image of the unit disk under $B(z)$? Does $B(z)$ appear to be univalent?

(c) What is the general formula for $B_{z_0}(0)$? What effect does $\arg(z_0)$ have on the location of $B_{z_0}(0)$? What effect does $|z_0|$ have on the image of the unit disk?

(d) Graph the image of the unit disk under $f_1(z) = (B_{0.5}(z))^2 = \left(\frac{0.5-z}{1-0.5z}\right)^2$. What is $f_1(0)$? What is the image of the unit disk under $f_1(z)$? Does $f_1(z)$ appear to be univalent? How does this compare to $f(z) = z^2$?

(e) Graph the image of the unit disk under $f_2(z) = B_{0.5}(z)B_{0.5i}(z)$, $f_3(z) = B_{0.5}(z)B_{0.2i}(z)$, and $f_4(z) = B_{0.5\exp(i\pi/4)}(z)B_{0.2i}(z)$. What are $f_2(0)$, $f_3(0)$, and $f_4(0)$? How do $f_2(0)$, $f_3(0)$, and $f_4(0)$ relate to the values of z_0 in the functions $B_{z_0}(z)$?

Try it out!

Definition 5.68. A *Blaschke factor* is $B_{z_0}(z) = \frac{z_0-z}{1-\overline{z_0}z}$, and a *finite Blaschke product* of order n is a product of n Blaschke factors, possibly multiplied by a constant ζ with $|\zeta| = 1$:

$$\zeta \prod_{k=1}^{n} \frac{z_k - z}{1 - \overline{z_k}z}.$$

Geometrically, the multiplication by ζ is a rotation.

Remark 5.69. This definition is non-standard. The standard definition of Blaschke product, as in Chapters 4 and 6, assumes that $|z_k| < 1$. We do not place that restriction on z_k to simplify our computations.

We use the result of Exercise 5.39 on page 290 to see that the dilatation of the harmonic polygonal functions to an n-gon is a Blaschke product of order at most $n - 2$. This was proved by Sheil-Small in [16], and is discussed in [5].

Let $f_k(z)$ denote the contribution to $f(z)$ that arises from applying the Poisson integral formula to the boundary correspondence for $t_{k-1} < t < t_k$. Then $f(z) = \sum_{k=1}^{n} f_k(z)$. On the interval $t_{k-1} < t < t_k$,

$$f_k(z) = \frac{v_k}{2\pi}(t_k - t_{k-1}) + \frac{v_k}{\pi} \arg\left(\frac{1 - ze^{-it_k}}{1 - ze^{-it_{k-1}}}\right)$$

by Definition 5.37. The canonical decomposition of $f_k(z)$ is $h_k(z) + \overline{g_k(z)}$. By Exercise 5.42, we have

$$h_k(z) = \frac{v_k}{2\pi}(t_k - t_{k-1}) + \frac{v_k}{2\pi i} \log\left(\frac{1 - ze^{-it_k}}{1 - ze^{-it_{k-1}}}\right)$$

$$= \frac{v_k}{2\pi}(t_k - t_{k-1}) + \frac{v_k}{2\pi i}\left(\log(1 - ze^{-it_k}) - \log(1 - ze^{-it_{k-1}})\right)$$

and

$$g_k(z) = \frac{\overline{v_k}}{2\pi i} \log\left(\frac{1 - ze^{-it_k}}{1 - ze^{-it_{k-1}}}\right)$$

$$= \frac{\overline{v_k}}{2\pi i}\left(\log(1 - ze^{-it_k}) - \log(1 - ze^{-it_{k-1}})\right).$$

The computations that follow give some rigor to our intuition: since h and g are sums of logarithms, their derivatives are sums of terms that have $1 - ze^{-it_k}$ in their denominators. We will combine the factors and write the derivatives as Blaschke products. From $h(z) = \sum_{k=1}^{n} h_k(z)$ we have:

$$h'(z) = \sum_{k=1}^{n} \frac{v_k}{2\pi i}\left(\frac{-e^{-it_k}}{1 - ze^{-it_k}} - \frac{-e^{-it_{k-1}}}{1 - ze^{-it_{k-1}}}\right)$$

$$= \sum_{k=1}^{n} \frac{v_k}{2\pi i}\left(\frac{1}{z - e^{it_k}} - \frac{1}{z - e^{it_{k-1}}}\right). \tag{5.14}$$

The function $g'(z)$ is identical except for having $\overline{v_k}$ instead of v_k:

$$g'(z) = \sum_{k=1}^{n} \frac{\overline{v_k}}{2\pi i}\left(\frac{-e^{-it_k}}{1 - ze^{-it_k}} - \frac{-e^{-it_{k-1}}}{1 - ze^{-it_{k-1}}}\right)$$

$$= \sum_{k=1}^{n} \frac{\overline{v_k}}{2\pi i}\left(\frac{1}{z - e^{it_k}} - \frac{1}{z - e^{it_{k-1}}}\right). \tag{5.15}$$

Combining like factors gives

$$h'(z) = \frac{1}{2\pi i}\sum_{k=1}^{n} \frac{v_k - v_{k+1}}{z - e^{it_k}} \quad \text{and} \quad g'(z) = \frac{1}{2\pi i}\sum_{k=1}^{n} \frac{\overline{v_k} - \overline{v_{k+1}}}{z - e^{it_k}}. \tag{5.16}$$

Because $v_{n+1} = v_1$, $\sum_{k=1}^{n}(v_k - v_{k+1}) = 0$, a result we use later.

Exercise 5.70. Prove that $\sum_{k=1}^{n}(v_k - v_{k+1}) = 0$, since $v_{n+1} = v_1$. Interpret the result geometrically. ***Try it out!***

We rely on the observations that

$$\overline{h'(1/\,\overline{z})} = z^2 g'(z) \qquad \text{or} \qquad \overline{g'(1/\,\overline{z})} = z^2 h'(z), \qquad (5.17)$$

which arise from:

$$
\begin{aligned}
\overline{h'(1/\,\overline{z})} - z^2 g'(z) &= \frac{-1}{2\pi i}\sum_{k=1}^{n}\frac{\overline{v}_k - \overline{v}_{k+1}}{1/z - e^{-it_k}} - \frac{z^2}{2\pi i}\sum_{k=1}^{n}\frac{\overline{v}_k - \overline{v}_{k+1}}{z - e^{it_k}}\\
&= \frac{-z}{2\pi i}\sum_{k=1}^{n}\frac{e^{it_k}(\overline{v}_k - \overline{v}_{k+1})}{e^{it_k} - z} - \frac{z^2}{2\pi i}\sum_{k=1}^{n}\frac{\overline{v}_k - \overline{v}_{k+1}}{z - e^{it_k}}\\
&= \frac{z}{2\pi i}\sum_{k=1}^{n}\frac{e^{it_k}(\overline{v}_k - \overline{v}_{k+1})}{z - e^{it_k}} - \frac{z^2}{2\pi i}\sum_{k=1}^{n}\frac{\overline{v}_k - \overline{v}_{k+1}}{z - e^{it_k}}\\
&= \frac{1}{2\pi i}\sum_{k=1}^{n}\frac{(\overline{v}_k - \overline{v}_{k+1})(ze^{it_k} - z^2)}{z - e^{it_k}}\\
&= \frac{1}{2\pi i}\sum_{k=1}^{n}\frac{(\overline{v}_k - \overline{v}_{k+1})(z)(e^{it_k} - z)}{z - e^{it_k}}\\
&= \frac{-1}{2\pi i}\sum_{k=1}^{n}(\overline{v}_k - \overline{v}_{k+1})(z)\\
&= 0.
\end{aligned}
$$

Exercise 5.71. We proved the first half of (5.17). Using that result, prove the second part of (5.17) with minimal computation. *Try it out!*

Exercise 5.72. Interpret the result geometrically. If $z_0 \in \mathbb{D}$ is such that $g'(z_0) = 0$, then what do we know about the zeros of h'? How are the locations of the zeros of h' related to the locations of the zeros of g'? Completing this exercise will give intuition about a later result. *Try it out!*

Exercise 5.73. Show that $h'(z)$ and $g'(z)$ of Exercise 5.66 satisfy (5.17). Do the conclusions of the last two exercises hold true for this example? *Try it out!*

We can already tell that if we have a common denominator for $h'(z)$ or $g'(z)$ that it is $\prod_{k=1}^{n}(z - e^{it_k})$, and we would guess that the ratio $g'(z)/h'(z)$ would be a product of rational functions. That is all we know now—it is not obvious that this product is a Blaschke product, although we may expect that from Exercise 5.66. The remainder of this section will be devoted to determining that it is, indeed, a Blaschke product, and we will find the order of the product.

Exploration 5.74. Based on the results of Exercise 5.66 and other examples, make a conjecture about the number of Blaschke factors that are in the dilatation of a harmonic function from $\mathbb{D}$ to an n-gon. *Try it out!*

The common denominator of h' and g' is $S(z) = \prod_{k=1}^{n} (z - e^{it_k})$, giving:

$$h'(z) = \frac{P(z)}{S(z)} \qquad \text{and} \qquad g'(z) = \frac{Q(z)}{S(z)}.$$

We want to find P and Q. Each term of the $P(z)$ has the form

$$(v_k - v_{k+1}) \prod_{j=1; j \neq k}^{n} (z - e^{it_j}),$$

a polynomial of degree at most $n-1$. The coefficient of the z^{n-1} term of $P(z)$ is $v_k - v_{k+1}$ for each piece of the sum in (5.14), so this coefficient is $\sum_{k=1}^{n} (v_k - v_{k+1})$, which is 0. Thus $P(z)$ has degree at most $n - 2$. The same argument works for $Q(z)$, because it has the same structure as $P(z)$.

Exercise 5.75. Show that $S(1/\overline{z}) = \left(\frac{1}{\overline{z}}\right)^n (-1)^n \overline{S(z)} \prod_{k=1}^{n} e^{it_k}$. ***Try it out!***

We could write this as

$$\overline{S(1/\overline{z})} = \left(\frac{1}{z}\right)^n (-1)^n S(z) \prod_{k=1}^{n} e^{-it_k}. \tag{5.18}$$

Exercise 5.76. Show that (5.18) holds for the denominator of the derivatives in Exercise 5.66, $2\pi i(1 - z^4)$. ***Try it out!***

We combine (5.17) and (5.18) to get a relationship between $P(z)$ and $Q(z)$. Substituting into $\overline{h'(1/\overline{z})} = z^2 g'(z)$, we see that

$$\overline{h'(1/\overline{z})} = z^2 g'(z)$$

$$\frac{\overline{P(1/\overline{z})}}{\overline{S(1/\overline{z})}} = z^2 \frac{Q(z)}{S(z)}$$

$$\frac{\overline{P(1/\overline{z})}}{\left(\frac{1}{z}\right)^n (-1)^n S(z) \prod_{k=1}^{n} e^{-it_k}} = z^2 \frac{Q(z)}{S(z)}.$$

This leads to

$$z^{n-2} \overline{P(1/\overline{z})} = (-1)^n Q(z) \prod_{k=1}^{n} e^{-it_k}. \tag{5.19}$$

Exercise 5.77. Using (5.19), show that

$$z^{n-2} \overline{Q(1/\overline{z})} = (-1)^n P(z) \prod_{k=1}^{n} e^{-it_k}. \tag{5.20}$$

Try it out!

Since f is orientation-preserving, we know that $h'(z) \neq 0$ in $\mathbb{D}$. This implies that $P(z) \neq 0$ in $\mathbb{D}$. In particular, $P(0) \neq 0$. Substituting 0 into (5.20), we find that the left-hand side is not zero, which forces the degree of Q to be at least $n - 2$. We determined that the degree of Q is at most $n - 2$, so its degree is precisely $n - 2$. Similarly, the degree of P is $n - 2$. Since the degree of Q is $n - 2$, we write

$$Q(z) = z^m \prod_{k=1}^{n-m-2} (z - z_k)$$

to show that Q may have m zeros at the origin and $n - m - 2$ zeros elsewhere (the z_k need not be distinct). Using (5.20), we write

$$z^{n-2} \overline{\left(\frac{1}{\overline{z}}\right)^m \prod_{k=1}^{n-m-2} \left(\frac{1}{\overline{z}} - z_k\right)} = (-1)^n P(z) \prod_{k=1}^{n} e^{-it_k}.$$

The left-hand side of the equation may be written as

$$z^{n-m-2} \frac{1}{z^{n-m-2}} \prod_{k=1}^{n-m-2} (1 - z\, \overline{z_k}),$$

so we see that since the zeros of Q are z_k, the zeros of P are the zeros of $\displaystyle\prod_{k=1}^{n-m-2} (1 - z\, \overline{z_k})$, which are precisely $1/\overline{z_k}$. Now we can see what the Blaschke product is.

Exercise 5.78. Find the relationship between the number of zeros of Q and the number of zeros of P. If Q has degree $n - 2$ with m zeros at the origin and $n - m - 2$ zeros away from the origin, then how many of the zeros of P are at the origin? How many of the zeros of P are away from the origin? *Try it out!*

We summarize the results of our work (as originally proved by Sheil-Small [16, Theorem 1]; see also [5])

Theorem 5.79. *Let f be the harmonic extension of the step function $\hat{f}(e^{it})$ in Definition 5.37. Then*

$$g'(z) = \frac{Q(z)}{S(z)} \qquad \text{and} \qquad h'(z) = \frac{P(z)}{S(z)},$$

where $Q(z)$, $P(z)$, and $S(z)$ are defined as above, and P and Q are polynomials of degree at most $n - 2$. Their ratio $\omega(z) = Q/P$ satisfies $|\omega(z)| = 1$ when $|z| = 1$, so is a Blaschke product of degree at most $n - 2$.

5.8 An Important Univalence Theorem

We examine a theorem of Sheil-Small that tells when the harmonic function in Definition 5.37 is univalent. It is enough to know the location of the zeros of the analytic dilatation $\omega(z) = \frac{g'(z)}{h'(z)}$.

Exploration 5.80. Open the *StarTool* applet. Check the box in front of **Show roots of** $\omega(z)$. You will see extra dots appear in the right-hand pane (the range of the function), as well as a unit circle for reference. The dots give the locations of the zeros of $\omega(z)$. Experiment with the values of p and r to see if there is a relationship between the roots of $\omega(z)$ and the univalence of the star. Do this for various values of n. Does your conjecture agree with what you learned in Exercise 5.66? *Try it out!*

The theorem below was proved by Sheil-Small, [17, Theorem 11.6.6].

Theorem 5.81. *Let f be a harmonic function of the form in Definition 5.37. It is the harmonic extension of a piecewise constant boundary function with values on the m vertices of a polygonal region Ω, so that, by Theorem 5.79, the dilatation of f is a Blaschke product with at most $m-2$ factors. Then f is univalent in $\mathbb{D}$ if and only if all zeros of ω lie in $\mathbb{D}$. In this case, f is a harmonic mapping of $\mathbb{D}$ onto Ω.*

Proof. Suppose that f is univalent in $\mathbb{D}$. If a Blaschke factor is $\varphi_{z_0}(z) = \frac{z_0 - z}{1 - \overline{z_0}z}$ with the constant z_0 not having modulus 1, then, since the dilatation ω is a product of a finite number of Blaschke factors, $\omega(z) \neq 0$ on the unit circle. This is because the zero of the Blaschke factor is z_0, and if $|z_0| = 1$, we get that $\varphi_{z_0}(z) = z_0$ for all z. If ω has a zero at z_0 outside of $\overline{\mathbb{D}}$, then it has a pole at $1/\overline{z_0} \in \mathbb{D}$. If it also has zeros in $\mathbb{D}$, then there are points in $\mathbb{D}$ where $|\omega(z)| < 1$ and other points where $|\omega(z)| > 1$. This implies that the Jacobian of f changes sign in $\mathbb{D}$, which would force the Jacobian to equal 0 at some point in $\mathbb{D}$, contradicting Lewy's theorem, which says that the Jacobian is non-zero since f is locally univalent. Thus there are two possibilities for a univalent f: Either all of the zeros of $\omega(z)$ lie in $\mathbb{D}$, or all lie outside $\overline{\mathbb{D}}$. If the zeros of ω lie outside of $\overline{\mathbb{D}}$, then $|\omega(z)| > 1$ in $\mathbb{D}$ and f has negative Jacobian, contradicting its construction as a sense-preserving boundary function. Therefore, all the zeros of ω lie in $\mathbb{D}$.

Conversely, assume all the zeros of ω lie within $\mathbb{D}$. By the mapping properties of Blaschke products, $|\omega(z)| < 1$ in $\mathbb{D}$. We use the argument principle to show that f is univalent in $\mathbb{D}$ and maps $\mathbb{D}$ onto Ω. Choose $w_0 \in \Omega$. Let C_ε be the path in $\overline{\mathbb{D}}$ consisting of arcs of the unit circle along with small circular arcs of radius ε about the points b_k (the points b_k are the arc endpoints in the domain disk), as shown in Figure 5.17.

If ε is sufficiently small, the image of C_ε will not go through w_0, and will have winding number $+1$ around w_0. Since $|\omega(z)| < 1$ inside C_ε, it follows from the argument principle for harmonic functions that $f(z) - w_0$ has one simple zero inside C_ε (or, put another way,

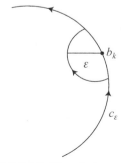

FIGURE 5.17. Tiny circle around b_k.

$f(z) = w_0$ has exactly one solution for $z \in \mathbb{D}$). Thus $\Omega \subset f(\mathbb{D})$. A similar construction with $w_0 \notin \Omega$ to shows that $w_0 \notin f(\mathbb{D})$. Thus f maps $\mathbb{D}$ univalently onto Ω. □

To apply the theorem to the star mappings of Section 5.6, we study the dilatation of the star mappings in detail.

5.9 The Dilatation for Star Mappings

We use Definition 5.63 of Section 5.6 as a starting point. We will build on the basic formula for the functions $h'(z)$ and $g'(z)$, and then will simplify the dilatation $\omega(z)$ as a Blaschke product. By doing so, we completely determine which star functions are univalent.

Using (5.14), (5.15), and Definition 5.63, we find that for the star functions:

$$
\begin{aligned}
h'(z) = {} & \frac{r\alpha^0}{2\pi i} \left(\frac{1}{z - \alpha^0 e^{ip\pi/n}} - \frac{1}{z - \alpha^0 e^{-ip\pi/n}} \right) \\
& + \frac{\alpha}{2\pi i} \left(\frac{1}{z - \alpha^2 e^{-ip\pi/n}} - \frac{1}{z - \alpha^0 e^{ip\pi/n}} \right) \\
& + \frac{r\alpha^2}{2\pi i} \left(\frac{1}{z - \alpha^2 e^{ip\pi/n}} - \frac{1}{z - \alpha^2 e^{-ip\pi/n}} \right) + \cdots \\
= {} & \frac{r}{2\pi i} \sum_{k=0}^{n-1} \alpha^{2k} \left(\frac{1}{z - \alpha^{2k} e^{ip\pi/n}} - \frac{1}{z - \alpha^{2k} e^{-ip\pi/n}} \right) \\
& + \frac{1}{2\pi i} \sum_{k=0}^{n-1} \alpha^{2k+1} \left(\frac{1}{z - \alpha^{2k+2} e^{-ip\pi/n}} - \frac{1}{z - \alpha^{2k} e^{ip\pi/n}} \right)
\end{aligned}
\tag{5.21}
$$

and

$$
\begin{aligned}
g'(z) = {} & \frac{r}{2\pi i} \sum_{k=0}^{n-1} \overline{\alpha^{2k}} \left(\frac{1}{z - \alpha^{2k} e^{ip\pi/n}} - \frac{1}{z - \alpha^{2k} e^{-ip\pi/n}} \right) \\
& + \frac{1}{2\pi i} \sum_{k=0}^{n-1} \overline{\alpha^{2k+1}} \left(\frac{1}{z - \alpha^{2k+2} e^{-ip\pi/n}} - \frac{1}{z - \alpha^{2k} e^{ip\pi/n}} \right).
\end{aligned}
\tag{5.22}
$$

Our goal is to express $\omega(z) = g'(z)/h'(z)$ as the ratio of Blaschke products guaranteed by Sheil-Small's work. We first establish some identities involving sums of the quantities of the type found in (5.21) and (5.22).

If ζ is a primitive mth root of unity, then

$$
\prod_{k=1}^{m} (z - \zeta^k) = z^m - 1.
\tag{5.23}
$$

Exercise 5.82. Prove (5.23). Interpret it geometrically. ***Try it out!***

Exercise 5.83. Let ζ be a primitive mth root of unity. Show that

$$
\prod_{k=0}^{m-1} (z - \zeta^k a) = z^m - a^m.
\tag{5.24}
$$

Hint: Replace z with z/a in (5.23). ***Try it out!***

Now the hard question: How do we add the sums in (5.21) and (5.22), given that their numerators are not 1? As a step toward an answer, we establish

Lemma 5.84. *If ζ is a primitive mth root of unity, then*

$$\sum_{k=0}^{m-1} \frac{\zeta^k}{z - \zeta^k a} = \frac{ma^{m-1}}{z^m - a^m}. \tag{5.25}$$

Exercise 5.85. Prove the lemma, using partial fraction decomposition:

(a) See how (5.24) relates to (5.25).

(b) Since we have n distinct linear factors in the denominator, we can expect to find that

$$\frac{ma^{m-1}}{z^m - a^m} = \sum_{k=0}^{m-1} \frac{a_k}{z - \zeta^k a}.$$

(c) We will find a_{k_0}. By setting $z = \zeta^{k_0} a$, establish that

$$a_{k_0} = \frac{ma^{m-1}}{\prod_{k \neq k_0} (\zeta^{k_0} a - \zeta^k a)} = \frac{m}{\prod_{k \neq k_0} (\zeta^{k_0} - \zeta^k)}.$$

(d) Show that $\displaystyle\prod_{k \neq k_0} (\zeta^{k_0} - \zeta^k) = m\zeta^{-k_0}$. Remember that ζ^{k_0} is an mth root of unity, so $\zeta^{k_0 m} = 1$.

(e) Conclude that $a_{k_0} = \zeta^{k_0}$.

(f) (5.25) follows.

Try it out!

From Exercise 5.64 on page 303 we know that

$$\alpha^{2k+2} e^{-ip\pi/n} = \alpha^{2k+1} e^{i(1-p)\pi/n} \qquad \text{and} \qquad \alpha^{2k} e^{ip\pi/n} = \alpha^{2k+1} e^{-i(1-p)\pi/n}.$$

Thus

$$\frac{1}{2\pi i} \sum_{k=0}^{n-1} \frac{\alpha^{2k+1}}{z - \alpha^{2k+2} e^{-ip\pi/n}} = \frac{-1}{2\pi i} \left(\frac{n e^{i(1-p)(n-1)\pi/n}}{z^n + e^{i(1-p)\pi}} \right)$$

and

$$-\frac{1}{2\pi i} \sum_{k=0}^{n-1} \frac{\alpha^{2k+1}}{z - \alpha^{2k} e^{ip\pi/n}} = \frac{1}{2\pi i} \left(\frac{n e^{-i(1-p)(n-1)\pi/n}}{z^n + e^{-i(1-p)\pi}} \right).$$

Exercise 5.86. Prove that

$$-\frac{1}{2\pi i} \sum_{k=0}^{n-1} \frac{\alpha^{2k+1}}{z - \alpha^{2k} e^{ip\pi/n}} = \frac{1}{2\pi i} \left(\frac{n e^{-i(1-p)(n-1)\pi/n}}{z^n + e^{-i(1-p)\pi}} \right).$$

Try it out!

Combining all these,

$$h'(z) = \frac{n}{2\pi i}\left(\frac{re^{i(\frac{n-1}{n})p\pi}}{z^n - e^{ip\pi}} + \frac{-re^{-i(\frac{n-1}{n})p\pi}}{z^n - e^{-ip\pi}}\right.$$

$$\left. + \frac{-e^{i(\frac{n-1}{n})(1-p)\pi}}{z^n + e^{i(1-p)\pi}} + \frac{e^{-i(\frac{n-1}{n})(1-p)\pi}}{z^n + e^{-i(1-p)\pi}}\right). \tag{5.26}$$

We know from Theorem 5.79 that $\frac{g'(z)}{h'(z)}$ can be written as a Blaschke product. To do this, we will find a common denominator and combine the four terms of $h'(z)$ to see the quotient. First, we find that we can write the common denominator more simply.

Exercise 5.87. Prove that $(z^n - e^{ip\pi})(z^n - e^{-ip\pi}) = (z^n + e^{i(1-p)\pi})(z^n + e^{-i(1-p)\pi})$, which we call $S_n(z)$. ***Try it out!***

Exercise 5.88. Using algebra to show that

$$h'(z) = \frac{n}{\pi S_n(z)}\left(z^n\left(r\sin\left(\frac{(n-1)p\pi}{n}\right) - \sin\left(\frac{(n-1)(1-p)\pi}{n}\right)\right)\right. \tag{5.27}$$

$$\left. + r\sin\left(\frac{p\pi}{n}\right) + \sin\left(\frac{(1-p)\pi}{n}\right)\right).$$

Hint: find a common denominator, simplify, and use properties of $z + \bar{z}$. ***Try it out!***

With similar methods, we can prove that

$$g'(z) = \frac{nz^{n-2}}{\pi S_n(z)}\left(z^n\left(r\sin\left(\frac{p\pi}{n}\right) + \sin\left(\frac{(n-1)(1-p)\pi}{n}\right)\right)\right. \tag{5.28}$$

$$\left. + r\sin\left(\frac{(n-1)p\pi}{n}\right) - \sin\left(\frac{(n-1)(1-p)\pi}{n}\right)\right).$$

If we let

$$c = \frac{\sin\frac{(n-1)(1-p)\pi}{n} - r\sin\frac{(n-1)p\pi}{n}}{r\sin\frac{p\pi}{n} + \sin\frac{(1-p)\pi}{n}} \tag{5.29}$$

we have

$$g'(z) = \frac{nz^{n-2}}{\pi}\left(r\sin\left(\frac{p\pi}{n}\right) + \sin\left(\frac{(1-p)\pi}{n}\right)\right)\frac{z^n - c}{S_n(z)}. \tag{5.30}$$

We combine the result of Theorem 5.81 with the dilatation. When the zeros of the dilatation are in the unit disk, then the harmonic function $f = h + \bar{g}$ that defines the star is univalent. By a straightforward computation, we find that the dilatation of f is

$$\omega(z) = \frac{z^{n-2}(z^n - c)}{1 - z^n c}. \tag{5.31}$$

Exercise 5.89. Verify that Theorem 5.81 holds for the star function. ***Try it out!***

Exploration 5.90. Using the fact that f is univalent when $|c| < 1$, do the following:

(a) Use the *StarTool* applet to explore graphically what relationship there is between n, p, and c.

(b) Given n, find the range of p-values that make $|c| < 1$.

(c) Given p, find the range of n-values that make $|c| < 1$.

Try it out!

Exercise 5.91. For a given n, consider the formula for c in (5.29) to be a function of p alone. Prove that any star configuration is possible; that is, prove that for any r, p can be found to make $|c| < 1$. What ranges of p makes this happen? Conversely, prove that if $r < \cos(\pi/n)$ or $r > \sec(\pi/n)$, p can be chosen to make the function not univalent. Why is this not true for $\cos(\pi/n) \le r \le \sec(\pi/n)$? For more information, see [7, Theorem 4]. *Try it out!*

Small Project 5.92. Refer to Chapter 2. For what values of c is the dilatation a perfect square? Find and describe the associated minimal surfaces. Can they be described as examples of other well-known surfaces? For more information, see [11].

5.10   Open Questions

Large Project 5.93. If you move one vertex of the star, do the same results hold for the relationship between n and p? (For example, take the vertex at r, and move it to $r + \varepsilon$ or $r - \varepsilon$. Is the star still univalent?)

Large Project 5.94. Can we map to any polygon univalently? The star takes full advantage of the symmetry. Once you lose that advantage it is harder to discover whether the zeros of the dilatation have modulus less than 1. This question is known as the *mapping problem*, proposed by Sheil-Small in [16].

Small Project 5.95. Look at a function f that is not univalent. Now look at the set $S \subset \mathbb{D}$ of points on which the function f is univalent. How do you find S? What is its shape? Is it starlike? Is it convex? Is it connected? Is it simply connected? Can $\mathbb{D} \backslash S$ be connected?

Small Project 5.96. We discussed one way of proving that a harmonic function is univalent by looking at zeros of the analytic dilatation $\omega(z)$. In Chapter 4, there is another set of criteria for univalence in Section 4.7. Connect the two. Does one imply the other? How does the work with stars generalize to the approach in Chapter 4? Are there results in this chapter that could not be found using the methods of Chapter 4?

5.11   Bibliography

[1] Lars Ahlfors, *Complex Analysis*, third edition, McGraw-Hill, New York, 1979.

[2] W. N. Bailey, *Generalized Hypergeometric Series*, Cambridge Tracts in Mathematics and Mathematical Physics, No. 32, Stechert-Hafner, Inc., New York, 1964.

[3] Tobin A. Driscoll and Lloyd N. Trefethen, *Schwarz-Christoffel Mapping*. Cambridge University Press, Cambridge, 2002.

[4] Kathy Driver Peter and Duren, Harmonic shears of regular polygons by hypergeometric functions, *J. Math. Anal. Appl.*, 239(1): 72–84, 1999.

[5] Peter Duren, *Harmonic Mappings in the Plane*, Cambridge Tracts in Mathematics, v. 156, Cambridge University Press, Cambridge, 2004.

[6] Peter Duren, Walter Hengartner, and Richard S. Laugesen, The argument principle for harmonic functions, *Amer. Math. Monthly*, 103(5): 411–415, 1996.

[7] Peter Duren, Jane McDougall, and Lisbeth Schaubroeck, Harmonic mappings onto stars. *J. Math. Anal. Appl.*, 307(1):312–320, 2005.

[8] Walter Hengartner and Glenn Schober, Harmonic mappings with given dilatation, *J. London Math. Soc. (2)*, 33(3): 473–483, 1986.

[9] Elgin Johnston, A "counterexample" for the Schwarz-Christoffel transform, *Amer. Math. Monthly*, 90(10): 701–703, 1983.

[10] Serge Lang, *Complex Analysis*, fourth edition, Graduate Texts in Mathematics, v. 103, Springer-Verlag, New York, 1999.

[11] Jane McDougall and Lisbeth Schaubroeck, Minimal surfaces over stars, *J. Math. Anal. Appl.*, 340(1): 721–738, 2008.

[12] Zeev Nehari, *Conformal Mapping*, Dover Publications Inc., New York, 1975.

[13] Earl D. Rainville, *Special Functions*, The Macmillan Co., New York, 1960.

[14] E. B. Saff and A. D. Snider, *Fundamentals of Complex Analysis for Mathematics, Science, and Engineering*, second edition, Prentice Hall, Upper Saddle River, NJ, 1993.

[15] Roland Schinzinger, and Patricio A. A. Laura, *Conformal Mapping: Methods and Applications*, Dover Publications Inc., New York, 2003.

[16] T. Sheil-Small, On the Fourier series of a step function, *Michigan Math. J.*, 36(3): 459–475, 1989.

[17] T. Sheil-Small *Complex Polynomials*, Cambridge Studies in Advanced Mathematics, v. 75, Cambridge University Press, Cambridge, 2002.

[18] Richard A. Silverman, *Complex Analysis with Applications*, Dover Publications Inc., New York, 1984.

[19] Dennis G. Zill and Patrick D. Shanahan, *A First Course in Complex Analysis with Applications*, Jones and Bartlett, Sudbury, MA, 2003.

6

Circle Packing

Kenneth Stephenson (text and software)

Complex analysis is in many ways the ultimate in continuous mathematics. It presents you with a smooth world: continuous variables, infinitely differentiable functions, smooth surfaces, backed up by a complex arithmetic with its power series, line integrals, and all sorts of handy formulas. In this chapter, however, we will develop a different view of the topic. Here the geometry behind analytic functions moves strongly to the fore as we see how one might discretize complex analysis.

An analogy is that of a mountain stream. We normally treat the stream as a continuous medium and use continuous variables and functions to understand it — pressure, velocity, vorticity — modeled, perhaps, by the Navier-Stokes and other partial differential equations. Yet we know that the stream is in fact a collection of individual water molecules, a discrete medium. The macro behavior, the waves, eddies, and currents, emerge from myriad local interactions among the discrete water molecules. So one might ask: what purely local rules of interaction among molecules could lead to the observed global behavior?

In this chapter, we address this issue for continuous analytic functions. Circles will be our molecules, and "packing" conditions will provide the local rules for their interaction. Circle packings, configurations of myriad individual circles, each interacting only with its neighbors, will manifest macro behavior that we will recognize as a version of analyticity. Later in the chapter we will also see that our discrete objects converge under appropriate refinement to the familiar continuous counterparts. In effect, then, circle packing provides a *quantum complex analysis*, one that is classical in the limit.

This is a topic ideal for visualization and experimentation. In fact, it is accessible only through the computer, so you will find yourself tethered to the software more closely than in the earlier chapters. Persevere and you may develop a new appreciation for complex analysis.

I want you to jump right in with an experiment from the foundations of the topic. The software and scripts are accessible through the book's web site. Download the Java application *CirclePack* as a **jar** file to your desktop and double click to start it. Using **File → Load script**, choose the file **Cookie.xmd** and run through it using the **NEXT** button. *Try it out!*

6.1 First Impressions

Circle packings were introduced to analysis by William Thurston in a talk in 1985 ([9]). The context was the conformal mapping of bounded, simply connected, plane regions Ω onto the unit disc $\mathbb{D}$. The central result is the famous Riemann mapping theorem stated in the Appendix, page 355. (Alternate statements: Theorem 3.33, p. 175, and Theorem 4.16, p. 209). For us, the preferred statement is this: Given Ω and designated points w_0, w_1 within Ω, there exists a unique one-to-one analytic mapping F from $\mathbb{D}$ onto Ω so that $F(0) = w_0$ and so that w_1 lies on the image of the positive real axis. Figure 6.1 therefore illustrates the *discrete* Riemann mapping theorem.

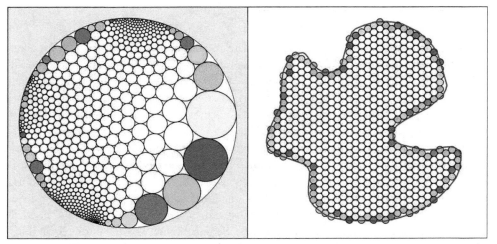

FIGURE 6.1. A conformal map display, domain (left) and range (right).

Figure 6.1 shows the mapping window as displayed by our experimental testbench, the software package *CirclePack*. In such side-by-side displays, the domain circle packing will be on the left, the range packing on the right. In this instance, the domain is $\mathbb{D}$ and the range is the given region Ω. The function itself can be interpreted in various ways. Each circle on the left corresponds to a circle on the right, so we can map circles to circles. Alternately, as we will see later, we can map the triangles formed by triples of circles on the left to those formed by the corresponding triples of circles on the right. The packings provide, in essence, a map from $\mathbb{D}$ onto Ω, a *discrete conformal map*. We will see the details shortly, but assuming you have run **Cookie** and have this example before you on the computer, try clicking the left mouse button on a circle in one packing: both it and the corresponding circle of the other packing will be highlighted (likewise, for faces, using the middle mouse button). ***Try it out!***

We say a few words about *CirclePack* before we go on. Figure 6.2 shows a typical screen. In addition to the mapping window with the two packings, it shows various frames and panels that you will use to manage *CirclePack*. Unlike the software in earlier chapters, *CirclePack* is a Java application rather than an applet: it accesses you computer's file system, it is larger, more comprehensive (and reconfigurable), and it calls shared C/C++ libraries for computationally intense experiments. It also has an important scripting feature:

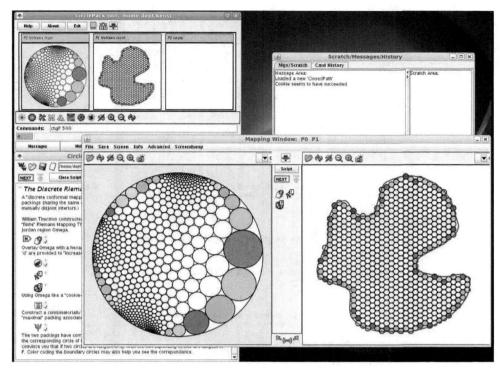

FIGURE 6.2. *CirclePack* in Mapping Window mode.

a *script* refers to a file with prepared sequences of commands. Scripts for this chapter are provided on the book's web site; they have file extension **.xmd** and are indicated in the text by **Name**. To run an experiment, then, start *CirclePack*, from within it load **Name.xmd**, then click through the prepared commands.

To recreate Figure 6.1, run the script **Cookie**. This script takes you through a very structured experiment: simply press the **NEXT** button (or press *enter* when the mouse is in the display canvas) to run through the demonstration steps. ***Try it out!*** (If things become jumbled in *CirclePack*— and they sometimes will! — it may be that you have outrun the computations: click the *up* icon to restart at the beginning of the script, or if things have really gone sour, exit and restart *CirclePack*.)

Even in the passive experimental mode of **Cookie** you can interact with the images by moving, focusing, clicking circles and faces, and so forth, as we describe later. However, it is hoped that as your experience and curiosity grow, you move through the following additional stages to gain more experimental independence:

(1) The **Open Script** button brings up the *CirclePack* Script Window. The commands of the script are laid out linearly and are typically accompanied by text explanations. If you didn't understand what the experiment was purporting to show, read along as you execute the commands. Figure 6.3 illustrates the Script Window, with text, icons that encapsulate the commands, included files, etc.

(2) In the script you can open a command icon by clicking its small "+" box to see the string of commands that it executes. For example.,

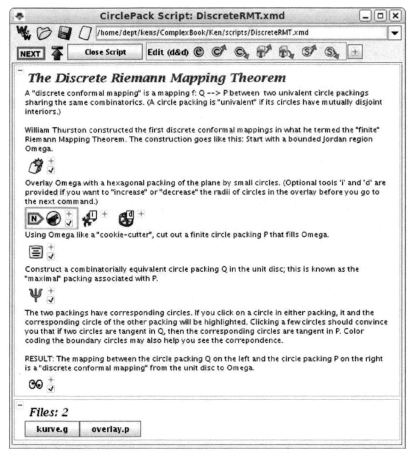

FIGURE 6.3. A typical *script* for *CirclePack*.

`infile_path kurve.g;cookie;disp -w -c -cf b`

tells *CirclePack* to read the path **`kurve.g`** from the script's data section, cookie-cut the portion of the resident packing inside that path, then clear the canvas and display the circles of the resulting packing, with the boundary circles filled. Modify these commands and you can change the action in the experiment. In many scripts, hints are given about alterations you might try. In this first experiment, for example, after overlaying Ω with a hexagonal circle packing, optional commands 'i' or 'd' will increase or decrease, respectively, the sizes of the circles in that overlay. After you have run through the experiment once, re-run it and try these adjustments to see what happens.

(3) If you want to dig deeper, click the **Advanced** button to bring up *CirclePack*'s main control frame, which gives access to its full functionality. The program holds up to three packings, has panels with messages, history, and rudimentary error feedback, **Commands:** lines where you can enter strings of commands, and various droppable icons: drag one onto a packing window and a prepared string of commands will be applied to that packing.

(4) For the ambitious, the Java code for CirclePack is available under the GNU open source license. It includes a **PackExtender** class that allows anyone (with enough work) to create specialized data structures and commands having full access to the core functionality of *CirclePack*. Several existing *PackExtender*s will be used in later experiments.

Figure 6.4 provides an annotated screen shot of *CirclePack* with some of its screens and auxiliary panels. (The screenshots are current as of publication, but *CirclePack* may change over time.)

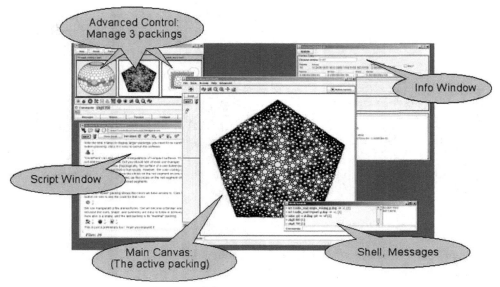

FIGURE 6.4. An annotated screenshot.

Bare-bones scripts are easy to write for yourself, but following, reading, and modifying those of others can be a challenge. The prepared scripts are built for ease of use, so the commands are often long and involved. For example, **disp -w -c** will clear a screen and draw the circles. However, **disp -w -ct3c20 i -cfc180 b** will clear the screen then draw the interior circles thicker and in blue and the boundary circles filled with red. Don't let the complications bother you — go ahead and start your own scripts with simple commands and small steps. When things go wrong, adjust the commands. As you learn *CirclePack* your scripts may become more involved and sophisticated, but also, more effective.

We now give the definitions and notations we will use in our discrete analytic function theory. Although static images are helpful, there is nothing like running experiments, even in automatic mode, to bring the topic to life. The conformal mapping experiment we did may help, and I hope it motivates the theory we will be investigating later. However, we have next a sequence of experiments tailored to display and explain the basics of the theory.

6.2   Basics of Circle Packing

Definition 6.1. A *circle packing* is a configuration of circles with a prescribed pattern of tangencies.

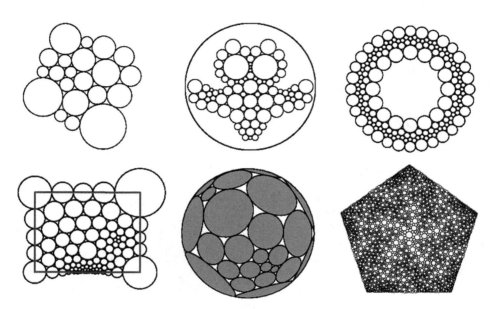

FIGURE 6.5. A small menagerie of circle packings.

Figure 6.5 illustrates a few circle packings from a tour you can take in **Menagerie**. *Try it out!* Among the things you might observe:

(a) The circles form mutually tangent *triples* and hence have triangular interstices.

(b) The packings can live in the Euclidean plane, $\mathbb{C}$, in the hyperbolic plane, $\mathbb{D}$ (the unit disc in $\mathbb{C}$), or on the Riemann sphere, $\overline{\mathbb{C}}$, represented as the unit sphere at the origin in $\mathbb{R}^3$. (See Appendix B.)

(c) The pattern of tangencies shows up in the *carrier* of the packing, the geometric triangles formed by connecting the centers of tangent circles.

(d) This pattern can be simply or multiply connected.

(e) There are *interior* circles, those completely surrounded by their neighbors, and *boundary* circles, those on the edge of the configuration.

(f) There are numerous manipulations, colors, and display options for displaying circle packings and their carriers which are invaluable in highlighting the structures you need to study.

We will have to make formal definitions shortly, but I suggest that you first play around with the CirclePack Owl in **OwlPlay**. You might run sequentially through the commands first, but by pressing the **Script** button you can open the Script Window and follow descriptions associated with the commands. There are suggestions along the way for changes that you can make to see how the packings react. *Try it out!*

Mathematical formalities are essential both for understanding what is going on and for establishing results that you and others can depend upon. Let me lay out the formal objects involved in circle packing. Follow along in **Objects**, where these are illustrated explicitly in turn, and read the descriptions in the script as you proceed. In other words, *Try it out!*

- **Complex, K:** The "prescribed pattern" of tangencies for a circle packing is represented as a triangulation K of a topological surface (technically, a simplicial 2-complex). This is a combinatorial object: K has *vertices* v, *edges* $\langle v, w \rangle$, and *faces* $\langle v, u, w \rangle$ formed by triples of vertices, but does have a geometric setting. It may be finite or infinite, may or may not have boundary, and may be simply or multiply connected. We always assume K is *oriented*, so in every face $\langle v, u, w \rangle$, the vertices are listed in positive (counterclockwise) order, and these orientations are globally consistent, meaning that if faces f and g share a directed edge $e = \langle v, w \rangle$, then e is positively oriented in one face and negatively oriented in the other. The *degree* of a vertex v is the number of edges from v (hence the number of *neighboring* vertices). As K is a "triangulation", every vertex will necessarily have finite degree.

  Of course, I cannot display an abstract complex, but **Objects** begins with a random geometric triangulation, one having no *a priori* connection with any circle packing — it is a simple graph that you might have sketched by hand and entered into a drawing program.

- **Circle Packing, P:** A circle packing for K is a configuration $P = \{c_v\}_{v \in K}$ of circles such that:

  (1) There is a circle c_v associated with every vertex v of K.

  (2) If vertices v and u are neighbors (denoted $v \sim u$ and meaning that $\langle v, u \rangle$ is an edge of K), then the circles c_v and c_u are (externally) tangent to one another.

  (3) If $\langle v, u, w \rangle$ is an oriented face of K, then the circles c_v, c_u, c_w form a triple of mutually tangent circles that is positively ordered.

  You will observe in **Objects** that the circles are indexed sequentially starting at 1. There is an icon tool in the script that displays the *flower* of a selected vertex; namely, the circle itself and the chain of neighboring *petal* circles. The script highlights other distinctions: *Interior* circles are those having closed flowers (their petals wrap completely around them), while the *boundary* circles are those on the edge, with flowers that do not close up. Likewise, boundary and interior edges and faces are illustrated. A circle packing P is said to be *univalent* if its circles have mutually disjoint interiors.

- **Carrier, $\mathrm{carr}(P)$:** The carrier of P is the union of triangles formed by connecting the centers of the triples of circles in P. In the script, the carrier serves to display the abstract complex. P is *locally univalent* if for every interior vertex v the faces in $\mathrm{carr}(P)$ corresponding to the flower of v have mutually disjoint interiors.

- **Label, R:** A label for K is a vector $R = \{r_v\}_{v \in K}$ of putative radii, one for each vertex of K. As you might imagine, circle radii are of central importance in circle packing. They are maintained in *CirclePack* and can be found using the **Info** button. (Radii and center data depend on which of the three geometries you are using.)

If you try to create a circle packing by hand, say by manipulating circle objects in some graphics program, you quickly learn how tightly choreographed their radii must be. The computational heart of circle packing is computing compatible radii; the neutral term "label" is used in place of "radii" because the algorithms typically start with values that do not fit together and then apply iterative adjustments to approach a solution.

- **Angle Sums, θ_R:** Given a complex K and an associated label R, how might you determine if circles with those radii would fit together in the pattern prescribed by K? The key lies with angle sums. For an interior vertex v let $\{v_1, v_2, \ldots, v_k\}$ be a list, in counterclockwise order, of its neighbors. Since the list is closed, it is convenient to write $v_{k+1} = v_1$.

  The faces containing v are $\langle v, v_j, v_{j+1} \rangle$, $j = 1, 2, \ldots, k$. Suppose r, r_j, r_{j+1} are the associated labels taken from R. For each face we could place circles c_v, c_j, c_{j+1} in the appropriate geometry to form a triple. The triangle formed by their centers would have edge lengths that are the sums of radii, and we could appeal to the law of cosines to compute the angle at v in the triangle. In Euclidean geometry, for instance, it would be

$$\alpha(r; r_j, r_{j+1}) = \arccos\left[\frac{(r + r_j)^2 + (r + r_{j+1})^2 - (r_j + r_{j+1})^2}{2(r + r_j)(r + r_{j+1})}\right].$$

  The sum of the angles at v in all the faces containing v is known as the *angle sum*:

$$\theta_R(v) = \sum_{\langle v, v_j, v_{j+1} \rangle} \alpha(r; r_j, r_{j+1}z).$$

  The neighbors of v will wrap precisely around v if and only if $\theta_R(v)$ is an integral multiple of 2π. This is illustrated in **Objects**, but note that we do not need to place any circles to compute an angle sum, as computations are done entirely with labels.

- **Packing Label:** A label R for K is known as a *packing label* if $\theta_R(v)$ is an integral multiple of 2π for every interior vertex v. If P is a circle packing for K (in one of our geometries), then the list R of its radii will necessarily be a packing label. The converse is true for simply connected complexes K: if R is a packing label for K, then one can lay out circles, using R for radii, to form a circle packing P for K.

CirclePack is a platform for creating, manipulating, analyzing, and displaying circle packings. Its computational core lies in computing packing labels, and although the packing algorithms themselves are interesting and revealing mathematics, they are not our target here. We will instead treat *CirclePack* as a black box computational engine so that we can concentrate on the underlying geometry. (Recall that there are several levels of usage for scripts; the more skill you develop, the more you will learn from experiments: (1) click **NEXT** to simply watch the action; (2) press **Open Script** to follow commands and descriptions; (3) modify the given script commands; (4) add new commands to a script; and (5) create new scripts from scratch.)

6.3   Circle Packing Manipulations

We base our introduction to circle packing on a sequence of scripts revealing the objects of study and their principal manipulations available within *CirclePack*.

1. Patterns concentrates on combinatorics. Starting with an empty canvas, a **seed** 7 command creates the simplest type of packing, namely a single flower — an interior circle surrounded by, in this case, seven petal circles. As you click through the script the first time, disregard the occasional repackings that are needed. The emphasis is on building and adjusting the combinatorics. You will see how to add generations of new neighbors, how to add and delete boundary circles, how to cut open or zip up a string of boundary circles, how to double a packing across a boundary segment, how to adjoin two packings, how to cookie-cut one packing from a larger one, and how the process of hex (hexagonal) refinement works. *Try it out!*

2. Geometries demonstrates that *CirclePack* operates in the three standard geometries: Euclidean, represented by the familiar complex plane $\mathbb{C}$; hyperbolic, represented by the unit disc $\mathbb{D} \subset \mathbb{C}$ with the Poincaré metric (density $|ds| = 2|dz|/(1 - |z|^2)$); and spherical, represented by the *Riemann sphere* $\overline{\mathbb{C}}$, namely, the unit sphere $\{(x, y, z) : x^2 + y^2 + z^2 = 1\}$ in $\mathbb{R}^3$. (See Appendix B.) The geometries are said to have *zero, negative,* or *positive* curvature, respectively.

The three geometries are nested, $\mathbb{D} \subset \mathbb{C} \subset \overline{\mathbb{C}}$, with this last inclusion *via* stereographic projection. Moreover, circles in one geometry are circles in the others, though with different centers and radii. The script illustrates changes in geometry, displays circles and faces in each, and shows the effects of automorphisms (i.e., Möbius transformations). The plane is the most comfortable setting for most of us. The sphere is also quite familiar — we live on one — but in circle packing it is by far the most challenging. Hyperbolic geometry is perhaps least familiar, but it is in many ways the richest. One detail to note in hyperbolic geometry is that we can have circles of infinite radius. Namely, a *horocycle* in $\mathbb{D}$ is a circle which is internally tangent to the unit circle. It can be treated in a natural way as a circle of infinite hyperbolic radius whose center is at the point of tangency with the unit circle. *Try it out!*

3. Layout may help you understand the repacking/layout process; though we treat them as black box operations in general, you should see a little about them. In the script, an embedding of a complex K is displayed without any reference to circles. A randomly chosen label R is generated (R will be different each time you run the script) and circles are drawn using the labels as radii. The **repack** command adjusts the labels iteratively based on their values and the neighbor relationships encoded in K. (One option for repacking lets you see the intermediate radii in stages.) Boundary radii (red circles) are fixed, interior radii (blue circles) change to meet the packing condition, but in the display the centers are temporarily fixed. Only when the radii are right (i.e., comprise a packing label), are we ready for the **layout** command to move the circles to consistent locations. First, the two circles for some edge are laid out to be tangent to one another; then the rest of the circles can be placed unambiguously in succession. The original combinatorial pattern is preserved, but now the locations for its vertices are determined by the circle centers. We have a packing! *Try it out!*

4. BdryControl illustrates manipulations critical to our later work: solutions of boundary value problems for radii and boundary angle sums. The first section emphasizes boundary angle sum manipulations. The script provides tools allowing you with the cursor and key presses to prescribe boundary angle sums of π, $\pi/2$, or $3\pi/2$ at various boundary vertices; other boundary vertices have unrestrained angle sums that adjust during repacking. The key lesson here is that while you control some things, the packing pushes back on others — there is a mixture of flexibility and rigidity. Here are some things to think about: Can you prescribe boundary angle sums arbitrarily? Do you encounter problem situations? What happens if you prescribe all but one of the boundary angle sums? In this case, can you compute that remaining one? Can you prescribe angles that will force the region to overlap itself? Some of these issues will come up in Exercise 6.5.

It is interesting to contemplate the boundary situation in the other geometries as well. The same script tools work to set boundary angle sums in hyperbolic geometry but you will see that the constraints of the geometry are different. Try building some quadrilaterals. *Try it out!* On the sphere, there is little we or *CirclePack* can do. As there is no packing algorithm known in spherical geometry, we can not generate or manipulate boundary values. (If you study spherical geometry, you can show that there are boundary constraints. For example, were you to build a spherical quadrilateral you would find that its boundary angle sums exceeded 2π.)

5. Type567 investigates connections between combinatorics and geometry using constant-degree packings, those in which every circle has the same degree (the same number of neighbors). The takehome message involves these intuitive associations: degree six circles, zero curvature (flat); degree five or less, positive curvature; degree seven or more, negative curvature. *Try it out!*

6. Maximal packings are particularly important to our work and are investigated in three scripts. A *maximal packing* for a complex K is a univalent circle packing whose carrier fills the underlying geometric space.

The theorem at the very foundation of circle packing is the Koebe-Andreev-Thurston (KAT) theorem. (We say K *is* a sphere when we mean that it is a *triangulation of a topological sphere*.)

Theorem 6.2 (KAT). *Let K be a sphere. Then there is univalent circle packing P_K for K living on the Riemann sphere, $\overline{\mathbb{C}}$. Moreover, P_K is unique up to Möbius transformations and inversions of $\overline{\mathbb{C}}$.*

We assume our complexes K and their associated packings P have the same orientation, so we will not allow inversions. Also, we say a circle packing for K is *essentially unique* if it is unique up to conformal automorphisms of the underlying space. In our terminology, then, the KAT Theorem states: *Every complex K which triangulates a sphere has a maximal packing P_K of the Riemann sphere, and P_K is essentially unique.*

If K is simply connected but does not triangulate a sphere, then it necessarily triangulates a topological disc, and the central case, both in theory and computations, is when K triangulates a *closed* topological disc (we will say that K *is* a closed disc). Specifically, the conditions are that K be finite, simply connected, and have one connected boundary component (a single chain of boundary edges and vertices).

Theorem 6.3. *If K is a closed disc, then there is a univalent packing P_K for K in the hyperbolic plane whose boundary circles are horocycles. This packing is unique up to Möbius transformations of $\mathbb{D}$.*

MaxPackHyp explores this theorem. As this is key to much of our later work, I recommend that you go through this script first. In addition to seeing the practical computations in action, you might also glean the intuition behind the formal proof.

MaxPackSph should be run next. It illustrates that the KAT Theorem and Theorem 6.3 are equivalent, each implies the other. Since packing computations cannot (yet) be carried out directly in spherical geometry, the hyperbolic computations are a prerequisite.

MaxPackInf is the third script. If K is simply connected but neither a sphere nor a closed disc, then it is necessarily infinite (possibly having boundary). This third script investigates some infinite maximal packings. These are only suggestive, of course, since one cannot compute with or display infinitely many circles. However, the intuition is theoretically sound. (See [1].)

These three scripts provide a quick course in maximal packing. *Try it out!* Since in experiments all simply connected complexes K are finite; *CirclePack* has a single command **max_pack** (or the droppable icon with the big maize 'M') which will automatically compute a normalized maximal packing P_K in the sphere or the unit disc, as appropriate. Maximal packings also exist for non-simply connected complexes; you will consider multiply-connected cases in Exercise 6.6 and triangulations of the torus in Exercise 6.23.

7. **Branching** illustrates a further source of control beyond the boundary conditions we have studied: namely, branching. A circle c_v is *branched* if it is an interior circle whose neighbors wrap more than once around it. For *simple* branching, the angle sum at v is 4π and the neighbors wrap twice around it. In general, if the angle sum is $2\pi n$, the neighbors wrap around n times and the circle is said to have *branch order* $n - 1$. Investigate the local geometry at branch circles in the script. *Try it out!*

8. **InputOutput** gives you a tour of *CirclePack* input and output operations. After you work with *CirclePack* for a while, you may want to read and write packings, load and save scripts, create PostScript images and *.jpg screendumps, etc. *Try it out!*

6.4   Discrete Function Theory

With experience from the previous scripts, you should be ready to investigate discrete analytic function theory. To see how *CirclePack* displays functions, repeat **Cookie**, which creates the discrete conformal mapping of Figure 6.1.

The script begins with a curve defining a Jordan region Ω. A regular hexagonal packing H_ε (i.e., all circles have the same radius ε) overlays Ω in the canvas on the right. Using Ω like a cookie cutter, a packing P of Ω is cut from H_ε. The associated complex K has a maximal packing $Q = P_K$ in $\mathbb{D}$, which *CirclePack* computes and displays in the canvas on the left. We now have, more or less, the situation as pictured in Figure 6.1. This is how we will represent our functions $f : Q \longrightarrow P$ in *CirclePack*: domain on the left, range on the right.

Definition 6.4. A *discrete analytic function* is a map $f : Q \longrightarrow P$ between circle packings Q and P that preserves tangency and orientation.

As a practical matter, our discrete analytic functions will have domain and range packings that share a common complex K (which may vary from one example to another). If v is a vertex of K with circle C_v in Q, then $c_v = f(C_v)$ is the circle for v in P. If $\langle v_1, v_2, v_3 \rangle$ is a positively-oriented face in K, then $\langle C_{v_1}, C_{v_2}, C_{v_3} \rangle$ and $\langle c_{v_1}, c_{v_2}, c_{v_3} \rangle$ form mutually tangent, positively-oriented triples of circles in Q and P, respectively.

Because of Theorem 6.3, we have already generated data for several discrete analytic functions. In particular, every packing P having a simply connected complex K is the range of a discrete analytic function: just build its associated maximal packing P_K, and violà, $f : P_K \longrightarrow P$ is a discrete analytic function. You might ask "What is the big deal about functions?" The function is not about the range (alone); it is about how the domain changes into the range — it is about the *mapping*. Which circles grow, which shrink? How does the boundary behave? Do the circles remain disjoint, or do some come to overlap? How does the geometry change under f? Which properties of a packing are preserved? which ones change?

Subsequent scripts will teach you how *CirclePack* can aid in generating mappings and investigating their properties. With domain and range as in Figure 6.1, you can click the left mouse button on a circle in one packing and both it and its counterpart in the other packing will be highlighted. Check a few pairs of tangent circles in Q to see that their counterparts are tangent in P. Also, the circle centered at 0 in Q corresponds to the circle of P that is also at 0: this is what we mean when we say $f(0) = 0$. Clicking the middle mouse button will highlight corresponding faces of Q and P. *CirclePack* provides various buttons for refreshing, zooming, etc., that you should experiment with. Your goal is to be able to probe the packings to discover properties of the mapping f.

6.4.1 Conformal Maps

The **cookie** method demonstrated in **Cookie** is a practical way to create *discrete conformal* mappings, so called because they parallel the conformal (i.e., one-to-one analytic) mappings of the classical theory.

There is a second approach, modeled on the Schwarz-Christoffel methods seen in Section 5.2. The aim is to find a conformal map of the upper half plane or the unit disc to the region bounded by a given polygonal path Γ.

The classical formula for the unit disc is (5.3). Having such a formula for f suggests a straightforward computation, but that is not the case. The β_j are known in advance from the corner angles of Γ, but the *prevertices*, the ζ_j in the formula, are the points mapping to those corners, and they are not known. Where does the turning occur? That is the real challenge with Schwarz-Christoffel. In the discrete version, one confronts this issue head on. Figure 6.6 illustrates a polygonal packing: the boundary edges of the carrier define a polygonal curve Γ. The problem, of course, is that in general you start with Γ and a complex K, not with a packing; you have to find the packing for K whose boundary edges most closely approximate Γ. See how it goes in the next exercise.

Exercise 6.5. First a little Euclidean geometry. Given a polygonal path Γ, its *turning angles* reside at the corners: when you travel in the positive direction along Γ and you come

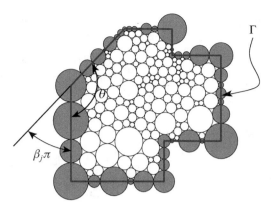

FIGURE 6.6. Packing of a polygonal region.

to a corner, the angle you turn through to start the next edge is by convention taken as positive if you turn left, negative if you turn right. Compute the algebraic total of all the turning angles for a simple closed polygonal path Γ.

In Schwarz-Christoffel notation, the jth turning angle is $\beta_j \pi$. We can obtain the boundary angle sum θ_j at the corner by observing (see Figure 6.6) that $\theta_j = \pi - \beta_j \pi$. Conditions on the total of turning angles therefore give conditions on the total of all boundary angle sums: this will constrain your experiments, so you should work this out before jumping into the script.

DiscreteSC starts with a worked out example for Γ a square. It then challenges you with a slightly more complicated Γ. You will have to open and read the script to see how the machinery has been set up and how to manipulate the packing with mouse and key presses. *Try it out!* .

Exercise 6.6. The Riemann mapping theorem says that every simply connected proper subdomain Ω of the plane is conformally equivalent to a particular model simply connected domain, namely, to the open unit disc. This has been extended to certain multiply connected domains Ω and various model multiply connected domains. For instance, suppose $\Omega \subset \mathbb{C}$ is a connected open set bounded by n disjoint Jordan curves (a *Jordan* curve is a closed curve without self-intersections.). It is known that Ω can be mapped conformally onto a *circle domain*, a plane domain bounded by a finite number of circles. In contrast to the simply connected case, where there are several numerical methods for computing the classical maps, such as Schwarz-Christoffel, the multiply connected case has few methods. In the discrete setting, however, maps to circle domains are quite elementary. See **CircleDomains**. *Try it out!*

6.4.2   Basic Function Theory

Let's look at scripts in which we construct and study additional types of functions. We start with maps of the unit disc into itself which are important in classical function theory and convenient for us since we can work in hyperbolic geometry. **Schwarz** starts with a somewhat random and small range packing P in $\mathbb{D}$. Incrementally increase the hyperbolic boundary radii to infinity (in practice, 5 is sufficiently large!) and P morphs into its maximal packing P_K. The map $f : P_K \longrightarrow P$ is a discrete analytic function of $\mathbb{D}$ into itself. We introduce some notation:

Definition 6.7. Let $f : Q \longrightarrow P$ be a discrete analytic function, where Q and P are packings with common complex K. The associated *ratio function* $f^\# : K \longrightarrow \mathbb{R}$ is defined by

$$f^\#(v) = \frac{\text{radius}(f(C_v))}{\text{radius}(C_v)}, \quad v \in K,$$

where C_v is the circle for v in Q and $f(C_v)$ is the corresponding circle in P.

$f^\#(v)$ is the stretch factor at v, the amount that the associated circle is stretched in going from Q to P. We can now illustrate our first piece of *discrete function theory*, the discrete Schwarz lemma.

Theorem 6.8 (Discrete Schwarz Lemma). *Let $f : P_K \longrightarrow P$ be a discrete analytic self-map of the unit disc with $f(0) = 0$. Then $f^\#(0) \leq 1$. Equality holds if and only if f is a rotation; that is, $P = \lambda P_K$ for some complex λ with $|\lambda| = 1$.*

Exercise 6.9. Run through **Schwarz** again, but more carefully. *Try it out!* Can you see the discrete Schwarz lemma in action? The ratio function plays the role in the discrete theory that the absolute value of the derivative, $|F'|$, plays in the classical theory, so the discrete Schwarz lemma precisely parallels its classical counterpart, even up to the equality statement. (See A.21.)

Exercise 6.10. There is a classical extension of Schwarz's lemma known as the Dieudonné-Schwarz lemma. If $F : \mathbb{D} \longrightarrow \mathbb{D}$ is analytic and $F(0) = 0$, then by Schwarz's lemma, $|F'(0)| \leq 1$. According to this extension, there is in fact a constant $\mathcal{C}, 0 < \mathcal{C} < 1$, independent of F, so that $|F'(z)| \leq 1$ whenever $|z| \leq \mathcal{C}$. **Dieudonne** sets up experiments so you can estimate the constant: you manipulate a range packing P in $\mathbb{D}$; with color coding, *CirclePack* shows you the circles in P_K whose (Euclidean) radii become smaller in P. How close can these circles come to the origin in P_K? $\mathcal{C}$ is larger than 1/2 but quite a bit smaller than 1. *Try it out!* (The correct value of $\mathcal{C}$ is given at the end of the script.)

The discrete Schwarz lemma is an example of a "monotonicity" result — if you decrease the boundary radii of Q to those of P, then the radius of the circle at the origin also decreases (after repacking). Details are given in [7], but the basic ideas can be seen in:

Exercise 6.11. Consider a mutually tangent triple of circles in Euclidean geometry as illustrated in Figure 6.7. Using the law of cosines, show that if the radius of circle C_v is increased, then the angle α at C_v will decrease while the angles β, γ at the other two circles will increase.

Exercise 6.12. Using the previous fact, prove:

Theorem 6.13 (Discrete Maximum Principle). *Let $f : Q \longrightarrow P$ be a discrete analytic function between Euclidean packings Q and P, and suppose Q is locally univalent. Then the ratio function $f^\#(v)$ attains it maximum at a boundary vertex v.*

Analogous results hold in hyperbolic geometry and are central to the computation of packing radii. The failure of monotonicity in spherical geometry may explain the lack of a

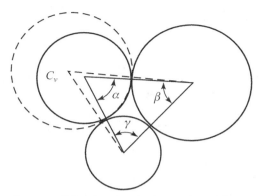

FIGURE 6.7. Monotonicity in Euclidean triples.

packing alogorithm there.

Exercise 6.14. Show with a computation in spherical geometry (using the spherical law of cosines) that the monotonicity of Exercise 6.11 can fail in spherical geometry. What causes the failure? *Try it out!*

The analogy between $f^{\#}$ and $|F'|$ is not perfect. While $|F'|$ is zero at branch points of F, $f^{\#}$ can never be zero. Nevertheless, the fact that discrete functions correctly realize the geometry underlying branching can be seen in the following Exploration.

Exploration 6.15. An important class of classical analytic self-maps of the disc, the *finite Blaschke products*, was introduced in Chapter 4. With the explicit product formula shown in (4.6), a Blaschke product $B : \mathbb{D} \longrightarrow \mathbb{D}$ has properties subject to direct computation. For instance, each factor in the formula accounts for one zero of B, and B extends to a continuous map on the unit circle, $\partial\mathbb{D}$, which maps $\partial\mathbb{D}$ n times around itself, where n is the number of factors in B.

However, there is a more geometric characterization of finite Blaschke products: they are precisely the *proper* analytic maps f of $\mathbb{D}$ onto itself. "Proper" means that whenever $\{z_j\}$ is a sequence of points in $\mathbb{D}$ converging to the unit circle, so $|z_j| \to 1$, then the images $\{f(z_j)\}$ also converge to the unit circle, so $|f(z_j)| \to 1$. By orientation, B wraps the boundary around itself some number n times, and from this the argument principle implies that B will have $n - 1$ branch points inside the disc. These geometric properties lead you to *discrete* finite Blaschke products, as you will see in **Blaschke**. *Try it out!* Zooming in to branch points gives you a chance to demystify their local geometric behavior.

Analytic functions present a fascinating mixture of rigidity within flexibility, which may be why they have attracted such attention for two centuries now. Though you can manipulate some features, you also have to live with the consequences. There are two notions that give some insight into their behavior: *harmonic measure* and *extremal length*.

6.4.3   Harmonic Measure

Let Ω be a Jordan region in the plane (one bounded by a Jordan curve) and let A be a subset of its boundary, $A \subseteq \partial\Omega$. There is a unique harmonic function h defined on Ω which goes to 1 as z approaches a point of A and goes to 0 as z approaches a point of $\partial\Omega \backslash A$. (We suppress technical details about measurability and continuity.) h is known as the *harmonic measure function* and written as $h(z) = \omega(z, A, \Omega)$. (Notations such as $\omega(z, A, \Omega)$ may be new to you: this denotes a function whose value depends on three quantities, the point $z \in \Omega$, the boundary arc A in $\partial\Omega$, and the region Ω itself. Hold any of these fixed, and it becomes a function of the others.)

We also say that $\omega(z, A, \Omega)$ is the *harmonic measure* of A at z with respect to Ω. This is called a measure because if you keep z and Ω fixed, then it is a probability measure on the variable set A. Observe that $\omega(z, A, \Omega) \in [0, 1]$, with $\omega(z, \emptyset, \Omega) = 0$ and $\omega(z, \partial\Omega, \Omega) = 1$, and that if A_1, A_2 are disjoint in $\partial\Omega$, then $\omega(z, A_1 \cup A_2, \Omega) = \omega(z, A_1, \Omega) + \omega(z, A_2, \Omega)$.

There are two nice properties of harmonic measure in the classical setting:

(1) It is sometimes easy to compute: If Ω is the unit disc $\mathbb{D}$ and z is the origin, then $\omega(0, A, \mathbb{D}) = \text{arclength}(A)/(2\pi)$. That is, harmonic measure gives the proportion of the unit circle that A occupies as seen from the origin.

(2) It is a *conformal invariant*; that is, it is invariant under conformal maps. More precisely, suppose $F : \Omega_1 \longrightarrow \Omega_2$ is a conformal map (i.e., a conformal homeomorphism) between two domains. Under quite weak conditions, it is known that F can be extended so it also maps the boundary onto the boundary, $F : \partial\Omega_1 \to \partial\Omega_2$. Given $A_1 \subseteq \partial\Omega_1$, define A_2 to be its image $A_2 = F(A_1)$ in $\partial\Omega_2$, and suppose $z_1 \in \Omega_1$ and $z_2 = F(z_1) \in \Omega_2$. Then $\omega(z_1, A_1, \Omega_1) = \omega(z_2, A_2, \Omega_2)$.

If you put these two properties together with the Riemann mapping theorem, you can, in theory, find the harmonic measure function for any simply connected region. How might this be useful? Here's one of many applications. (See Section 3.11.) Suppose a homogeneous thin metal plate is cut out in the shape of Ω. Suppose a segment A of the boundary is held at 100 degrees centigrade, while the remainder of the boundary is held at 0 degrees. After the plate has had time to reach thermal equilibrium, so it has a steady state temperature, what is the temperature at a point z in the interior of the plate? It is evidently somewhere between 0 and 100, it would approach 100 as z comes closer to A, and 0 as z gets closer to other parts of the boundary. But what is the temperature at z? The answer is $100 \cdot \omega(z, A, \Omega)$.

In **HarmMeasure** we investigate the discrete analogy. I can give the definition now, since it is motivated directly by property (1). Suppose K triangulates a closed topological disc, A is a chain of edges in ∂K, and $v \in K$ is an interior vertex of K. Let P_K be the maximal packing for K in $\mathbb{D}$ that has the circle for v centered at the origin. An edge $e \in A$ corresponds to an arc $\alpha_e \subset \partial\mathbb{D}$ between the centers of the two horocycles forming e.

Definition 6.16. Given K, A, and v as described, the *discrete harmonic measure* of A at v with respect to K is

$$\omega(v, A, K) = \sum_{e \in A} \text{arclength}(\alpha_e)/(2\pi).$$

So discrete harmonic measure is defined directly on K. The invariance of property (2) above in the discrete setting is merely due to the definition. In the script, however, you will see its geometric nature exhibited, much as you would see the temperature distributed about a plate in the application. Experiments are quite convincing that the same geometry is at work in both the discrete and the classical settings: in **HarmMeasure** we discuss harmonic measure and exit probabilities, comparing the classical and discrete versions for a region, and we illustrate the *extension of domain principle*.

6.4.4 Extremal Length

Extremal length is a more nuanced measure of conformal shape. Start with a Jordan region Ω, but this time suppose we are given disjoint closed subarcs A and B of its boundary; the triple $\mathcal{Q} = \{\Omega; A, B\}$ is known as a *conformal quadrilateral*.

A classical theorem states that there exists a conformal mapping $F : \Omega \longrightarrow R$, where R is a Euclidean rectangle, so that F extends continuously to the boundary and so that the images $F(A)$, $F(B)$ are opposite ends of the boundary of R. F is unique up to Euclidean similarities. (You may be reminded of the Schwarz-Christoffel situation, as investigated earlier, but this is different: the boundary points mapping to corners are specified, but the size of the image rectangle is not set in advance.)

Figure 6.8 illustrates the analogous discrete result you will see in **ExtLength**. By a rigid motion, image rectangles can be positioned as shown here, with the images of A and B on the left and right, respectively. The *aspect ratio* of R is width over height, $\text{Aspect}(R) = W/H$.

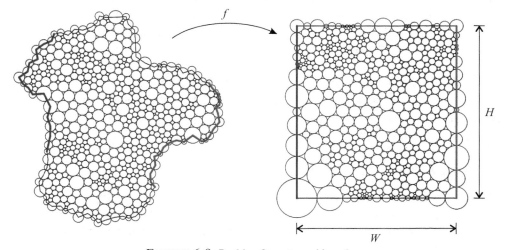

FIGURE 6.8. Packing for extremal length.

Continuing our discussion of the classical setting, $\text{Aspect}(R)$ is known as the *extremal distance* from A to B, relative to Ω, or the *extremal length* of the quadrilateral $\mathcal{Q}$, denoted $\text{EL}(\mathcal{Q})$. As with harmonic measure, there are two key properties:

(i) Extremal length is sometimes easy to compute: If Ω is a Euclidean rectangle width W and height H, and A and B are its ends, then $\text{EL}(\{\Omega, A, B\}) = W/H$.

(ii) Extremal length is a *conformal invariant*; that is, if $Q_1 = \{\Omega_1; A_1, B_1\}$ and $Q_2 = \{\Omega_2; A_2, B_2\}$ are conformal quadrilaterals and $F : Q_1 \longrightarrow Q_2$ is a conformal mapping such that $F(A_1) = A_2$ and $F(B_1) = B_2$, then $\mathrm{EL}(Q_1) = \mathrm{EL}(Q_2)$

As with harmonic measure, there are physical interpretations. Suppose Ω is a uniform conducting metal plate in the plane, the boundary segment A is attached to a 1 volt power supply, the boundary segment B is grounded, and the remaining segments of the boundary are insulated. Then the current flow between A and B in amperes will be proportional to $\mathrm{EL}(\{\Omega, A, B\})$, the constant depending on the properties of the metal.

Discrete analytic mappings to rectangles having designated boundary vertices as corners are easy to compute by methods you have already seen in **Layout**: you specify that boundary vertices have angle sum π except for the four that are to be the corners, and these you set to $\pi/2$. **ExtLength** illustrates the construction of Figure 6.8 and explains how *CirclePack* reports the extremal length when you have a rectangle. You are asked to modify the constructions to answer a question about current flow. You might then test your intuition with an extremal length analogue of the extension of domain principle we saw with harmonic functions.

One last script for this section, **MiscFtnTheory**, illustrates the range of miscellaneous behaviors you might encounter with discrete analytic functions. They may provide additional insights you can carry back to the classical setting.

6.5   Function Construction

We have hands-on control of our discrete functions, but this is not the formula-driven and algebraic control available in the classical setting. In the discrete world, we have limited ability to add or multiply functions, and essentially no way to consider compositions of the type studied in Chapter 1. Discrete functions require some geometric hook. We have seen examples: to get maximal packings in $\mathbb{D}$, we send boundary radii to infinity; to build Euclidean polygons, we control their boundary angle sums; to create Blaschke products, we designate branch circles. The next sequence of scripts give additional examples, some involving techniques, others relying on special circumstances or outright tricks. There is plenty of room for your creativity here.

6.5.1   Discrete Exponentials

Doyle challenges you to adjust parameters in a family of intriguing circle packings known as Doyle spirals. With patience you can build a range packing that clearly mimics the classical exponential function. *Try it out!* Read the script for guidance, but then experiment on your own. Examples of good parameter values are provided at the end of the script.

6.5.2   Discrete Rational Functions

A classical *rational* function has the form $F(z) = P(z)/Q(z)$, where $P(z)$ and $Q(z)$ are complex polynomials with no common factors. Projecting stereographically to the sphere in both domain and range, F may be considered as a map from the Riemann sphere to itself, $F : \overline{\mathbb{C}} \longrightarrow \overline{\mathbb{C}}$, and that will be our view. (See Definition B.11.) If $P(z) = 0$, then

$F(z) = 0$, while if $Q(z) = 0$, then $F(z) = \infty$. The behavior at ∞ in the domain depends on relative degrees: if $\deg(P) > \deg(Q)$, then $F(\infty) = \infty$, if $\deg(P) < \deg(Q)$, then $F(\infty) = 0$, and if $\deg(P) = \deg(Q)$, $F(\infty)$ is a non-zero complex number.

The sphere is compact, and $F : \overline{\mathbb{C}} \longrightarrow \overline{\mathbb{C}}$ is an open continuous map, so key properties of rational maps (including polynomials) are purely topological: F has a constant *valence*, $N > 0$; that is, for a point $p \in \overline{\mathbb{C}}$, $F^{-1}(p)$ has N points (counting multiplicities at branch points, which may include infinity). The valence N satisfies $N = \max\{\deg(P), \deg(Q)\}$. There are precisely $2(N - 1)$ branch points. As you build examples you should watch for these properties directly.

Polynomial gives a general method for creating discrete polynomials in the plane. Treating them as rational maps, however, suggests a broader notion: A map of the form $F = \phi \circ P$, where P is a polynomial and ϕ is a Möbius transformation, is said to be a *polynomial-like* rational map. These are characterized by having half of their branching at a single point, and this is exploited in **PolyLike** to build examples.

To get a taste for more typical rational maps, **Rational** illustrates an example constructed using *Schwarz triangles*, with valence seven and with twelve simple branch points. This is the discrete analogue of the classical rational function

$$F(z) = z^2(z^5 - 7)/(7z^5 + 1).$$

[This formula was found by Edward Crane; the one given in [2] and [7] is not correct.] Lovely as this discrete example is, it highlights a key capability that is missing in our discrete theory: we have no algorithm for computing packing labels directly in spherical geometry. See Exercise 6.25 at the end of the chapter for more on the geometry behind this Schwarz triangle construction.

6.5.3   Ratio Function Constructions

If $f : Q \longrightarrow P$ is a discrete analytic function, then the associated local scaling, the analogue of $|F'|$ for analytic functions, is given by its ratio function $f^\#$. (See Definition 6.7.) If we wish to build a discrete analogue f for some classical function F, we might begin from this differential level, mimicking $|F'|$.

RatioBuild builds a discrete analogue for the classical function $F(z) = (1+z)^2$ on the unit disc. You should compare the mapping image the script generates to Figure 4.12, page 208, created using the applet *ComplexTool*. Modify the script to build the other analytic examples illustrated in Chapter 4.

Ratio function constructions also give an approach to entire functions. We illustrate in **Erf** a function central to statistics, the *error function* $\mathrm{erf}(z)$ defined by

$$E(z) = \mathrm{erf}(z) = \frac{2}{\sqrt{\pi}} \int_0^z e^{-\zeta^2}\, d\zeta.$$

When z is real, $z = x$, the derivative of $E(z)$ is the Gaussian distribution, the famous bell-shaped curve. You may recall that this function has no closed form anti-derivative, hence the integral representation.

The construction of the discrete error function illustrates several construction issues.

(1) E is an entire function defined on the whole complex plane. Discrete entire functions would likewise be defined on infinite complexes, as the discrete exponential we defined earlier on the regular hexagonal packing. Experiments and computations, however, are restricted to finite complexes. Our approach relies on the method of exhaustion: we mimic E using finite packings that take up successively larger, but finite, portions of $\mathbb{C}$.

(2) E has infinite valence: for every $w \in \mathbb{C}$, $E^{-1}\{w\}$ has infinitely many points. In fact, its mapping behavior is fascinating to watch. Early in the construction, it may appear that the image omits two points, w_1, w_2, much as the exponential omits the image value 0. That, however, would contradict the little Picard theorem, which tells us that a nonconstant entire function can omit at most one value in the complex plane. Watch more stages in the construction to learn how our growing function manages to cover everything. Though not omitted, w_1 and w_2 are what is known as *asymptotic values*, that is, there are paths in the domain whose images under E converge to w_1 or w_2. Can you find such paths in the *CirclePack* experiments? ***Try it out!***

(3) Does the sequence of packings in the method of exhaustion ultimately converge to an infinite circle packing? Each stage of construction is finite, so this is a convergence issue that remains unresolved. The script has both positive and negative indications. What do you think? If there is a limit packing, this construction would answer in the negative one of the earliest questions in circle packing:

Question 1. Are the regular hexagonal packings and the Doyle spirals (see Section 6.5.1) the only locally univalent hexagonal circle packings in the plane?

6.5.4   Harmonic Functions

According to Theorem 4.31, harmonic mappings F on the unit disc may be written as $F(z) = H(z) + \overline{G}(z)$ where H and G are analytic in $\mathbb{D}$. While the addition occurring here is not a geometric feature which can be carried directly by our circles, there is a discrete fact that captures the local behavior that we will discuss in a moment. The upshot is that if one just jumps in with discrete functions, the fundamental global phenomena appear to persist. In this section we start with the mechanics, see some examples parallel to ones in Chapter 4, then set up a few lines of open experimentation for you.

A run through **HGintro** shows you how to create *discrete harmonic functions*. ***Try it out!*** The way *CirclePack* operates, h and g will be discrete analytic functions on the disc, so within each example they will share a complex K and have its maximal packing P_K as their common domain. There will be no need to view P_K, but in **HGintro** h is taken as the identity function, so P_K itself appears in **p0** as the range of h. ("**p0**" refers to **Pack 0** in *CirclePack*.) The image of g appears in **p1**. On execution of **h_g_bar**, the sum $h + \overline{g}$ is displayed in **p2**. *CirclePack* requires that the range packings for h and g be Euclidean, and a key feature in **HGintro** is that their boundary radii are identical. The script starts with g also the identity, so the initial f is boring. However, if you press the letter **b** while the cursor is on an interior circle of h, you will cause the circle to become branched for g. The image of f is immediately more interesting. With additional branch points you will observe some of the behavior described in Chapter 4. Since the boundary radii are identical for h and g, the complex dilatation function $w(z)$ is a discrete finite Blaschke product (see Sections 4.6 and 5.7).

The *CirclePack* mechanics behind **h_g_bar** are straightforward: for each vertex, the center in **p0** is added to the conjugate of the center in **p1**, giving a location that is plotted in **p2**. With the centers in place, the edges are drawn in **p2** to give the triangulation. In **FtnAddition** we display, strictly for purposes of comparison, the analogous addition $h + g$ (no conjugation bar on g). The result $f = h + g$ in **p2** will approximate the classical sum of these analytic functions. However, there is no sensible way to assign radii in **p2** and there is no packing. So in this algebraic situation, the circles are no longer doing the work for you.

A last script, **HGtrials**, sets up various adjustments you can implement with mouse and keys for more open-ended experiments. You can also add actions of your own devising. Certain keys will increase boundary radii of g, which introduces you to another feature: faces of f (in **p2**) are colored blue if f reverses their orientation. We know from the classical theory that this can happen only if $|G'|/|H'| > 1$ on some part of $\partial \mathbb{D}$. In the discrete setting, absolute values of derivatives are replaced by ratio functions; if you return to the definition of the ratio function and recall that h and g share the same domain packing, you see that the discrete condition becomes $g^{\#}/h^{\#} > 1$ on some vertices of ∂K. Therefore, by manipulating the relative sizes of the boundary circles in **p0** and **p1**, we affect the orientation of $f = h + \overline{g}$.

Exercise 6.17. Verify this bit of local geometry, which may account for the properties of these discrete harmonic functions. Consider Euclidean triangles T and t with vertices located at $\{Z_1, Z_2, Z_3\}$ and $\{z_1, z_2, z_3\}$, respectively. Assume that each is positively oriented. Form a new triangle Δ with vertices $\{w_1, w_2, w_3\}$, where $w_j = Z_j + \overline{z_j}$, $j = 1, 2, 3$. (The bar denotes complex conjugation of z_j.) The issue is whether Δ is positively oriented.

In general, the answer is "no", but here is a pretty fact that you should try to prove: If T and t happen to be formed by mutually tangent triples of circles $\{C_1, C_2, C_3\}$ and $\{c_1, c_2, c_3\}$, and if radius$(c_j)/$radius$(C_j) \leq 1$, $j = 1, 2, 3$, then Δ is indeed positively oriented. In other words, the classical boundary condition for preserving orientability of $H + G$, namely $|G'|/|H'| < 1$, carries over to the discrete case at the local face level.

Use your favorite computer math package to build examples. Show first that Δ can have reverse orientation in general. Even more is true: show that even if you arrange that the side lengths of t are individually less than the corresponding side lengths of T, Δ may still have reverse orientation. Mysteriously, when the side lengths are sums of radii, and the radii for t are individually less than those for T, then Δ remains positively oriented!

6.6 Convergence

We have introduced the local geometry of circle packings, moved on to defining mappings between packings, and investigated the emergent global geometric properties. The claim has been made — with convincing evidence, I hope — that these mappings deserve the name of discrete analytic function. Can we move beyond mere parallels?

The answer is a resounding yes. Though granularity prevents discrete analytic functions from actually being analytic functions, with a suitable refinement process for improving resolution they can be made to approximate analytic functions. Circle packing began with exactly this approximation issue with the celebrated Rodin-Sullivan theorem explained below. Let's see it in action first.

DiscreteRMT picks up from our earlier script **Cookie**. It builds a range packing P using **cookie** to cut a region Ω from a regular hexagonal packing. The maximal packing Q having the same combinatorics becomes the domain. A standard *CirclePack* normalization puts the circle for the α vertex at the origin in both Q and P, and a γ vertex on the imaginary axis in each. The map $f : Q \longrightarrow P$ is then a discrete conformal map.

At issue: How well do the discrete mappings approximate the classical conformal mapping $F : \mathbb{D} \longrightarrow \Omega$ guaranteed by the Riemann mapping theorem? This script provides options to change the radii in the hexagonal overlay. Making those circles smaller — hence using more of them — is an example of *refinement*. The effectiveness of such refinements is the content of the foundational theorem by Burt Rodin and Dennis Sullivan, paraphrased here.

Theorem 6.18 (Rodin and Sullivan, [6]). *Let Ω be a bounded simply connected domain in the plane and for each $\varepsilon > 0$ let f_ε be the discrete conformal mapping created as in the script, based on a regular hexagonal packing with circles of radius ε. Let $F : \mathbb{D} \to \Omega$ be the classical conformal mapping with the same normalization. Then the functions f_ε converge to F uniformly on compact subsets of $\mathbb{D}$ as $\varepsilon \to 0$. Moreover, the ratio functions $f^\#$ converge uniformly on compact subsets of D to the modulus of the classical derivative, $|F'|$.*

The script illustrates the quality of approximation in two ways. One is self explanatory, imposing a spoke-and-wheel grid on the domain disc and having the discrete analytic function carry it forward to Ω. This result is compared in Figure 6.9 with the analogous result computed using Toby Driscoll's Schwarz-Christoffel software ([5]). (These markings are related to the *flow* lines employed in Chapter 3.)

A second test of quality is illustrated in the script by using color-coding. Treat both Q and P as Euclidean packings and pay attention to their carriers. We can think of $f : Q \longrightarrow P$ as a more familiar type of function; namely, as a *piecewise affine* function

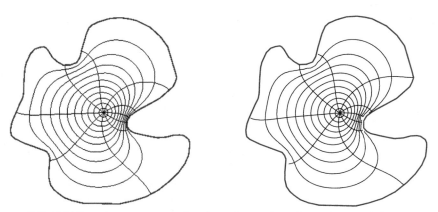

FIGURE 6.9. Grid lines from two approximations of a conformal mapping: the left was computed using Schwarz-Christoffel and the right using *CirclePack*.

$f : \operatorname{carr}(Q) \longrightarrow \operatorname{carr}(P)$ mapping each face of $\operatorname{carr}(Q)$ affinely onto the corresponding face of $\operatorname{carr}(P)$, with the vertices (circle centers) mapped to corresponding vertices. The map may be defined using barycentric coordinates: if T_1 is a face of $\operatorname{carr}(Q)$ and t_1 the corresponding face of $\operatorname{carr}(P)$, then for appropriate real values a, b, c, d, x_0, y_0, f has the form

$$f(x, y) = (ax + by + x_0, cx + dy + y_0), \ (x, y) \in T_1.$$

This is *quasiconformal* on T_1 with *dilatation* given by $k = w + \sqrt{w^2 - 1}$, where $w = (a^2 + b^2 + c^2 + d^2)/(2(ad - bc))$. Note that $k \geq 1$; moreover, f is conformal (analytic) on T_1 if and only if $k = 1$, that is, if and only if T_1 and t_1 are similar triangles. The larger k becomes, the further f is from being conformal. The experiments in **DiscreteRMT** color-code the faces in the range based on k. Faces are white when $k = 1$ and become more red as k gets larger up to some cutoff that you can set. Faces with dilatation larger than the cutoff are blue. In **DiscreteRMT**, commands have been set up to display both grids and the color coding: run the script with increasing refinement levels to see what you can say about the affects on k.

The convergence of discrete objects to their classical counterparts seen in the Rodin-Sullivan theorem holds more generally, indeed, in essentially all the settings we have discussed in this chapter. First, a few cautionary words on increasing the complexity:

(a) Computational penalty: repacking and display times can increase dramatically.

(b) Roundoff error: errors accumulate, interfering with layout, display, and other operations.

(c) Display crowding: objects and indices are harder to see, harder to pick out with the mouse, and display updates are slower.

Nevertheless, *CirclePack* can handle packings with hundreds of thousands of circles, so give the following scripts a try, noting that a few ill-advised mouse clicks can push you into computational neverland — but of course, you can always kill *CirclePack* and restart!

Exercise 6.19. RandomRefine illustrates the use of random packings in place of regular hexagonal packings. You can create your own simply connected region Ω and then apply the commands developed in the script to create and investigate the resulting discrete conformal mappings. Refinement is accomplished by increasing the number of circles in the random packing.

Exercise 6.20. MultiRefine illustrates maps between multiply connected regions, as in **CircleDomains**. If you want to modify the region Ω, note that each component of $\partial\Omega$ requires its own **PATH** file.

Exercise 6.21. Run **ErfRefine** to pursue hex refined packings for the error function encountered in **Erf** (see [7, §14.4]). One method for judging how closely the results mimic the classical $erf(z)$ is to compare the images of the unit circle under the discrete map and the classical map. Methods are given in the script to display the former. However, you should use a math package to generate the latter; the script explains how to read and display a path given as a list of x, y coordinates.

Exercise 6.22. This exercise sets a simple task: estimate the complex number $B(3i/4)$ for a classical finite Blaschke product B. Here is what you know about B: $B(0) = 0$ and $B(1) = 1$; B has three zeros (counting multiplicities); and B branches (i.e., B' vanishes) at $z_1 = (1 + i)/2$ and $z_2 = -1/2$. It is known that appropriate discrete finite Blaschke products b_n will converge uniformly on compact subset of $\mathbb{D}$ to B (an extension of the Rodin-Sullivan theorem to branched packings). **BlaschkeRefine** allows you to create discrete functions b_n (at various refinement levels) based solely on this type of data. You will have to tailor the construction and normalizations to match this particular B. Then you are asked to apply two methods for estimating $B(3i/4)$: (1) Use the values $b_n(3i/4)$ for the b_n you constructed. (2) Locate the zeros of the b_n, use those to build a classical finite Blaschke product $\widehat{B}$ close to B, then compute $\widehat{B}(3i/4)$ in a math package. The point is that classical Blaschke products are best constructed from their zeros, while their discrete analogues are best constructed from their branch points.

Nearly every script from Sections 6.4 and 6.5 involves packings that can be refined as we discussed, and invariably the discrete objects will converge under the refinement to their classical counterparts. The reader should pick a topic and create a script that implements appropriate refinement: discrete exponentials, discrete rational maps, discrete ratio function constructions, discrete harmonic functions — each project will have its distinct character.

6.7   Wrapup

I hope you have enjoyed this chapter. Perhaps it has given you a new way to look at and think about analytic functions. As discrete theories go, circle packing is arguably unique in the breadth, depth, and fidelity with which it captures its subject. The essential phenomena appear with even the simplest of packings, and you seem to glimpse the heart of analyticity. As complexity increases, the fidelity only grows, to the point that you can reliably investigate — perhaps even discover new phenomena — in the classical theory. So it is a full-featured quantum theory, classical in the limit, with experimental and visualization capabilities to boot.

This chapter may also have introduced you to a topic of interest in its own right. There is a lot of room for growth in circle packing: packings on surfaces, packings of random triangulations, graph embedding, and a growing list of applications. Wherever circle packing goes, there, too, will be complex analysis!

6.8   Additional Exercises

Here are three additional exercises to stretch your understanding of circle packing. The first challenge in each is understanding the mechanisms being illustrated in the scripts. After that you may see new directions for experiments, and you are encouraged to set your own path of discovery.

Exercise 6.23. Seeing the connection between triangulations and circle packings may have piqued your curiosity about more general triangulations. Until now, most of our complexes have been simply connected, but **Torus** is different: K triangulates a torus. A torus is topologically equivalent to the surface of an inner tube, so you would rightly guess that we cannot realize a circle packing for K on the plane or on the sphere. Yet the script presents you with the "packing" P shown on the left in Figure 6.10.

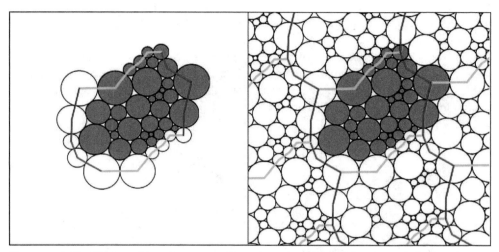

FIGURE 6.10. Circle packing of a torus and its universal covering with color-coded side pairings.

Interpretation of P rests on the color coded side-pairings: The 32 shaded circles form the packing, but additional tangencies exist on the torus. Since we cannot realize them in the plane, we create ghost circles (the unshaded ones on the left): the ghost circles centered along one colored edge are identified with corresponding circles along the other edge sharing that color. You may be able to see the matchings in the image either from the combinatorics or the radii, but the script should help you explore. The image in the right of Figure 6.10 presents this in another fashion; translated copies of the packing tile the plane.

It is up to you to come to terms with the representation for circle packings of tori. It may help to create your own triangulated torus and insert it into the script data section in place of the initial packing. Encoding of complexes is described in the **Formats** tab of the Help Frame of *CirclePack*. Let me offer one specific challenge: create a combinatorial torus K having just seven vertices — the smallest number possible for a proper triangulation (that is, one in which two faces are disjoint, share a single vertex, or share an edge and its end vertices). This may be a little harder than you expect.

Exercise 6.24. An experimental capability of *CirclePack* that we touched on briefly in **Patterns** involves the **adjoin** operation for triangulations. Observe that when two simply connected triangulations K_1 and K_2 have the same number of boundary vertices, they can be pasted together along their boundaries. The result is a new triangulation K of the sphere. A packing P for K in $\overline{\mathbb{C}}$ then provides simultaneous packings for K_1 and K_2 and gives an embedding for their common boundary curve, the so-called *welding curve*, on the sphere.

The basics are demonstrated with Owl in **OwlWelding**. Interest lies in how the shape of the welding curve is determined by our choices in attaching K_1 and K_2. There is a

connection with conformality related to a classical topic called *conformal welding* (see
[10]). Learn about the **double** and **adjoin** operations, and let the experiments begin.

Exercise 6.25. The *CirclePack* algorithms for computing circle packings work in Euclidean and hyperbolic geometry. Unfortunately, there is (as yet) no general algorithm
that works directly in spherical geometry, and the spherical packings in previous scripts
have been stereographic projections from the plane or disc. All is not lost, however. At
least a smattering of examples can be created by hand using basic spherical geometry in
conjunction with combinatorial symmetries. Figure 6.11 illustrates an example with the
combinatorics of the soccer ball: the left side is the maximal packing, while the right is
a branched packing having the same combinatorics. The 12 shaded circles are the branch
circles. Investigate this further in **SoccerPack**. *Try it out!* .

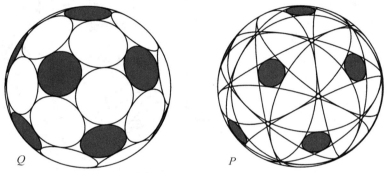

FIGURE 6.11. Circle packings with soccer ball combinatorics.

Both packings in Figure 6.11 are constructed using the geometry of Schwarz triangles.
The univalent packing Q shows the symmetry of the regular dodecahedron: twelve symmetrically distributed circles each with five neighbors, the remaining twenty circles each
with six neighbors. (This packing can also be created by standard methods in *CirclePack*;
see **MaxPackSph**). The packing P shares the same combinatorics, but all degree five circles are branch circles, so P defines a 7-fold covering of the sphere. The map $f : Q \to P$
is thus a discrete rational function with twelve simple branch points and valence seven.

Both Q and P can be created using the spherical triangles of Figure 6.12, two of
several triangles known as *Schwarz triangles*. The Schwarz $\langle 2, 3, 5 \rangle$ triangle t (i.e., one
having angles $\pi/2, \pi/3$, and $\pi/5$) is pictured in Figure 6.12(a). A second triangle, the

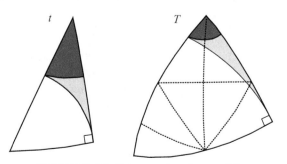

FIGURE 6.12. Schwarz triangles t and T.

$\langle 2, 3, 5/2 \rangle$ Schwarz triangle T, is shown in (b) (with a different scale). Dashed lines on T show how it is built from seven copies of t. A pattern of two circular arcs is imprinted on t in (a) and the analogous pattern is imprinted on T in (b).

A Schwarz triangle can be reflected across any of its edges to form a copy of itself with reverse orientation. Repeated reflections will tile the sphere; for t this involves 120 non-overlapping copies. Carrying along the imprinted arcs during the reflection process gives us a univalent circle packing with 32 circles, our maximal packing Q. The identical reflection process, starting with T, gives 120 copies which tile the sphere seven layers deep and whose imprinted arcs give our packing P. In any copy of T, it takes ten reflections around the vertex with angle $2\pi/5$ to close up, and this is what happens at each of the twelve branch points of P. The script **SoccerPack** lets you check a branch circle visually to see the five (huge) neighboring circles wrap twice around it.

Schwarz triangles were first used to generate branched circle packings of the sphere in [2] (see Appendix H of [7]). However, the example of Figure 6.13 and other equally beautiful examples were created by Samantha Corvino, one of my REU (Research Experience for Undergraduates) students.

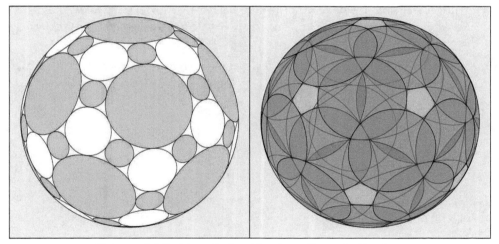

FIGURE 6.13. A branched packing due to Samantha Corvino.

There are a handful of additional Schwarz triangles and other options for circular markings, so the challenge to you is to create more examples. For each, you will need to encode the combinatorics K for *CirclePack*, which can then generate the associated maximal packing. However, *CirclePack* cannot find the radii and centers for the branched packing. For this you will need to apply some spherical trigonometry, record the results in a packing file along with K, and have *CirclePack* display them. To get you started, I've put another of Samantha's examples with more details in **Tri432**. *Try it out!*

6.9 Bibliography

[1] Alan F. Beardon and Kenneth Stephenson, Circle packings in different geometries, *Tohoku Math. J.*, **43** (1991), 27–36.

[2] Philip L. Bowers and Kenneth Stephenson, A branched Andreev-Thurston Theorem for circle packings of the sphere, *Proc. London Math. Soc.*, **73** (1996), 185–215.

[3] Charles R. Collins and Kenneth Stephenson, A circle packing algorithm, *Computational Geometry: Theory and Applications* **25** (2003), 233–256.

[4] Tobin A. Driscoll and Lloyd N. Trefethen, *Schwarz-Christoffel Mapping*, Camb. Univ. Press, New York, 2002.

[5] Tobin Driscoll, *Schwarz-Christoffel Toolbox* (for **Matlab**), http://www.math.udel.edu/~driscoll/software/SC/.

[6] Burt Rodin and Dennis Sullivan, The convergence of circle packings to the Riemann mapping, *J. Differential Geometry* **26** (1987), 349–360.

[7] Kenneth Stephenson, *Introduction to Circle Packing: the Theory of Discrete Analytic Functions*, Camb. Univ. Press, New York, 2005.

[8] Kenneth Stephenson, Circle packing: A mathematical tale, *Notices Amer. Math. Soc.* **50** (2003), no. 11, 1376–1388.

[9] William Thurston, The finite Riemann mapping theorem, Invited talk, *An International Symposium at Purdue University in celebration of de Branges' proof of the Bieberbach conjecture*, March 1985 (unpublished).

[10] G. Brock Williams, Discrete conformal welding, *Indiana Univ. Math. J.*, **53** (2004), 765–804.

Background

This appendix summarizes some major topics and theorems from a standard undergraduate complex analysis course. Since it is a review, we do not prove results; for more details you can look to reference texts such as [1] and [3], and occasionally to the more advanced text [2]. A few advanced topics are also summarized here, including the Riemann mapping theorem, the open mapping theorem, and Schwarz's lemma. The extended notion of analyticity on the Riemann sphere $\overline{\mathbb{C}}$ is addressed separately in Appendix B.

A.1 Functions of a Complex Variable as Mappings

Let $\mathbb{R}$ denote the real numbers, i the imaginary number $\sqrt{-1}$, and $\mathbb{C} = \{x+iy : x, y \in \mathbb{R}\}$ the complex numbers. (We employ standard notations which the reader can find in the Index of Notation on page 365.)

A complex-valued function f defined on a set $\Omega \subseteq \mathbb{C}$ is denoted $f : \Omega \to \mathbb{C}$. We often denote the domain set of f by $domain(f)$, and express $z = x + iy = (x, y)$ and $f(z) = f(x, y)$ with the implicit understanding that $x = \text{Re}(z)$ and $y = \text{Im}(z)$. The term "mapping", synonymous with "function", is often used in geometric settings such as the complex plane, and tempting as it may be to treat f as mapping two real variables to two real variables, it is far better to work with it as a complex function of a single complex variable. The set $f(\Omega) = \{f(z) : z \in \Omega\}$ is called the *range* of f. If $A \subseteq \mathbb{C}$, we denote the *inverse image* (or preimage) of A under f by $f^{-1}(A) = \{z \in \Omega : f(z) \in A\}$.

A.1.1 Linear Functions

We can examine the linear mappings $f(z) = az+b$, for constants $a, b \in \mathbb{C}$, by considering the simpler functions $h(z) = az$ and $g(z) = z + b$.

Writing $z = re^{i\theta}$ and $a = |a|e^{i\alpha}$ in polar form, we can express the image point as $h(z) = az = r|a|e^{i(\theta+\alpha)}$. Hence, h maps z to the point $h(z)$, which geometrically is a stretching or contraction of the modulus by $|a|$ and a rotation about the origin by the angle α. Follow this with the action of $g(z) = z + b$, which translates a point by $|b|$ units in the direction of the *vector* b. The result is the geometric action of the composition $f(z) = az + b = g(h(z))$. See Figure A.1 for an example.

A.1.2 Power Functions

Let $f(z) = z^n$ for fixed $n \in \mathbb{N}$, where $\mathbb{N}$ denotes the natural numbers. Writing $z = re^{i\theta}$ in polar form, we see that $f(z) = (re^{i\theta})^n = r^n e^{in\theta}$; that is, under the action of f the point

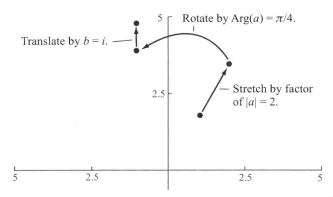

FIGURE A.1. An illustration of the geometric action of $f(z) = az + b$ for $a = 2e^{i\pi/4}$ and $b = i$.

z has its modulus raised to the nth power and its argument multiplied by n. See Figure A.2 for an example.

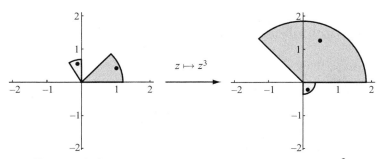

FIGURE A.2. An illustration of the geometric action of $z \mapsto z^3$.

Roots of unity

For a positive integer n, we call any of the n solutions to the equation $z^n = 1$ an *nth root of unity*. These have the form $\omega_k = e^{2k\pi i/n}$ for $k = 0, \ldots, n - 1$, and written as $1, \omega_1, \omega_1^2, \ldots, \omega_1^{n-1}$, we see they are equally spaced points on the unit circle $|z| = 1$.

A.1.3   The Exponential Function

The definition of the exponential function arises naturally out of Euler's formula $e^{i\theta} = \cos\theta + i\sin\theta$.

Definition A.1. If $z = x + iy$, then

$$e^z = e^x(\cos y + i\sin y).$$

The definition has two important consequences:

(a) The exponential function maps $\mathbb{C}$ onto $\mathbb{C}\backslash\{0\}$.

(b) The exponential function is periodic with period $2\pi i$.

A.1.4 The Logarithm Function

The logarithm function provides an inverse of the exponential function. However, since the exponential function is not one-to-one, special consideration must be taken to understand in what sense the logarithm is an inverse. We begin by first understanding $\arg(z)$.

Definition A.2. For a complex number $z \neq 0$, the argument of z, denoted $\arg(z)$, is the set of all $\theta \in \mathbb{R}$ such that $z = |z|e^{i\theta}$. The *principal value* of the argument, denoted $\text{Arg}(z)$, is the unique such angle θ with $-\pi < \theta \leq \pi$.

Definition A.3. The logarithm function $\log(z)$ is the multiple-valued function

$$\log(re^{i\theta}) = \ln(r) + i\theta,$$

where $\ln(r)$ denotes the real-valued logarithm. We also write

$$\log(z) = \ln|z| + i \arg(z).$$

Since $\log(z)$ is multiple valued (because $\arg(z)$ is multiple valued), it is not a function in the traditional sense. To make it a function, we can use the *principal value* of the argument, $\text{Arg}(z)$.

Definition A.4. The *principal value* of the logarithm is $\text{Log}(z)$ where

$$\text{Log}(z) = \ln|z| + i\,\text{Arg(z)}.$$

It has domain set $\mathbb{C}\backslash\{0\}$ and range $-\pi < \text{Im}(w) \leq \pi$, but fails to be continuous at any point of the negative real axis.

Technology Note: We use $\text{Log}(z)$ and $\text{Arg}(z)$ to denote the principal values of the logarithm and argument functions, but the applet *ComplexTool* uses $\log(z)$ and $\arg(z)$ to express them.

Remark A.5. While the principal branch of the logarithm is the one most widely used, it is possible to define other branches of the logarithm function that are continuous, for example, along the negative real axis. Any branch of the logarithm must involve a branch cut, a line of discontinuities running from infinity to 0. More information is in [1] and [3].

A.1.5 Trigonometric Functions

The trigonometric functions of a complex number z are defined in terms of the exponential function.

Definition A.6. Given a complex number z, we define

$$\sin z = \frac{e^{iz} - e^{-iz}}{2i} \qquad \text{and} \qquad \cos z = \frac{e^{iz} + e^{-iz}}{2}.$$

We can show that the image of the infinite vertical strip $-\pi/2 \leq \text{Re}\, z \leq \pi/2$ under either map is the entire plane, so these trigonometric functions are unbounded in $\mathbb{C}$.

A.2 Continuity and Analyticity in $\mathbb{C}$

Definition A.7. Let $\Omega \subseteq \mathbb{C}$ and consider a complex-valued function $f : \Omega \to \mathbb{C}$.

(a) We say that f is *continuous* at $z_0 \in \Omega$ if $\lim_{z \to z_0} f(z) = f(z_0)$; i.e., for each $\varepsilon > 0$ there exists $\delta > 0$ such that $|f(z) - f(z_0)| < \varepsilon$ whenever $z \in \Omega$ and $|z - z_0| < \delta$. We say f is continuous on a set $U \subseteq \Omega$ if it is continuous at each point of U.

(b) We say that f is *differentiable* at z_0 when its derivative $f'(z_0) = \lim_{z \to z_0} \frac{f(z) - f(z_0)}{z - z_0}$ exists.

(c) When a function is differentiable at all points of an open set Ω, it is said to be *analytic* on Ω. (Some authors refer to such a function as *holomorphic* in Ω.)

(d) The term *domain* refers to an open connected subset of the plane (or of the Riemann sphere), since these are the natural domain sets for analytic functions.

A.2.1 Cauchy-Riemann Equations

Writing $f(z) = f(x, y) = u(x, y) + iv(x, y)$ with $u(x, y) = \operatorname{Re} f(x, y)$ and $v(x, y) = \operatorname{Im} f(x, y)$, we have the partial derivatives

$$\frac{\partial f}{\partial x} = \frac{\partial u}{\partial x} + i\frac{\partial v}{\partial x} \qquad \text{and} \qquad \frac{\partial f}{\partial y} = \frac{\partial u}{\partial y} + i\frac{\partial v}{\partial y}. \tag{A.1}$$

When f is differentiable at z_0, we have $f'(z_0) = \frac{\partial f}{\partial x}(z_0) = -i\frac{\partial f}{\partial y}(z_0)$, which by equating real and imaginary parts yields the Cauchy-Riemann equations

$$\frac{\partial u}{\partial x} = \frac{\partial v}{\partial y} \qquad \text{and} \qquad \frac{\partial u}{\partial y} = -\frac{\partial v}{\partial x}. \tag{A.2}$$

Theorem A.8. *A function $f(z) = u(x, y) + iv(x, y)$ is analytic in an open set Ω if and only if the first partial derivatives u_x, u_y, v_x, v_y exist, are continuous, and satisfy the Cauchy-Riemann equations.*

Using the Cauchy-Riemann equations (A.2), we can show that if $f(z) = u(x, y) + iv(x, y)$ is analytic in an open set Ω, then the component functions u and v are *harmonic* in Ω, that is, $u_{xx} + u_{yy} = 0$ and $v_{xx} + v_{yy} = 0$.

A.2.2 Conformal Mappings

A function $f : \Omega \to \mathbb{C}$ is called *univalent* if it is one-to-one; i.e., if $z_1 \neq z_2$ in Ω, then $f(z_1) \neq f(z_2)$. A function that is both analytic and univalent on an open set is said to be *conformal* there. Geometrically, conformal means that the function is locally angle-preserving, preserving both the magnitude and sense of the angle between intersecting curves. You can get a feel for this by graphing a univalent analytic function with the *ComplexTool* applet, using either a rectangular or circular grid, and zooming in on points in the range to see that the right angles where grid lines intersect are preserved. An analytic map f is *locally univalent* at $z_0 \in \Omega$ when its restriction to some neighborhood of z_0 is univalent. As discussed in Section A.6.1, this is the case if and only if $f'(z_0) \neq 0$.

A.3   Contour Integration

Contour integration is a form of line integration and involves expressions of the form $\int_C f(z)\,dz$. Here C is a curve in the plane and f is a complex function defined on C.

A.3.1   Planar Contours

A path in the plane is defined as a continuous map $\gamma : [a, b] \to \mathbb{C}$ where $[a, b]$ is a parameter interval on the real line. It suffices to consider piecewise differentiable functions γ, and since the parameterization is generally irrelevant in computations, the term *contour* (also, *curve*) will be used in place of path for the trace C (i.e., the image of γ).

Definition A.9.   Suppose $\gamma : [a, b] \to \mathbb{C}$ defines a piecewise smooth contour C in the plane.

(a) Points $z_0 = \gamma(a)$ and $z_1 = \gamma(b)$ are known as the *initial* and *terminal* points of C; the orientation of the contour is from z_0 to z_1.

(b) C is a *closed contour* if its initial and terminal points coincide.

(c) C is a *simple closed contour* if it is a closed contour that does not cross itself. It is *positively oriented* if the inside is on the left as you travel around the contour.

See Appendix B.3 for information regarding topology on both the plane and the Riemann sphere.

A.3.2   Examples

The value of a contour integral on C is independent of the continuous function γ which is used to parameterize C. In practice, of course, some parameterization $\gamma : [a, b] \to C$ must be chosen, then the integral is calculated as

$$\int_C f(z)\,dz = \int_{t=a}^{b} f(\gamma(t))\gamma'(t)\,dt. \tag{A.3}$$

When C is closed and positively oriented, the special integral sign $\oint$ is often used to emphasize that fact.

Here are two examples.

(a) The line segment C from z_0 to z_1 can be parameterized by $\gamma(t) = (1 - t)z_0 + tz_1$, where $0 \le t \le 1$. Thus,

$$\int_C f(z)\,dz = \int_0^1 f((1 - t)z_0 + tz_1)\,(z_1 - z_0)\,dt.$$

(b) The contour C defined as wrapping once counterclockwise around the circle of radius R centered at a can be parameterized as $\gamma(t) = a + R\,e^{it}$, where $0 \le t \le 2\pi$. Hence,

$$\oint_C f(z)\,dz = \int_0^{2\pi} f(a + Re^{it})\,Ri\,e^{it}\,dt.$$

Though the value of a contour integral on C does not depend on the parameterization γ, it would seem to depend on the geometry of C and on its endpoints if C is not closed. In the study of analytic functions, however, the following results are central.

A.3.3   Four Integral Theorems

Theorem A.10 (Cauchy's Theorem). *Let $f(z)$ be analytic everywhere on and inside a simple closed contour C. Then*

$$\oint_C f(z)\, dz = 0.$$

One corollary is that if $f(z)$ is analytic on and inside the region between two contours C_1 and C_2 which share the same initial and same terminal points, then $\int_{C_1} f(z)\, dz = \int_{C_2} f(z)\, dz$. Another is that if f is analytic on and in between positively-oriented simple closed contours K_1 and K_2 with K_1 interior to K_2, then $\int_{K_1} f(z)\, dz = \int_{K_2} f(z)\, dz$. These are both aspects of the *principle of deformation of contours*.

Theorem A.11 (Cauchy's Integral Formula). *Let $f(z)$ be analytic everywhere on and inside a simple closed positively-oriented contour C. Let a be a point inside C. Then*

$$\oint_C \frac{f(z)}{z - a}\, dz = 2\pi i\, f(a).$$

Theorem A.12 (Cauchy's Generalized Integral Formula). *Let $f(z)$ be analytic everywhere on and inside a simple closed positively-oriented contour C. Let a be a point inside C. Then, for $n \in \mathbb{N}$,*

$$\oint_C \frac{f(z)}{(z - a)^{n+1}}\, dz = \frac{2\pi i\, f^{(n)}(a)}{n!}.$$

If C is not a simple closed contour, then these integral theorems do not apply. Fortunately, we may still be able to avoid direct use of (A.3) by an application of the fundamental theorem of calculus.

Theorem A.13. *Let C be a contour from z_0 to z_1. Suppose f has an antiderivative in some open set Ω containing C; that is, there exists a complex function $F : \Omega \to \mathbb{C}$ so that $F'(z) = f(z)$ for all $z \in \Omega$. Then $\int_C f(z)\, dz = F(z_1) - F(z_0)$.*

A.4   Taylor Series and Laurent Series

A.4.1   Taylor Series

The theory of Taylor series carries over from the theory in real variables. However, it is even better here, as the following theorem shows:

Theorem A.14. *Let $f(z)$ be an analytic function in a domain Ω and let $z_0 \in \Omega$. Then*

(a) $f(z)$ can be represented by a power series which converges in the disk $|z - z_0| < R$, where R is the distance from z_0 to the nearest singularity of $f(z)$. If $f(z)$ is entire, i.e., is analytic on all of $\mathbb{C}$, then $R = \infty$ and the power series converges everywhere in the plane.

(b) The series representing $f(z)$ is its Taylor series; namely,

$$f(z) = \sum_{n=0}^{\infty} a_n (z - z_0)^n, \text{ where } a_n = \frac{f^{(n)}(z_0)}{n!}.$$

Here $f^{(n)}(z_0)$ denotes the nth derivative of f at z_0. The a_n are also given by the contour integrals

$$a_n = \frac{1}{2\pi i} \oint_C \frac{f(z)}{(z - z_0)^{1+n}} \, dz,$$

where the contour C is any positively-oriented circle centered at z_0 of radius $r < R$.

(c) The Taylor series is the unique power series representing f at z_0.

Power series can be added, subtracted, multiplied and divided. They can also be differentiated and integrated term by term to obtain new series. Of particular note is the fact that the derivative of an analytic function is itself analytic.

A.4.2 Laurent Series

Laurent series are a generalization of Taylor series. Suppose f is analytic throughout an annular domain $\Omega = \{z : R_1 < |z - z_0| < R_2\}$. Then f has a unique representation as a *Laurent series*

$$f(z) = \sum_{n=0}^{\infty} a_n (z - z_0)^n + \sum_{n=1}^{\infty} \frac{b_n}{(z - z_0)^n}, z \in \Omega. \tag{A.4}$$

Both the values $R_1 = 0$ and $R_2 = +\infty$ are allowed. The coefficients a_n are as given in the integral representation above and the b_n by

$$b_n = \frac{1}{2\pi i} \oint_C \frac{f(z)}{(z - z_0)^{1-n}} \, dz,$$

where the contour C is any positively-oriented circle centered at z_0 of radius r, $R_1 < r < R_2$.

A.4.3 Isolated Singularities

A function f that is analytic on $\Delta(z, \varepsilon) \setminus \{z_0\}$, that is, on a deleted ε neighborhood of z_0, can be expressed by its Laurent series as in (A.4). This leads to three types of singularities:

Definition A.15.

(a) When the coefficients $b_n = 0$ for every $n = 1, 2, \ldots$, then $f(z)$ is said to have a *removable singularity* at z_0. Then $f(z) = \sum_{n=0}^{\infty} a_n (z - z_0)^n$, and setting $f(z_0) = a_0$ will make f analytic in the full neighborhood $\Delta(z, \varepsilon)$ (we have removed the singularity).

(b) When there exists $N \in \mathbb{N}$ such that $b_N \neq 0$ and $b_n = 0$ for all $n > N$, then f has a *pole of order N* at z_0. We can express f in a factored form $f(z) = (z - z_0)^{-N} h(z)$ where $h(z) = b_N + b_{N-1}(z - z_0) + \cdots$ is analytic on $\Delta(z, \varepsilon)$ and $h(z_0) = b_N \neq 0$.

(c) When infinitely many b_n are nonzero, we say f has an *essential singularity* at z_0.

A.4.4   Residue Theorem

The Laurent expansion is very useful in complex integration. When $f(z)$ is expressed by a Laurent series as in (A.4) on a punctured disk ($R_1 = 0$) with center z_0, then the *residue* of f at z_0 is the coefficient of $\frac{1}{(z-z_0)}$ in the Laurent expansion, that is, $\mathrm{Res}(f, z_0) = b_1$.

Theorem A.16 (Residue Theorem). *Let C be a simple closed positively-oriented contour in $\mathbb{C}$. If a function f is analytic on and inside C except for a finite number of isolated singularities z_k, $k = 1, \ldots, n$, inside C, then*

$$\oint_C f(z)\,dz = 2\pi i \sum_{k=1}^{n} \mathrm{Res}(f, z_k).$$

Definition A.17. A function f is *meromorphic* in a domain Ω if it is analytic at every point of Ω except possibly at poles.

A.5   Key Theorems

A.5.1   Maximum Modulus Theorem

Two key results about the maximum modulus of an analytic function are the following.

Theorem A.18 (Maximum Modulus Theorem). *If a function f is analytic and not constant in a domain $\Omega \subseteq \mathbb{C}$, then $|f(z)|$ has no maximum value in Ω. That is, there is no point $z_0 \in \Omega$ such that $|f(z)| \le |f(z_0)|$ for all $z \in \Omega$.*

Corollary A.19. *Suppose that a function f is continuous on a closed and bounded region $E \subset \mathbb{C}$ and is analytic and non constant in the interior of E. Then the maximum value of $|f(z)|$ in E, which is always attained, occurs on the boundary of E and not in the interior.*

A.5.2   Argument Principle

The argument principle lets us count the number of zeros and poles inside a contour.

Theorem A.20 (Argument Principle). *Let C be a positively-oriented simple closed contour. Suppose f is meromorphic on an open set Ω which contains C and the open set bounded by C. Suppose f has no poles or zeros on C. Then*

$$\frac{1}{2\pi i} \oint_C \frac{f'(z)}{f(z)}\,dz = N_0 - N_p,$$

where N_0 is the number of zeros of f inside C and N_p is the number of poles of f inside C. In determining N_0 and N_p, zeros and poles are counted according to their orders (or multiplicities).

A.5.3   Schwarz's Lemma

We have the following geometric result regarding functions on the open unit disk $\mathbb{D}$.

Theorem A.21 (Schwarz's Lemma). *Suppose $f : \mathbb{D} \to \overline{\mathbb{D}}$ is analytic and $f(0) = 0$. Then $|f'(0)| \le 1$ and $|f(z)| \le |z|$ for all $z \in \mathbb{D}$. If $|f'(0)| = 1$ or if $|f(z)| = |z|$ for any $z \in \mathbb{D} \setminus \{0\}$, then f is of the form $f(z) = e^{i\theta} z$ for some $\theta \in \mathbb{R}$ (i.e., f is a rotation).*

A.6 More Advanced Results

A.6.1 Local Properties of Analytic Maps

The existence of power series representations of analytic maps is a powerful tool for understanding their local, and sometimes global, behavior. If an analytic function f is defined in a domain $\Omega \subseteq \mathbb{C}$, then for any $z_0 \in \Omega$, Theorem A.14 tells us that

$$f(z) = a_0 + a_1(z - z_0) + a_2(z - z_0)^2 + \ldots \tag{A.5}$$

on any Euclidean disk $\Delta(z_0, r) = \{z \in \mathbb{C} : |z - z_0| < r\}$ which is contained in Ω.

Suppose for the moment that $f'(z_0) = a_1 \ne 0$. Then for z very close to z_0 we see that the linear contribution $a_0 + a_1(z - z_0)$ dominates the rest of the series in (A.5). Omitting the details, we can say that $f(z) \approx A_1(z) = a_0 + a_1(z - z_0)$ for $z \in \Delta(z_0, \varepsilon)$ when $\varepsilon > 0$ is small. Since $A_1(z)$ is a linear map, with well understood properties, we can expect that $f(z)$ will have the same properties as $A_1(z)$ when $z \in \Delta(z_0, \varepsilon)$. In particular, for every w near $a_0 = f(z_0)$, there is exactly one value z_1 near z_0 that maps to w. This holds for A_1, and for f. Thus f is *locally one-to-one* (also called *locally univalent*) at z_0 when $f'(z_0) \ne 0$. See Figure A.3.

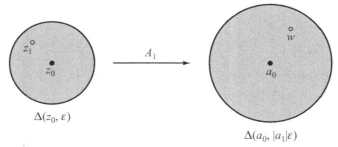

$$\Delta(z_0, \varepsilon) \qquad \Delta(a_0, |a_1|\varepsilon)$$

FIGURE A.3. An illustration of the map $A_1(z) = a_0 + a_1(z - z_0)$ mapping $\Delta(z_0, \varepsilon)$ onto $\Delta(a_0, |a_1|\varepsilon)$ in a one-to-one fashion.

Similarly, we can use approximations to understand the local behavior of $f(z)$ when $a_1 = 0$, i.e., $f'(z_0) = 0$. If f is not constant,

$$f(z) = a_0 + a_k(z - z_0)^k + \ldots \tag{A.6}$$

where a_k is the first non-zero coefficient (other than possibly a_0) in (A.5). Then for z very close to z_0, the terms $a_0 + a_k(z - z_0)^k$ dominate the rest of the series. Omitting the details, we can say that $f(z) \approx A_k(z) = a_0 + a_k(z - z_0)^k$ for $z \in \Delta(z_0, \varepsilon)$ when $\varepsilon > 0$ is small. Note that A_k is a composition $h_3 \circ h_2 \circ h_1$ of the simple maps $h_1(z) = z - z_0$ (translation), $h_2(z) = z^k$ (power function), and $h_3(z) = a_0 + a_k z$ (linear function). As such, $A_k(z)$ has well understood properties, and so we can expect that $f(z)$ will have the

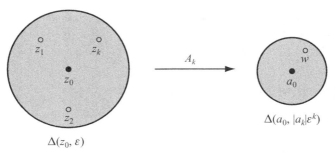

FIGURE A.4. An illustration of the map $A_k(z) = a_0 + a_k(z - z_0)^k$ for $k = 3$ mapping $\Delta(z_0, \varepsilon)$ onto $\Delta(a_0, |a_k|\varepsilon^k)$ in a k-to-one fashion.

same properties when $z \in \Delta(z_0, \varepsilon)$. In particular, for every w near $a_0 = f(z_0)$, there are k values $z_1, \dots, z_k$ symmetrically arranged near z_0 that map to w. This holds for A_k, and for f, though since we have only an approximation the k values may not be exactly symmetrically arranged about z_0. See Figure A.4. Thus we say that f is *locally k-to-one* at z_0 when $f'(z_0) = f''(z_0) = \dots = f^{(k-1)}(z_0) = 0$, but $f^{(k)}(z_0) \neq 0$. We also describe this situation by saying that z_0 maps to $f(z_0)$ with *degree* (or *order* or *multiplicity* or *valency*) k.

When z_0 maps to $f(z_0)$ with multiplicity k, we can rewrite (A.6) as $f(z) = a_0 + (z - z_0)^k[a_k + a_{k+1}(z - z_0) + \dots]$. Because $a_k + a_{k+1}(z - z_0) + \dots$ determines an analytic map on D, we have the following.

Lemma A.22. *Let f be a function analytic at z_0 with multiplicity k. Then there is a map h that is analytic at z_0 such that $h(z_0) \neq 0$ and $f(z) = f(z_0) + (z - z_0)^k h(z)$. In particular, if $f(z_0) = 0$, that is, if f has a zero of order k at z_0, then $f(z) = (z - z_0)^k h(z)$.*

Open Mapping Theorem

A fact not usually introduced in standard undergraduate complex variables courses is that analytic maps are open maps, which we define more carefully below. Some chapters in this text use this and other such results, and so we present them here. Proofs are in more advanced texts such as [2, p. 344-348].

Consider a non-constant function f that is analytic at z_0. Section A.6.1 shows that, whether or not $f'(z_0) = 0$, the image of a small neighborhood of z_0 covers a small neighborhood of $f(z_0) = a_0$. This is enough to assert the following.

Theorem A.23 (Open Mapping Theorem). *If f is a function on a domain $\Omega \subseteq \mathbb{C}$ that is analytic and non-constant, then the range $f(\Omega)$ is an open set.*

Since *open maps* are those for which the image of an open set is an open set, non-constant analytic maps are open maps. An application is the following.

Theorem A.24 (Inverse Function Theorem). *Suppose that $\Omega \subseteq \mathbb{C}$ is a domain and that $f : \Omega \to \mathbb{C}$ is a univalent analytic map. Then the inverse function $f^{-1} : f(\Omega) \to \Omega$ is analytic.*

A use for the inverse function theorem is that it shows that analytic locally one-to-one functions have local analytic inverse functions. Let f be an analytic map defined on a domain Ω, where f need not be one-to-one on all of Ω. If $f'(z_0) \neq 0$ for some $z_0 \in \Omega$, then there is a small disk $D = \Delta(z_0, \varepsilon)$ on which f is one-to-one. Hence by the inverse function theorem, there is a map $g : f(D) \to D$, the *local inverse* of f, defined on the open set $f(D)$ that is analytic and satisfies $(g \circ f)(z) = z$ for all $z \in D$.

Riemann Mapping Theorem

A very important theorem in complex analysis states that all simply connected domains (other than the whole plane) are conformally equivalent to the unit disk:

Theorem A.25 (Riemann Mapping Theorem). *Suppose that Ω is a simply connected domain in the complex plane, $\Omega \neq \mathbb{C}$, and that $z_0 \in \Omega$. Then there exists a unique conformal mapping f of Ω onto the unit disk $\mathbb{D}$ satisfying $f(z_0) = 0$ and $f'(z_0) > 0$.*

A proof is in [2, p. 420].

A.7   Bibliography

[1] Ruel V. Churchill and James Ward Brown, *Complex Variables and Applications*. McGraw-Hill Book Co., New York, eighth edition, 2009.

[2] Bruce P. Palka, *An Introduction to Complex Function Theory: Undergraduate Texts in Mathematics*. Springer-Verlag, New York, 1991.

[3] Edward B. Saff and Arthur David Snider, *Fundamentals of Complex Analysis with Applications to Engineering, Science, and Mathematics*. Prentice Hall, New Jersey, third edition, 2003.

The Riemann Sphere

This appendix describes how "infinity" is incorporated in complex analysis via the Riemann sphere and how complex arithmetic extends slightly beyond the complex plane. As points z in $\mathbb{C}$ move arbitrarily far from the origin we informally say that they go to infinity. To make that precise, we adjoin a point denoted ∞ to $\mathbb{C}$ to get the *Riemann sphere* (or *extended complex plane*), denoted here as $\overline{\mathbb{C}} = \mathbb{C} \cup \{\infty\}$.

We extend the concepts of continuity and analyticity from $\mathbb{C}$ to $\overline{\mathbb{C}}$ by using stereographic projection, the intrinsic *spherical metric* σ, and the corresponding topology. We even give an answer to the division-by-zero problem. We do not prove results or even give a full exposition; our goal is to provide enough background so the reader can grasp the material in chapters that require it. More details can be found in the reference texts [1], [3], and the more advanced text [2].

B.1 Stereographic Projection and Spherical Geometry

We model the Riemann sphere $\overline{\mathbb{C}}$ by identifying points in $\mathbb{C}$ with points in the unit sphere $\mathbb{S} = \{(x_1, x_2, x_3) : x_1^2 + x_2^2 + x_3^2 = 1\} \subset \mathbb{R}^3$ through *stereographic projection*. See Figure B.1. Label $N = (0, 0, 1)$ the north pole on $\mathbb{S}$. Then for any $z = x + iy = (x, y, 0)$ in $\mathbb{C}$, consider the line in $\mathbb{R}^3$ between z and N. The line intersects $\mathbb{S}$ at two points, N and a second point we call $Z = (x_1, x_2, x_3)$. We then associate the points z and Z. Formally, we define a map $\pi : \mathbb{C} \to \mathbb{S}$ given by $\pi(z) = Z$. Omitting the details (which can be found in [2, p. 351]), the formula for the map is $Z = \pi(z) = \pi(x + iy) = \left(\frac{2x}{|z|^2+1}, \frac{2y}{|z|^2+1}, \frac{|z|^2-1}{|z|^2+1} \right)$. However, we rarely use the formula since it is the idea and the picture that provide the necessary understanding.

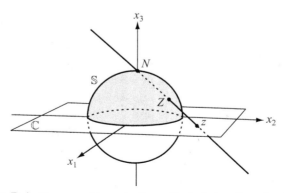

FIGURE B.1. Stereographic projection with shaded northern hemisphere.

The only point on $\mathbb{S}$ that is not associated with a complex number is the north pole N and the association between $\mathbb{C}$ and $\mathbb{S} \setminus \{N\}$ is bijective (that is, both one-to-one and onto). We now decide how to associate a point with N.

If $|z|$ is large (i.e., z is very far from the origin), then $Z \in \mathbb{S}$ is close to N. It is therefore natural to say that a point associated with N must be infinitely far from the origin. It is for this reason that we adjoin ∞ to $\mathbb{C}$ and extend the definition of π by defining $\pi(\infty) = N$. We now have an identification, i.e., a bijection, between all of $\overline{\mathbb{C}} = \mathbb{C} \cup \{\infty\}$ and all of $\mathbb{S}$. We can think of $\overline{\mathbb{C}}$ in two ways: as the complex plane with an added point ∞ infinitely far from the origin, or as the sphere $\mathbb{S}$ with N representing ∞. Both views are useful.

Stereographic projection identifies the origin $0 \in \mathbb{C}$ with the south pole $(0, 0, -1)$ in $\mathbb{S}$, the unit disk $\mathbb{D} = \{z : |z| < 1\}$ with the southern hemisphere, the unit circle $|z| = 1$ with the equator, and the real axis with a great circle through the two poles and the points $\pi(1) = (1, 0, 0)$ and $\pi(-1) = (-1, 0, 0)$. In the plane there are many ways a point can go to infinity. On the real axis, for example, it is normal to say that a point moving off arbitrarily far to the left is going to $-\infty$ and one moving to the right, to $+\infty$. On $\mathbb{S}$, however, these point come together at N. There is only one ∞ on $\mathbb{S}$.

As an exercise you are asked to match the curves in $\mathbb{C}$ with their projections onto $\overline{\mathbb{C}}$ given in Figure B.2.

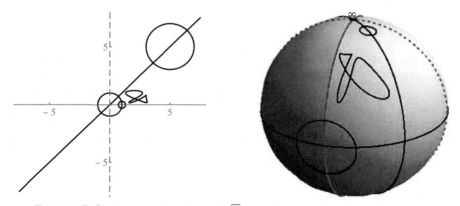

FIGURE B.2. Curves in $\mathbb{C}$ projected to $\overline{\mathbb{C}}$; ∞ is the point N on top of the sphere.

B.2   The Spherical Metric σ

The standard way we measure the distance between two points z and w in $\mathbb{C}$ is by $|z - w|$, the Euclidean metric on $\mathbb{C}$. Is there a natural way to define the distance between points on $\overline{\mathbb{C}}$ using the sphere model? The answer is yes. We use the natural spherical metric on $\mathbb{S}$ and transfer that back to $\overline{\mathbb{C}}$. Here's how.

The spherical distance between two points Z and W on $\mathbb{S}$ is defined to be the arclength of the shortest path on $\mathbb{S}$ that connects Z and W, which is the shorter arc of the great circle that runs through Z and W. We denote this distance by $d(Z, W)$. For example, the distance between $(1, 0, 0)$ and $N = (0, 0, 1)$ is $\pi/2$, a quarter of the circumference of a great circle with radius 1.

We can now transfer the metric from $\mathbb{S}$ to $\overline{\mathbb{C}}$ using the map $\pi : \overline{\mathbb{C}} \to \mathbb{S}$ by defining the distance between two points $z, w \in \overline{\mathbb{C}}$ to be $\sigma(z, w) = d(Z, W) = d(\pi(z), \pi(w))$. In other words, project z and w to the sphere $\mathbb{S}$ and measure the distance there. The new *metric* (distance function) on $\overline{\mathbb{C}}$ does not treat ∞ as special; the point ∞ plays the same role as any other point of $\overline{\mathbb{C}}$.

There is a formula for $\sigma(z, w)$. We do not need it, but a few observations about σ and a few examples may be helpful. The largest possible spherical distance is π, and for $r \in (0, \pi)$, $\triangle_\sigma(a, r) = \{z \in \overline{\mathbb{C}} : \sigma(z, a) < r\}$ is the *spherical disc* of radius r and center a. For example, $\triangle_\sigma(0, \pi/2)$ is the southern hemisphere of the Riemann sphere and corresponds to the unit disc $\mathbb{D}$, while $\triangle_\sigma(\infty, \pi/2)$ is the northern hemisphere and corresponds to $\overline{\mathbb{C}} \setminus \overline{\mathbb{D}}$. In general, spherical discs are spherical caps of the sphere cut off by planes in $\mathbb{R}^3$. As suggested in Figure B.2, circles in $\mathbb{C}$ project under π to circles in the metric σ (though the center in $\mathbb{C}$ does not generally project to the center in $\overline{\mathbb{C}}$). Straight lines in $\mathbb{C}$, all of which are regarded as passing through ∞, project to circles in $\overline{\mathbb{C}}$ passing through ∞. The extended real axis $\overline{\mathbb{R}} = \mathbb{R} \cup \{\infty\}$, for instance, projects to a great circle through ∞, as does every line in $\mathbb{C}$ through the origin. In general, a spherical disc D corresponds under π to a halfplane of $\mathbb{C}$ if $\infty \in \partial D$, to the outside of a Euclidean circle if $\infty \in D$, and to the inside of a Euclidean circle otherwise.

B.3   Topology in $\mathbb{C}$ and $\overline{\mathbb{C}}$

With the spherical metric in hand, we can now define the corresponding topological concepts on $\overline{\mathbb{C}}$. We begin by defining the interior, closure, and boundary of a set $E \subseteq \overline{\mathbb{C}}$ as follows: The *interior* of E, denoted $\mathrm{Int}(E)$, is the set of points $z \in E$ for which there exists $r > 0$ so that $\triangle_\sigma(z, r) \subseteq E$. The *closure* of E, denoted $\overline{E}$ or closure(E), is the set that contains all points $z \in \overline{\mathbb{C}}$ such that for any $r > 0$ we have $\triangle_\sigma(z, r) \cap E \neq \emptyset$. The *boundary* of E, denoted ∂E, is the set that contains all points $z \in \overline{\mathbb{C}}$ such that for any $r > 0$ we have $\triangle_\sigma(z, r) \cap E \neq \emptyset$ and $\triangle_\sigma(z, r) \cap (\overline{\mathbb{C}} \setminus E) \neq \emptyset$. Hence, $\partial E = \overline{E} \cap \overline{\overline{\mathbb{C}} \setminus E}$ and $\overline{E} = E \cup \partial E$.

Example B.1. Readers should convince themselves that $\mathrm{Int}(\mathbb{D}) = \mathbb{D}$, that $\partial \mathbb{D}$ is the unit circle $|z| = 1$, and that $\overline{\mathbb{D}} = \{z \in \mathbb{C} : |z| \leq 1\}$. In general, $\partial\triangle(a, r) = C(a, r)$, where $C(a, r) = \{z \in \mathbb{C} : |z - a| = r\}$ is the *Euclidean circle* of radius $r > 0$ and center $a \in \mathbb{C}$. Likewise, for $a \in \overline{\mathbb{C}}$ and $r \in (0, \pi)$, $\partial\triangle_\sigma(a, r) = C_\sigma(a, r)$, the *spherical circle*. What happens when $r = \pi$? Show that $\{z \in \overline{\mathbb{C}} : \sigma(a, z) > r\}$ is a spherical disc and give its center.

Exercise B.2. Identify the interior, closure, and boundary of the following subsets of $\overline{\mathbb{C}}$. No proofs are required. ***Try it out!***

(a) $A = \emptyset$

(b) $B = \{z \in \mathbb{C} : |z| \leq 1\}$

(c) $C = \{z \in \mathbb{C} : |z| = 1\}$

(d) $D = \{z \in \mathbb{C} : |z| > 1\}$

(e) $E = \mathbb{R}$

(f) $F = \mathbb{C} \setminus \mathbb{R}$

(g) $G = \{x + iy \in \mathbb{C} : x, y \in \mathbb{Q}\}$, where $\mathbb{Q}$ is the set of rational numbers in $\mathbb{R}$

Definition B.3.

(a) A set $A \subseteq \overline{\mathbb{C}}$ is called *open in* $\overline{\mathbb{C}}$ if $\text{Int}(A) = A$, i.e., if for each point $z \in A$, there exists $r > 0$ such that $\triangle_\sigma(z, r) \subseteq A$.

(b) A set $A \subseteq \mathbb{C}$ is called *open in* $\mathbb{C}$ if for each point $z \in A$, there exists $r > 0$ such that $\triangle(z, r) \subseteq A$.

Definition B.4.

(a) A set $A \subseteq \overline{\mathbb{C}}$ is called *closed in* $\overline{\mathbb{C}}$ if its complement $\overline{\mathbb{C}} \setminus A$ is open in $\overline{\mathbb{C}}$.

(b) A set $A \subseteq \mathbb{C}$ is called *closed in* $\mathbb{C}$ if its complement $\mathbb{C} \setminus A$ is open in $\mathbb{C}$.

Remark B.5. Often we refer to open or closed sets without making explicit whether we are working in $\mathbb{C}$ or in $\overline{\mathbb{C}}$. Context will in general make the meaning clear. These facts may also help. Take a moment to convince yourself of their validity:

(a) A set $A \subseteq \mathbb{C}$ is open in $\overline{\mathbb{C}}$ if and only if it is open in $\mathbb{C}$.

(b) A set $E \subset \mathbb{C}$ is *bounded* if there exists some $R > 0$ such that $|z| \leq R$ for all $z \in E$. A bounded subset E of $\mathbb{C}$ is closed in $\mathbb{C}$ if and only if it is closed in $\overline{\mathbb{C}}$.

(c) There exist sets $B \subseteq \mathbb{C}$ which are closed in $\mathbb{C}$ but not closed in $\overline{\mathbb{C}}$. For example, $\mathbb{R}$ or $\mathbb{C}$ itself.

An open set $A \subseteq \overline{\mathbb{C}}$ is called *connected* if given any two points $z, w \in A$ there exists a polygonal line in A that connects z to w. A *polygonal line*, is the union of a finite number of line segments joined end to end. If U is an open set and $z_0 \in U$, then the set $U(z_0)$ of all points $z \in U$ such that there is a polygonal line in U which connects z_0 to z is called the *(connected) component of U containing z_0*. Two facts regarding an open set U can be confirmed, proofs are in [2]:

(a) Components $U(z)$ and $U(w)$ are equal exactly when there is a polygonal path in U which connects z to w.

(b) An open set U is equal to the union of its components, i.e., $U = \cup_{z \in U} U(z)$.

A *domain* is a non-empty open connected set in $\overline{\mathbb{C}}$. (The *domain set* of a function is not necessarily a domain in this sense.) A domain $D \subseteq \overline{\mathbb{C}}$ is *simply connected* if every closed curve in D can be shrunk continuously to a point while remaining in D. Informally, this means that D has no "holes" in it. For example, the unit disc $\mathbb{D}$ is a simply connected domain, while the set $\mathbb{D} \setminus \{0\}$ is not. A *Jordan curve* in $\overline{\mathbb{C}}$ is a simple closed curve, that is, one which does not intersect itself. The Jordan curve theorem tells us that removing a Jordan curve Γ from the Riemann sphere leaves two simply connected domains. The domain inside Γ is the one on its left as you traverse Γ, and this is known as a *Jordan domain*. Observe that $\mathbb{D}$ is a Jordan domain, whereas the slit disc, $\mathbb{D} \setminus [0, 1)$, though simply connected, is not a Jordan domain.

A *neighborhood* of a point $z \in \overline{\mathbb{C}}$ is an open set that contains z. A *deleted neighborhood* of z is a set $U \setminus \{z\}$ where U is a neighborhood of z. When a neighborhood U is of the form $\triangle_\sigma(z, \varepsilon)$ or $\triangle(z, \varepsilon)$ we refer to it as an *ε-neighborhood* of z.

B.3.1   Convergence in the Riemann Sphere

With the spherical metric σ on $\overline{\mathbb{C}}$ and the notion of ε-neighborhood, we can now define convergence of sequences in $\overline{\mathbb{C}}$:

Definition B.6. A sequence of points $\{z_n\}_{n=1}^{\infty}$ in $\overline{\mathbb{C}}$ is said to *converge* to $z \in \overline{\mathbb{C}}$ if $\sigma(z_n, z) \to 0$ as $n \to \infty$, in which case we write $z_n \to z$.

The statement $z_n \to z$ when $z \in \mathbb{C}$ (i.e., $z \neq \infty$) conforms to the standard notion of convergence in $\mathbb{C}$. The next proposition records this fact and more that relate the Euclidean metric in $\mathbb{C}$ to the spherical metric σ on $\overline{\mathbb{C}}$.

Proposition B.7. *For points $z_n, z,$ and w in $\mathbb{C}$, we have:*

(a) $\sigma(z_n, z) \to 0$ *if and only if* $|z_n - z| \to 0$.

(b) $|z| > |w|$ *if and only if* $\sigma(z, \infty) < \sigma(w, \infty)$.

(c) $z_n \to \infty, |z_n| \to +\infty,$ *and* $\sigma(z_n, \infty) \to 0$ *are equivalent.*

(d) $z_n \to 0, |z_n| \to 0, 1/|z_n| \to +\infty,$ *and* $1/z_n \to \infty$ *are equivalent.*

B.4   Continuity on the Riemann Sphere

Let Ω be a subset of $\overline{\mathbb{C}}$ and consider an extended complex-valued function $f : \Omega \to \overline{\mathbb{C}}$. We write $\lim_{z \to z_0} f(z) = L$ and we say the function f *approaches* L as z *approaches* z_0, if for each $\varepsilon > 0$ there exists $\delta > 0$ such that $\sigma(f(z), L) < \varepsilon$ whenever $z \in \Omega$ and $0 < \sigma(z, z_0) < \delta$. We say that f is *continuous* at z_0 if $z_0 \in \Omega$ and $\lim_{z \to z_0} f(z) = f(z_0)$. We say f is continuous on a set U in its domain set if it is continuous at each point of U. From Proposition B.7, this notion of continuity conforms with our usual notion of continuity in $\mathbb{C}$ (see Section A.2).

Example B.8. We consider the map $\phi : z \mapsto 1/z$ and its relation to the division-by-zero problem. Using complex arithmetic, we can see that ϕ is a continuous function from $\overline{\mathbb{C}}$ to itself, except at $z = 0$ and $z = \infty$ where it is not defined. Use Proposition B.7 to show that $\phi(0) = \infty$ is the only way to extend the definition of ϕ so that it becomes continuous at 0. Thus, in this context $1/0 = \infty$. Show as well that $\phi(\infty) = 0$ is the only way to make ϕ continuous at ∞, meaning $1/\infty = 0$.

Example B.9. Mappings from $\overline{\mathbb{C}}$ to itself can also be treated geometrically by identifying $\overline{\mathbb{C}}$ with the sphere $\mathbb{S}$. Consider the x_1-axis of $\mathbb{S}$ in $\mathbb{R}^3$, which pierces $\mathbb{S}$ at the points $(1, 0, 0)$ and $(-1, 0, 0)$, corresponding to 1 and -1 in $\mathbb{C}$. What is the effect of rotating $\mathbb{S}$ by $180° = \pi$ radians about this axis? Observe that the rotation carries $\mathbb{S}$ one-to-one onto itself, fixes each of the points 1 and -1, carries the equator (the unit circle) onto itself, carries the real axis (a great circle) onto itself, and interchanges the north and the south poles. Readers should convince themselves that this rotation is precisely the map $\phi : z \mapsto 1/z$.

You can use $\phi : z \mapsto 1/z$ for addressing continuity (and later analyticity) on the sphere by the judicious interchange of ∞ and 0.

Theorem B.10. *Let $z_0, w_0 \in \mathbb{C}$ and let $f(z)$ be a complex-valued function defined in a deleted neighborhood of z_0. Then,*

(a) $\lim_{z \to z_0} f(z) = \infty$ if and only if $\lim_{z \to z_0} \dfrac{1}{f(z)} = 0$.

(b) $\lim_{z \to \infty} f(z) = w_0$ if and only if $\lim_{z \to 0} f\left(\dfrac{1}{z}\right) = w_0$.

(c) $\lim_{z \to \infty} f(z) = \infty$ if and only if $\lim_{z \to 0} \dfrac{1}{f(1/z)} = 0$.

By pre- or post-composing f with the rotation $\phi : z \mapsto 1/z$, the above theorem converts each statement on the left (involving ∞) to the corresponding statement on the right (which avoids ∞). By considering the topology of the Riemann sphere $\overline{\mathbb{C}}$ near ∞ and the continuity of ϕ, the details of the proof can easily be checked (and they can also be found in [1, p. 51]).

The simplest and most useful class of maps that are continuous on all of $\overline{\mathbb{C}}$ is the class of rational maps.

Definition B.11. *A quotient of two polynomials is called a rational function.*

Let $f(z) = \frac{P(z)}{Q(z)}$ be a rational function in reduced form (i.e., polynomials $P(z)$ and $Q(z)$ have no common zeros). Although this formula for f is defined only for complex values where Q is not zero, we can regard f (as a mapping into $\overline{\mathbb{C}}$) as being defined and continuous on all of $\overline{\mathbb{C}}$. If $Q(a) = 0$, we set $f(a) = \infty$. We also set $f(\infty) = \lim_{z \to \infty} f(z)$. We leave it to the reader to check that this gives f the desired continuity properties. We illustrate this in the next examples.

Example B.12. Let $f(z) = z^2 + 5i$ and $g(z) = \frac{3z^2-5}{z^2+2z}$. Then $f(\infty) = \infty$, $g(\infty) = 3$, $g(-2) = \infty$, and $g(0) = \infty$. Defining the maps this way, we see that f and g are continuous at every point of $\overline{\mathbb{C}}$. For instance, $f(\infty) = \lim_{z \to \infty} f(z) = \infty$ because

$$\lim_{z \to 0} \frac{1}{f(1/z)} = \lim_{z \to 0} \frac{1}{(1/z)^2 + 5i} = \lim_{z \to 0} \frac{z^2}{1 + 5iz^2} = 0.$$

The reader should use Theorem B.10 to check the continuity of g at $-2, 0$, and ∞.

Example B.13. Special among rational maps are the *Möbius transformations*. These are of form $M(z) = \frac{Az+B}{Cz+D}$ where A, B, C, and D are complex numbers satisfying $AD - BC \neq 0$. The inversion map $z \to 1/z$ discussed earlier is one example. Show that in general $M(-D/C) = \infty$ and $M(\infty) = A/C$ and explain the interpretations when $C = 0$. For more on Möbius transformations, see Sections B.1 and 4.2.

Example B.14. In contrast to the case for rational functions, the exponential function $f(z) = e^z$ cannot be defined at ∞ to make it continuous there. You can see this by noting that on the real line $\lim_{x \to +\infty} e^x = \infty$, but $\lim_{x \to -\infty} e^x = 0$, which means that $\lim_{z \to \infty} e^z$ does not exist.

B.5 Analyticity on the Riemann Sphere

As the rotation $z \mapsto 1/z$ was used in Theorem B.10 to explain the notion of continuity on the Riemann sphere $\overline{\mathbb{C}}$, so too can we use it to extend our notion of *analyticity*.

Definition B.15. Let $\Omega \subseteq \overline{\mathbb{C}}$ be an open set and suppose $f : \Omega \to \overline{\mathbb{C}}$ is continuous:

(a) If $z_0 \in \Omega \cap \mathbb{C}$ and $f(z_0) = \infty$, then f is *analytic* at z_0 if $\frac{1}{f(z)}$ is analytic at z_0.

(b) If $\infty \in \Omega$ and $f(\infty) \neq \infty$, then f is *analytic* at ∞ if $f(\frac{1}{z})$ is analytic at 0.

(c) If $\infty \in \Omega$ and $f(\infty) = \infty$, then f is *analytic* at ∞ if $\frac{1}{f(\frac{1}{z})}$ is analytic at 0.

One way to summarize the above is to say that a map is analytic in the extended sense if after moving each instance of ∞ to 0 (by pre- or post-composing with the map $z \mapsto 1/z$), we get a map that is analytic in the usual sense.

Exercise B.16. Show that $z \mapsto \sin \frac{1}{z}$ is analytic at ∞, but $z \mapsto \sin z$ is not.

Remark B.17. When $z_0 \in \mathbb{C}$ is a pole of f of order k, then by Definition A.15 we know that we can write $f(z) = (z - z_0)^{-k} h(z)$ where h is analytic at z_0 (in the usual sense of Section A.2) and $h(z_0) \neq 0$. Thus, $\frac{1}{f(z)} = (z - z_0)^k / h(z)$ is analytic at z_0, which means we can say f is analytic at z_0 by Definition B.15(a), in the extended sense. Since the word "analytic" is used in this text in the usual sense of Section A.2 and in the extended sense, the reader must always be careful to use context to decide in which sense it is being used. Since the context is usually clear, no confusion should arise.

A map f with an isolated singularity at $z_0 \in \mathbb{C}$ (see Definition A.15) is analytic in the usual sense for a removable singularity and analytic in the extended sense for a pole. Thus, only if z_0 is an essential singularity does f fail to be analytic in any sense.

Example B.18. Let $f(z) = \frac{1}{1+z^2}$. Since $f(\frac{1}{z}) = \frac{1}{1+(\frac{1}{z})^2} = \frac{z^2}{z^2+1}$ is analytic at 0 (and has the value 0 there), we can say that f is analytic at ∞ (and $f(\infty) = 0$). Similarly, since $1/f(z) = 1 + z^2$ is analytic at $\pm i$, f is analytic at the poles $\pm i$, a fact stated more generally in Remark B.17.

Example B.19. Let $f(z) = 3z - \frac{1}{z}$ so $f(\infty) = \infty$. Since

$$k(z) = \frac{1}{f(\frac{1}{z})} = \frac{1}{3(\frac{1}{z}) - z} = \frac{z}{3 - z^2}$$

is analytic at zero we say that $f(z)$ is analytic at ∞. Because $f'(z) = 3 + \frac{1}{z^2}$, $f'(\infty) = 3$. We also have $k'(z) = \frac{3+z^2}{(3-z^2)^2}$ yielding $k'(0) = 1/3 = 1/f'(\infty)$. The relationship between $k'(0)$ and $f'(\infty)$ is an important and general property:

Lemma B.20. *Suppose f is analytic at ∞ with $f(\infty) = \infty$. If k is defined by $k(z) = \frac{1}{f(\frac{1}{z})}$, then $f'(\infty) = 1/k'(0)$. If $k'(0) = 0$, then by our conventions $f'(\infty) = \infty$.*

Proof. Assuming f is analytic at ∞ means k is analytic at 0; also, $f(\infty) = \infty$ means $k(0) = 0$. We let N denote the multiplicity of the zero of k at 0. There exists $r > 0$ such that k is analytic on $\Delta(0, r)$. Therefore, $f(z) = \frac{1}{k(\frac{1}{z})}$ is analytic on $\{z \in \mathbb{C} : |z| > 1/r\}$. Hence f may be represented by a Laurent series

$$f(z) = \sum_{n=0}^{\infty} a_n z^n + \sum_{n=1}^{\infty} b_n z^{-n}$$

on $\{z \in \mathbb{C} : |z| > 1/r\}$. Since

$$\frac{1}{k(z)} = f\left(\frac{1}{z}\right) = \sum_{n=0}^{\infty} a_n z^{-n} + \sum_{n=1}^{\infty} b_n z^n$$

has a pole at 0 of order N, we have $a_N \neq 0$ and $a_n = 0$ for all $n > N$. Hence $f(z) = z^N h(z)$ where $h(z) = \cdots + a_{N-1} z^{-1} + a_N$.

Let $g(z) = h(\frac{1}{z}) = a_N + a_{N-1} z + \cdots$. Then

$$k(z) = \frac{1}{f(\frac{1}{z})} = \frac{1}{z^{-N} h(\frac{1}{z})} = \frac{z^N}{g(z)}.$$

Thus $k'(z) = \dfrac{N z^{N-1} g(z) - g'(z) z^N}{[g(z)]^2}$ and so

$$k'(0) = \begin{cases} 1/a_N & \text{when } N = 1 \\ 0 & \text{when } N > 1. \end{cases}$$

Because $h'(z) = \cdots - a_{N-1} z^{-2}$, as $z \to \infty$ we have

$$f'(z) = N z^{N-1} h(z) + z^N h'(z) \to \begin{cases} a_N & \text{when } N = 1 \\ \infty & \text{when } N > 1. \end{cases}$$

Thus $f'(\infty) = 1/k'(0)$ as claimed. $\square$

B.6   Bibliography

[1] Ruel V. Churchill and James Ward Brown, *Complex Variables and Applications*, McGraw-Hill Book Co., New York, eighth edition, 2009.

[2] Bruce P. Palka, *An Introduction to Complex Function Theory: Undergraduate Texts in Mathematics*, Springer-Verlag, New York, 1991.

[3] Edward B. Saff and Arthur David Snider, *Fundamentals of Complex Analysis with Applications to Engineering, Science, and Mathematics*, Prentice Hall, New Jersey, third edition, 2003.

Index

Index of Notation

Index of Terms

About the Authors

Michael Brilleslyper was raised in Southern California. He earned a bachelor's degree in applied mathematics and a master's degree in mathematics from Arizona State University. He obtained his Ph.D. from the University of Arizona in 1994, under the direction of Doug Pickrell. He spent the first six years of his career at Arizona State working extensively with their introductory and calculus courses. For two of those years he served as coordinator for their First Year Mathematics Program. In 2000, he came to the United States Air Force Academy in Colorado Springs, CO. At the Academy, Dr. Brilleslyper enjoys teaching at all levels of the curriculum. He has worked on numerous curricular development projects involving writing, technology, and fundamental skills. He is extremely active in the MAA, having been in the second cohort of Project NExT fellows. More recently he has served as Rocky Mountain section chairman, he has twice been a program chair, he has served as chair of the professional development committee, and he currently serves as Governor of the Rocky Mountain section. He lives in Colorado Springs with his wife and their two daughters.

Michael Dorff is a professor of mathematics at Brigham Young University. He was born in Duluth, Minnesota and grew up in southern California. After teaching high school for four years, he earned an MS degree at the Univ. of New Hampshire and in 1997 a PhD from the Univ. of Kentucky in complex analysis. He was a professor at the Univ. of Missouri-Rolla before accepting a position in 2000 at BYU. He has published over 20 research papers and has given talks at over 100 different conferences, universities, and colleges. He founded the BYU mathematics REU and in 2007 he founded CURM, the national Center of Undergraduate Research in Mathematics, which promotes, trains, and supports professors across the U.S. in doing research with undergraduate students. He is a member of the MAA, AMS, CUR, and Project NExT, and has served in many positions including governor of the MAA Intermountain section, chair of several MAA committees, member of the Executive Board of CUR, and member of the editorial boards of the journals American Math Monthly and Involve. He was a Fulbright Scholar in Poland and has received numerous teaching awards including the MAA's Deborah and Franklin Tepper Haimo Award in 2010. He is married with 5 daughters. His interests include reading (Dostoyevsky and Dickens through Stegner and Saramago), traveling (invite him to visit you!), running (even at 3 am on the streets in Utah), music (classical, Norah Jones), and soccer.

Jane McDougall received her PhD from Northwestern University in 1996, where she studied functions of one complex variable. She has been a member of the Department of Mathematics and Computer Science at Colorado College since 1997, during which time her research interests have grown into geometric function theory and harmonic mappings.

Jim Rolf was born in Nashville, TN but moved to his adopted home state of Texas at age two. He earned his Bachelor of Science degree in mathematics in his home town of Waco, TX while attending Baylor University. This was followed by a Master of Divinity degree from Southwestern Baptist Theological Seminary in Fort Worth, TX. In 1997 he completed his doctoral work in numerical analysis at Duke University under the direction of William K. Allard. After three years on the faculty at West Point, Jim moved to Colorado Springs, CO and the math department at the United States Air Force Academy. In the fall of 2012, Jim will join the math department at Yale University. In addition to his professional interests, Jim enjoys cycling, gardening, and woodworking.

Lisbeth Drews Schaubroeck was born in the Netherlands, but spent most of her childhood in Geneseo, IL. She earned her bachelor's degree in Mathematics and Mathematics Education at Wartburg College in Waverly, IA. In 1998, she completed her doctoral work in complex analysis at the University of North Carolina at Chapel Hill under the direction of John Pfaltzgraff. Currently, Beth is a professor at the United States Air Force Academy in Colorado Springs, CO. There she enjoys teaching all levels of undergraduate mathematics to future Air Force officers. She is active in the Rocky Mountain Section of the MAA and has been on the awards committee and the student activities coordinator. Beth especially enjoys her work in faculty development in a wide range of venues. She mentors Air Force officers who are new to teaching at the Air Force Academy, new PhDs through Project NExT, and seasoned faculty from civilian universities and community colleges through an NSF-funded PREP workshop. She lives in Colorado Springs with her husband and two sons.

Richard L. Stankewitz was born in Royal Oak, Michigan. He earned a bachelor's degree with distinction in mathematics from the University of Michigan at Ann Arbor in 1991. He obtained a Ph.D. in 1998 under the direction of his advisor Aimo Hinkkanen at the University of Illinois at Urbana in the field of complex dynamics. Dr. Stankewitz held positions at Texas A&M University at College Station and Penn State University at Erie before coming to Ball State University in 2002, where in 2010 he attained his current position of Professor of Mathematics. He is a member of the American Mathematical Society and has a long time interest in supporting efforts to encourage undergraduate research in mathematics. In addition to publishing over 15 research articles, he has given talks in over 50 national and international conferences, universities, and colleges.

Ken Stephenson was born in South Haven, Michigan, in 1945. He received a BS in mathematics at Michigan and an MS in mathematics at Wisconsin before serving 3 years as a Naval Officer. He returned to his studies at Wisconsin, receiving his PhD under Walter Rudin in 1976. From Wisconsin he joined the faculty at the University of Tennessee, where he continues his research and teaching.

Complex function theory and the associated conformal geometry have remained Dr. Stephenson's core mathematical interest, but in 1985 a fascinating talk by Bill Thurston on "circle packing" profoundly changed his outlook. He began contributing to the development of circle packing and its associated "discrete" conformal geometry through publications and software, culminating in his 2005 book *Introduction to Circle Packing: the Theory of Discrete Analytic Functions*, Cambridge University Press, and his software package "CirclePack". This is a new type of experimentally driven mathematics, and his principal goal now is promoting circle packing and its applications.

Dr. Stephenson has been supported throughout his years of research by grants from the National Science Foundation, the Tennessee Science Alliance, and currently the Simons Foundation. He has spoken on circle packing at numerous national and international conferences and has held visiting positions at the University of Hawaii, the Open University (England), the University of Cambridge, Florida State University, the Technical University of Berlin, the Free University of Berlin, Oak Ridge National Laboratory, and the Mathematical Sciences Center (Tsinghua University, Beijing). He is a member of the American Mathematical Society and the Mathematical Association of America, and is a Fellow of AAAS (American Association for the Advancement of Science).